APPROXIMATE CALCULATION OF INTEGRALS

Vladimir Ivanovich Krylov

Translated by
Arthur H. Stroud

DOVER PUBLICATIONS, INC.
Mineola, New York

Bibliographical Note

This Dover edition, first published in 2005, is an unabridged republication of the Arthur H. Stroud translation, originally published by The Macmillan Company, New York, in 1962. The text has been translated from the Russian book *Priblizhennoe Vychislenie Integralov,* Gos. Izd. Fiz.-Mat. Lit., Moscow, 1959.

Library of Congress Cataloging-in-Publication Data

Krylov, V. I. (Vladimir Ivanovich), 1902–
 [Priblizhennoe vychislenie integralov. English]
 Approximate calculation of integrals / Vladimir Ivanovich Krylov ; translated by Arthur H. Stroud.
  p. cm.
 Originally published: New York : Macmillan Co., 1962.
 Includes index.
 ISBN 0-486-44579-8 (pbk.)
 1. Integrals. 2. Approximation theory. I. Title.

QA311.K713 2005
515'.43—dc22

2005051789

Manufactured in the United States of America
Dover Publications, Inc., 31 East 2nd Street, Mineola, N.Y. 11501

PREFACE

The author attempts in this book to introduce the reader to the principal ideas and results of the contemporary theory of approximate integration and to provide a useful reference for practical computations.

In this book we consider only the problem of approximate integration of functions of a single variable. We almost completely ignore the more difficult problem of approximate integration of functions of more than one variable, a problem about which much less is known. Only in one place do we mention double and triple integrals in connection with their reduction to single integrals.

But even for single integrals the author has omitted many interesting considerations. Problems not touched upon are, for example, methods of integration of rapidly oscillating functions, the calculation of contour integrals of analytic functions, the application of random methods, and others. The book is devoted for the most part to methods of mechanical quadrature where the integral is approximated by a linear combination of a finite number of values of the integrand.

The contents of the book are divided into three parts. The first part presents concepts and theorems that are met with in the theory of quadrature, but are at least partially outside of the programs of higher academic institutions.

The second part is devoted to the problem of calculation of definite integrals. Here we consider, in essence, three basic topics: the theory of the construction of mechanical quadrature formulas for sufficiently smooth integrand functions, the problem of increasing the precision of quadratures, and the convergence of the quadrature process.

In the third part of the book we study methods for the calculation of indefinite integrals. Here we confine ourselves for the most part to a study of methods for constructing computational formulas. In addition we indicate stability criterions and the convergence of the computational process.

My colleagues in this work, M. K. Gavurin and I. P. Mysovskich, examined a large part of the manuscript and I am very thankful for their remarks and advice.

Academy of Sciences of the
Byelorussian Socialist Soviet Republic   V. I. KRYLOV

TRANSLATOR'S PREFACE

This book provides a systematic introduction to the subject of approximate integration, an important branch of numerical analysis. Such an introduction was not available previously. The manner in which the book is written makes it ideally suited as a text for a graduate seminar course on this subject.

A more exact title for this book would be *Approximate Integration of Functions of One Variable*. As in many aspects of the theory of functions the theory developed here for functions of one variable is very difficult to extend to functions of more than one variable, and the corresponding results are mostly unknown. Several years from now, after methods for integration of functions of more than one variable have been investigated more thoroughly, a book entirely devoted to this subject will be needed.

As a source of reference for other topics concerning approximate integration see "A Bibliography on Approximate Integration," *Mathematics of Computation* (vol. 15, 1961, pp. 52–80), which was compiled by the translator. This is a reasonably complete bibliography, particularly for papers published during the past several decades.

The only significant change in this translation from the original is the inclusion in the appendices of slightly more extensive tables of Gaussian quadrature formulas. The formulas in Appendix A for constant weight function are taken from a memorandum by H. J. Gawlik and are published with the permission of the Controller of Her Britannic Majesty's Stationery Office, and the British Crown copyright is reserved.

I wish to thank Dr. V. I. Krylov for the assistance he provided in furnishing a list of corrections to the original edition. I am also indebted to Professor G. E. Forsythe for the interest he expressed on behalf of the Association for Computing Machinery in having this book published in the present monograph series. Finally I am indebted to James T. Day for his interest in this book and for his assistance in reading parts of the manuscript.

University of Wisconsin
Madison, Wisconsin   A. H. STROUD

CONTENTS

Part One

PRELIMINARY

INFORMATION

Part One of this book presents certain selected results from the following special mathematical topics: Bernoulli numbers and Bernoulli polynomials, orthogonal polynomials, interpolation, linear operators and convergence of sequences of such operators. These topics are needed to construct the theory of approximate integration and are presented only to the extent required to understand the other chapters. The results developed here can be found in special literature, but we think it is useful to present them in this book to free the reader from the inconvenience of looking up literature references.

CHAPTER 1

Bernoulli Numbers and Bernoulli Polynomials

1.1. BERNOULLI NUMBERS

Bernoulli polynomials and Bernoulli numbers are needed in later chapters (Sections 6.3 and 11.3) to construct Euler-Maclaurin formulas and other similar formulas which serve to increase the accuracy of approximate quadrature.

The Bernoulli numbers can be defined by means of the following generating function. Let t be a complex parameter. Consider the function

$$g(t) = \frac{t}{e^t - 1}. \tag{1.1.1}$$

For k, an integer, the points $t = 2k\pi i$ are zeros of the denominator. All of these are simple zeros because the derivative of the denominator is e^t and is different from zero for all finite t. The point $t = 0$ is not a singular point of $g(t)$ because $\lim\limits_{t \to 0} \dfrac{t}{e^t - 1} = 1$.

The function $g(t)$ is holomorphic in the circle $|t| \leq 2\pi$ and thus can be expanded there in a power series in t. We write the expansion in the form

$$\frac{t}{e^t - 1} = \sum_{n=0}^{\infty} \frac{B_n}{n!} t^n, \qquad |t| < 2\pi. \tag{1.1.2}$$

The numbers B_n defined by this equation are called *Bernoulli numbers*.

If both sides of (1.1.2) are multiplied by $e^t - 1 = \sum\limits_{\nu=1}^{\infty} \dfrac{t^{\nu}}{\nu!}$, then we obtain the equation

$$\left(\frac{t}{1!} + \frac{t^2}{2!} + \frac{t^3}{3!} + \cdots\right) \sum_{n=0}^{\infty} \frac{B_n}{n!} t^n = t,$$

valid for all t in the circle $|t| < 2\pi$. After multiplying out the power series on the left side of this equation there must remain only the first power of t with coefficient of unity. Thus the powers of t higher than the first must all become zero: $B_0 = 1$ and for $n = 2, 3, \ldots$ we must have

$$\frac{B_0}{n!} + \frac{B_1}{(n-1)!1!} + \frac{B_2}{(n-2)!2!} + \cdots + \frac{B_{n-1}}{1!(n-.1)!} = 0.$$

This last equation permits us to sequentially calculate all of the Bernoulli numbers. We can obtain other forms which are more convenient for some purposes. Multiplying the last equation by $n!$ and adding B_n to both sides we obtain

$$\sum_{k=0}^{n} \frac{n!}{k!(n-k)!} B_k = B_n.$$

Comparing this equation with the binomial expansion we see that it can be written in the form

$$(B + 1)^n = B_n \qquad (1.1.3)$$

if we interpret this equation to mean that after raising the binomial $B + 1$ to the nth power the kth power of B is the Bernoulli number B_k $(k = 0, 1, \ldots, n)$.

We can easily verify that all Bernoulli numbers with odd indices, greater than unity, are equal to zero:

$$B_{2k+1} = 0, \qquad k > 0. \qquad (1.1.4)$$

In order to show this replace t by $-t$ in (1.1.2):

$$\frac{-t}{e^{-t} - 1} = \sum_{n=0}^{\infty} \frac{(-1)^n B_n}{n!} t^n.$$

On the other hand

$$\frac{-t}{e^{-t} - 1} = \frac{te^t}{e^t - 1} = t + \frac{t}{e^t - 1} = t + \sum_{n=0}^{\infty} \frac{B_n}{n!} t^n,$$

and therefore we must have

$$t + \sum_{n=0}^{\infty} \frac{B_n}{n!} t^n = \sum_{n=0}^{\infty} \frac{(-1)^n B_n}{n!} t^n.$$

Comparing the coefficients of t^n, for $n > 1$, gives

$$B_n = (-1)^n B_n.$$

When n is an odd integer $2k + 1$ $(k > 0)$ we have

$$B_{2k+1} = -B_{2k+1},$$

which is equivalent to (1.1.4).

The values of the nonzero Bernoulli numbers for $n \leq 30$ are:

$B_0 = 1$ $\qquad B_{10} = \dfrac{5}{66}$ $\qquad B_{22} = \dfrac{854513}{138}$

$B_1 = -\dfrac{1}{2}$ $\qquad B_{12} = -\dfrac{691}{2730}$ $\qquad B_{24} = -\dfrac{236364091}{2730}$

$B_2 = \dfrac{1}{6}$ $\qquad B_{14} = \dfrac{7}{6}$ $\qquad B_{26} = \dfrac{8553103}{6}$

$B_4 = -\dfrac{1}{30}$ $\qquad B_{16} = -\dfrac{3617}{510}$ $\qquad B_{28} = -\dfrac{23749461029}{870}$

$B_6 = \dfrac{1}{42}$ $\qquad B_{18} = \dfrac{43867}{798}$ $\qquad B_{30} = \dfrac{8615841276005}{14322}$

$B_8 = -\dfrac{1}{30}$ $\qquad B_{20} = -\dfrac{174611}{330}$

The Bernoulli numbers with even indices are related to sums of even negative powers of the natural numbers by the following remarkable identity:

$$B_{2k} = \frac{(-1)^{k-1}(2k)!}{2^{2k-1}\pi^{2k}}(1 + 2^{-2k} + 3^{-2k} + 4^{-2k} + \cdots). \qquad (1.1.5)$$

From this it is seen that for increasing k the Bernoulli numbers B_{2k} will increase in size and for large k will asymptotically approach

$$B_{2k} \approx 2(-1)^{k-1}(2k)!(2\pi)^{-2k}.$$

Equation (1.1.5) follows at once from the expansion (1.3.1) which we will obtain for Bernoulli polynomials in a trigonometric series on the segment $[0, 1]$.

1.2. BERNOULLI POLYNOMIALS

Bernoulli polynomials can be defined by various methods, but for our purpose it is convenient to define them by means of a generating function. We introduce the function

$$g(x, t) = e^{xt} \frac{t}{e^t - 1}. \tag{1.2.1}$$

This differs from (1.1.1) by the factor e^{xt} which does not vanish, so $g(x, t)$ has the same singular points as $g(t)$. In particular it is holomorphic in the circle $|t| < 2\pi$ and can be expanded there in a power series in t:

$$g(x, t) = e^{xt} \frac{t}{e^t - 1} = \sum_{n=0}^{\infty} \frac{B_n(x)}{n!} t^n. \tag{1.2.2}$$

In the next paragraph we will see that the functions $B_n(x)$ are polynomials of degree n. They are called the *Bernoulli polynomials*.

If in $g(x, t)$ the factor e^{xt} is replaced by the series $\sum_{\nu=0}^{\infty} \frac{x^\nu t^\nu}{\nu!}$ and $\frac{t}{e^t - 1}$ is replaced by the expansion (1.1.2), then we obtain the identity

$$\sum_{\nu=0}^{\infty} \frac{x^\nu t^\nu}{\nu!} \sum_{k=0}^{\infty} \frac{B_k}{k!} t^k = \sum_{n=0}^{\infty} \frac{B_n(x)}{n!} t^n \qquad |t| < 2\pi.$$

Comparing the coefficients of t^n leads to the equation

$$\frac{B_n(x)}{n!} = \frac{x^n B_0}{n!} + \frac{x^{n-1} B_1}{(n-1)!1!} + \cdots + \frac{B_n}{n!}.$$

After multiplying by $n!$ we obtain the following expression for $B_n(x)$

$$B_n(x) = \sum_{k=0}^{n} \frac{n!}{k!(n-k)!} B_{n-k} x^k, \tag{1.2.3}$$

which shows that $B_n(x)$ is indeed a polynomial of degree n. The expression (1.2.3) can be written in a simpler form

$$B_n(x) = (x + B)^n \tag{1.2.4}$$

if we agree to consider that after raising the binomial $x + B$ to the nth power that the kth power of B is taken to be the kth Bernoulli number B_k.

We will need to be familiar with certain properties of the Bernoulli polynomials; these will now be developed.

1. The value of a Bernoulli polynomial.

For $x = 0$ the value of a Bernoulli polynomial is the corresponding Bernoulli number:

$$B_n(0) = B_n \tag{1.2.5}$$

which we see from (1.2.3).

2. Differentiation and integration of $B_n(x)$.

Differentiating (1.2.2) with respect to x gives

$$te^{xt}\frac{t}{e^t - 1} = \sum_{n=0}^{\infty} \frac{B_n'(x)}{n!} t^n.$$

The lefthand side of this equation is different from $g(x, t)$ only by the factor t and therefore must be

$$te^{xt}\frac{t}{e^t - 1} = \sum_{n=0}^{\infty} \frac{B_n(x)}{n!} t^{n+1}.$$

The power expansions of the two previous equations must be identically equal and thus

$$\frac{B_n'(x)}{n!} = \frac{B_{n-1}(x)}{(n-1)!}$$

or

$$B_n'(x) = nB_{n-1}(x). \tag{1.2.6}$$

From this and from (1.2.5) we immediately have the following relationship for the integration of Bernoulli polynomials

$$B_n(x) = B_n + n\int_0^x B_{n-1}(t)\,dt. \tag{1.2.7}$$

3. Multiplication of the argument by a constant.

Let m by any positive integer

$$e^{mxt}\frac{t}{e^t - 1} = \sum_{n=0}^{\infty} \frac{B_n(mx)}{n!} t^n.$$

By a very simple transformation we can obtain another expansion

$$e^{mxt} \frac{t}{e^t - 1} = \frac{1}{m} e^{mxt} \left[\frac{mt(1 + e^t + \cdots + e^{(m-1)t})}{e^{mt} - 1} \right] =$$

$$= \frac{1}{m} \sum_{s=0}^{m-1} \frac{e^{\left(x + \frac{s}{m}\right)mt}}{e^{mt} - 1} mt = \frac{1}{m} \sum_{s=0}^{m-1} \sum_{n=0}^{\infty} \frac{m^n B_n\left(x + \frac{s}{m}\right)}{n!} t^n.$$

From these two expansions we deduce the relationship for multiplication of the argument by a constant factor:

$$B_n(mx) = m^{n-1} \sum_{s=0}^{m-1} B_n\left(x + \frac{s}{m}\right). \tag{1.2.8}$$

4. Representations for the polynomials $B_n(x)$.

In order to study the behavior of $B_n(x)$ it is convenient to replace the variable x by a new variable $z = x(1 - x)$. We will show the validity of the following assertions concerning representations for Bernoulli polynomials in the variable z.

Each polynomial $B_n(x)$ of even order $n = 2k$ can be expanded in powers of z:

$$(-1)^k [B_{2k}(x) - B_{2k}] = \sum_{\nu=0}^{k-2} F_{k,\nu} z^{k-\nu} \tag{1.2.9}$$

where $F_{k,0} = 1$ and $F_{k,\nu} > 0$ $(\nu = 1, 2, \ldots, k - 2)$. Each Bernoulli polynomial of odd order $n = 2k - 1$ can be represented in the form:

$$(-1)^k B_{2k-1}(x) = (1 - 2x) \sum_{\nu=0}^{k-2} H_{k,\nu} z^{k-\nu-1} \tag{1.2.10}$$

where all the coefficients $H_{k,\nu}$ $(\nu = 0, 1, \ldots, k - 2)$ are positive.

Let us verify the first of these assertions concerning the polynomials $B_{2k}(x)$ of even order. To simplify the discussion we introduce the auxiliary variable ξ, setting $x = 1/2 + \xi$. The variables ξ and z are related by

$$z = x(1 - x) = \frac{1}{4} - \xi^2.$$

In order to see that $B_{2k}(x)$ is a polynomial in the variable z it is sufficient to establish that the expansion of $B_{2k}(x)$ in powers of ξ will contain only even powers of ξ.

Differentiating the function (1.2.1) with respect to the variable ξ gives the following expression

$$g(x,\ t)\ =\ e^{\left(\frac{1}{2}+\xi\right)t}\ \frac{t}{e^t-1}\ =\ e^{\xi t}\ \frac{te^{\frac{1}{2}t}}{e^t-1}\ =$$

$$=\ e^{\xi t}\ \frac{t}{e^{\frac{1}{2}t}-e^{-\frac{1}{2}t}}\ =\ e^{\xi t}\ \frac{t/2}{\sinh\ t/2}\ .$$

$\dfrac{B_{2k}(x)}{(2k)!}$ is the coefficient of t^{2k} in the expansion of $g(x,\ t)$ in powers of t. The factor $\dfrac{t/2}{\sinh\ t/2}$ is an even function of t and its power series in t will contain only even powers of t. After multiplication of this series by

$$e^{\xi t}\ =\ \sum_{\nu=0}^{\infty}\ \frac{\xi^{\nu}t^{\nu}}{\nu!}$$

in order to obtain the term in t^{2k} we must take from the series for $e^{\xi t}$ only terms with even powers of t. But all of these also contain only even powers of ξ, and thus $B_{2k}(x)$ will contain only even powers of ξ.

For $x = 0$ we also have $z = 0$, and hence the difference $B_{2k}(x) - B_{2k}$ will be a polynomial in z without a constant term and must have the form

$$(-1)^k[B_{2k}(x)\ -\ B_{2k}]\ =\ \sum_{\nu=0}^{k-1}\ F_{k,\nu}z^{k-\nu}.$$

There remains only to verify the assertion about $F_{k,\nu}$. The coefficient in $B_{2k}(x)$ of the highest degree (that is the coefficient of x^{2k}) is equal to unity, and therefore we must have $F_{k,0} = 1$. In addition the coefficient of x in $B_{2k}(x)$ is $2kB_{2k-1} = 0$ and because the first power of x on the righthand side can be only contained in the term corresponding to $\nu = k - 1$, then $F_{k,k-1} = 0$. We may construct a recursion relation to find the remaining $F_{k,\nu}$. Let us calculate the second derivative with respect to x of both sides of (1.2.9). Because

$$B_{2k}''(x)\ =\ 2k(2k\ -\ 1)B_{2k-2}(x)$$

and because the operators of differentiation with respect to x and z are related by

$$\frac{d}{dx} = \frac{dz}{dx}\frac{d}{dz} = (1 - 2x)\frac{d}{dz},$$

$$\frac{d^2}{dx^2} = (1 - 2x)^2\frac{d^2}{dz^2} - 2\frac{d}{dx} = (1 - 4z)\frac{d^2}{dz^2} - 2\frac{d}{dz}$$

then we obtain:

$$(-1)^k 2k(2k - 1)B_{2k-2}(x) = \sum_{\nu=1}^{k-1} F_{k,\nu-1}(k - \nu + 1)(k - \nu)z^{k-\nu-1} -$$

$$- \sum_{\nu=0}^{k-2} F_{k,\nu}(2k - 2\nu)(2k - 2\nu - 1)z^{k-\nu-1}.$$

Comparing this with (1.2.9) for $B_{2k-2}(x)$, namely with

$$(-1)^{k-1}[B_{2k-2}(x) - B_{2k-2}] = \sum_{\nu=0}^{k-2} F_{k-1,\nu}z^{k-\nu-1},$$

we obtain the desired recursion relation for $F_{k,\nu}$

$$(2k - 2\nu)(2k - 2\nu - 1)F_{k,\nu} =$$

$$= 2k(2k - 1)F_{k-1,\nu} + (k - \nu + 1)(k - \nu)F_{k,\nu-1}.$$

Hence knowing $F_{k,0} = 1$ and $F_{k,k-1} = 0$ $(k = 1, 2, \ldots)$ we can sequentially find $F_{k,\nu}$ $(k = 3, 4, \ldots; \nu = 1, 2, \ldots, k - 2)$, and all of them turn out to be positive.

To establish the representation for $B_{2k-1}(x)$ it suffices to differentiate both sides of (1.2.9) with respect to x:

$$(-1)^k 2kB_{2k-1}(x) = \sum_{\nu=0}^{k-2} F_{k,\nu}(1 - 2x)(k - \nu)z^{k-\nu-1}.$$

Hence we see that (1.2.10) is valid with

$$H_{k,\nu} = \frac{(k - \nu)F_{k,\nu}}{2k} > 0.$$

5. Symmetry of $B_n(x)$.

Consider the point $x = 1/2$ on the x axis. The points x and $1 - x$ are symmetrically situated with respect to this point. The parameter $z = x(1 - x)$ does not change in value if we replace x by $1 - x$. Thus from

(1.2.9) we obtain

$$B_{2k}(1 - x) = B_{2k}(x); \tag{1.2.11}$$

the graph of $B_{2k}(x)$ is symmetric with respect to the line $x = 1/2$.

The factor $\sum_{\nu=0}^{k-2} H_{k,\nu} z^{k-\nu-1}$ in (1.2.10) has the same value at the points x and $1 - x$. The factor $(1 - 2x)$ has the same absolute value but opposite sign at these points. Therefore

$$B_{2k-1}(1 - x) = -B_{2k-1}(x). \tag{1.2.12}$$

Thus the graph of $B_{2k-1}(x)$ is centrally symmetric with respect to the point $x = 1/2$.

From (1.2.5) and (1.2.11) we obtain

$$B_{2k}(1) = B_{2k},$$

and from (1.2.12) for $k \geq 2$ we obtain

$$B_{2k-1}(1) = -B_{2k-1}.$$

Thus each Bernoulli polynomial, except $B_1(x)$, has equal values at the ends of the segment [0, 1].

$$B_n(1) = B_n(0) = B_n. \tag{1.2.13}$$

6. The behavior of the Bernoulli polynomials on the segment [0, 1].

We will need to know the value $B_n(1/2)$ which can be easily calculated from (1.2.8). If in (1.2.8) we substitute $m = 2$ and $x = 1/2$ we obtain

$$B_n(1) = 2^{n-1}\left[B_n\left(\frac{1}{2}\right) + B_n(1)\right].$$

But since

$$B_n(1) = B_n \quad (n > 1),$$

then for every n

$$B_n\left(\frac{1}{2}\right) = -(1 - 2^{-n+1})B_n. \tag{1.2.14}$$

We will also need some properties of the polynomials

$$\gamma_n(x) = B_n(x) - B_n$$

which are essentially the same as $B_n(x)$, but which are more convenient for some purposes. Consider, at first, the polynomial of even order $n = 2k$, which by (1.2.9) is

$$(-1)^k y_{2k}(x) = \sum_{\nu=0}^{k-2} F_{k,\nu} z^{k-\nu}. \qquad (1.2.15)$$

The points $x = 0$ and $x = 1$ are zeros of $y_{2k}(x)$:

$$y_{2k}(0) = B_{2k}(0) - B_{2k} = B_{2k} - B_{2k} = 0$$

$$y_{2k}(1) = B_{2k}(1) - B_{2k} = B_{2k} - B_{2k} = 0.$$

It is easily seen that for $k \geq 2$ both of these points are zeros of multiplicity two; for example for $x = 0$

$$y'_{2k}(0) = 2k B_{2k-1}(0) = 0$$

$$y''_{2k}(0) = 2k(2k-1) B_{2k-2}(0) = 2k(2k-1) B_{2k-2} \neq 0.$$

By $(1.2.11)$ the same holds true for $x = 1$. For $0 < x < 1$ the parameter z will lie within the limits $0 < z \leq 1/4$, and since $F_{k,\nu} > 0$

$$(-1)^k y_{2k}(x) > 0 \qquad \text{for } 0 < x < 1.$$

In the open segment $0 < x < 1$ the polynomial $y_{2k}(x)$ has no zeros and has the same sign as $(-1)^k$.

When x varies from zero up to $1/2$, the function $z = x(1-x)$ will increase from zero up to $1/4$, and as x varies from $1/2$ up to 1, the function z will decrease from $1/4$ to zero.

As can be seen from $(1.2.15)$ as x varies from zero up to $1/2$ the polynomial $(-1)^k y_{2k}(x)$ will increase from zero up to $(-1)^k y_{2k}(1/2) = |B_{2k}(1/2) - B_{2k}| = (2 - 2^{-2k+1})|B_{2k}|$. When x varies from $1/2$ up to 1, the polynomial $(-1)^k y_{2k}(x)$ will decrease again to zero. Each value a in the range $0 < a < (2 - 2^{-2k+1})|B_{2k}|$ will be taken on twice by $y_{2k}(x)$, on the segment $(0, 1)$, at two points which are symmetrically located with respect to $x = 1/2$.

Let us consider now a polynomial $y_n(x)$ of odd order $n = 2k - 1$. If we take $k \geq 2$ then

$$y_{2k-1}(x) = B_{2k-1}(x)$$

and

$$(-1)^k y_{2k-1}(x) = (1 - 2x) \sum_{\nu=0}^{k-2} H_{k,\nu} z^{k-\nu-1}. \qquad (1.2.16)$$

The points $x = 0$ and $x = 1$ will be zeros of $y_{2k-1}(x)$, and we can see that they both will be zeros of multiplicity one. In fact

$$y'_{2k-1}(0) = y'_{2k-1}(1) = (2k-1) B_{2k-2}(0) \neq 0.$$

In addition, from $(1.2.16)$ and from $H_{k,\nu} > 0$, we see that $x = 1/2$ is a simple zero of $y_{2k-1}(x)$ and these are the only zeros of this polynomial on the closed segment $0 \leq x \leq 1$. The sign of $y_{2k-1}(x)$ is given by

$$(-1)^k y_{2k-1}(x) > 0 \qquad \text{for } 0 < x < \frac{1}{2},$$

$$(-1)^k y_{2k-1}(x) < 0 \qquad \text{for } \frac{1}{2} < x < 1.$$

Here we give a table of the first ten Bernoulli polynomials.

$B_0(x) = 1$
$B_1(x) = x - 1/2$
$B_2(x) = x^2 - x + 1/6$
$B_3(x) = x^3 - 3/2\, x^2 + 1/2\, x$
$B_4(x) = x^4 - 2\, x^3 + x^2 - 1/30$
$B_5(x) = x^5 - 5/2\, x^4 + 5/3\, x^3 - 1/6\, x$
$B_6(x) = x^6 - 3\, x^5 + 5/2\, x^4 - 1/2\, x^2 + 1/42$
$B_7(x) = x^7 - 7/2\, x^6 + 7/2\, x^5 - 7/6\, x^3 + 1/6\, x$
$B_8(x) = x^8 - 4\, x^7 + 14/3\, x^6 - 7/3\, x^4 + 2/3\, x^2 - 1/30$
$B_9(x) = x^9 - 9/2\, x^8 + 6\, x^7 - 21/5\, x^5 + 2\, x^3 - 3/10\, x$
$B_{10}(x) = x^{10} - 5\, x^9 + 15/2\, x^8 - 7\, x^6 + 5\, x^4 - 3/2\, x^2 + 5/66.$

Figure 1 illustrates the behavior of the Bernoulli polynomials $B_n(x)$ on the segment $[0, 1]$.

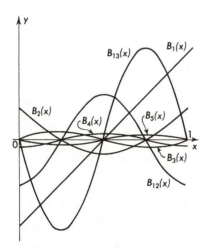

Figure 1. Bernoulli polynomials on the segment (0,1).

1.3. PERIODIC FUNCTIONS RELATED TO BERNOULLI POLYNOMIALS

To study certain questions connected with Bernoulli polynomials we introduce the functions $B_n^*(x)$, of period one, defined by the conditions

$$B_n^*(x) = B_n(x), \qquad 0 \le x < 1$$

$$B_n^*(x + 1) = B_n^*(x).$$

$B_0^*(x)$ is a constant equal to 1; $B_1^*(x)$ is a discontinuous function with a jump of -1 at each integer; for $n > 1$, $B_n^*(x)$ is a continuous function.

Let us construct the trigonometric Fourier series for $B_n^*(x)$. For this purpose we construct the Fourier series for the generating function

$$g(x, t) = e^{xt} \frac{t}{e^t - 1}$$

for $0 \le x < 1$. To do this we expand $g(x, t)$ in an exponential series

$$g(x, t) = \sum_{m=-\infty}^{+\infty} C_m e^{i2\pi m x};$$

$$C_m = \int_0^1 g(x, t) e^{-i2\pi m x} dx = \frac{t}{e^t - 1} \int_0^1 e^{xt} e^{-i2\pi m x} dx =$$

$$= \frac{t}{e^t - 1} \left[\frac{e^{x(t - i2\pi m)}}{t - i2\pi m} \right]_0^1 = \frac{t}{t - i2\pi m}.$$

By singling out the summand $C_0 = 1$ and combining the terms in the series corresponding to the indices m and $-m$ we obtain

$$g(x, t) = 1 + \sum_{m=1}^{\infty} \left[\frac{t}{t - i2\pi m} e^{i2\pi m x} + \frac{t}{t + i2\pi m} e^{-i2\pi m x} \right].$$

It can be shown that for any value of x on the segment $0 \le x \le 1$ the series on the right hand side of this equation will converge for all t distinct from $i2k\pi$ ($k = 0, \pm1, \pm2, \dots$). To prove this we take any bounded part σ of the plane t and exclude from the series the terms at the beginning which have poles in this part of the plane; the terms which remain will converge uniformly relative to t in σ. From these remarks it is easy to justify the change of order of the summation which we will make below in the construction of a power series for $g(x, t)$.

If we consider $|t| < 2\pi$ and expand the right side of the last equation in powers of t then the coefficient of t^n will be a trigonometric series for $B_n(x)/n!$. It will also give a representation for $B_n^*(x)/n!$ for all x.

$$\frac{t}{t - i2\pi m} = -\frac{t}{i2\pi m} \left(\frac{1}{1 - \frac{t}{i2\pi m}} \right) = -\sum_{n=1}^{\infty} \left(\frac{t}{i2\pi m} \right)^n$$

$$\frac{t}{t + i2\pi m} = \sum_{n=1}^{\infty} (-1)^{n-1} \left(\frac{t}{i2\pi m}\right)^n$$

$$g(x, t) = 1 + \sum_{m=1}^{\infty} \sum_{n=1}^{\infty} \left[\frac{(-1)^{n-1}}{(i2\pi m)^n} t^n e^{-i2\pi m x} - \frac{1}{(i2\pi m)^n} t^n e^{i2\pi m x}\right] =$$

$$= 1 + \sum_{n=1}^{\infty} \frac{t^n}{(i2\pi)^n} \sum_{m=1}^{\infty} \left[\frac{(-1)^{n-1}}{m^n} e^{-i2\pi m x} - \frac{1}{m^n} e^{i2\pi m x}\right]$$

thus, for $n > 1$,

$$B_n^*(x) = \frac{n!}{(2\pi i)^n} \sum_{m=1}^{\infty} \left[\frac{(-1)^{n-1}}{m^n} e^{-i2\pi m x} - \frac{1}{m^n} e^{i2\pi m x}\right].$$

For even and odd orders the calculations give the following results

$$B_{2k}^*(x) = \frac{(-1)^{k-1}(2k)!}{2^{2k-1}\pi^{2k}} \sum_{m=1}^{\infty} \frac{\cos 2\pi m x}{m^{2k}}, \tag{1.3.1}$$

$$B_{2k+1}^*(x) = \frac{(-1)^{k-1}(2k+1)!}{2^{2k}\pi^{2k+1}} \sum_{m=1}^{\infty} \frac{\sin 2\pi m x}{m^{2k+1}}. \tag{1.3.2}$$

From this we obtain, for $x = 0$, the series (1.1.5) for the Bernoulli numbers.

1.4. EXPANSION OF AN ARBITRARY FUNCTION IN BERNOULLI POLYNOMIALS

Theorem 1. *If f has a continuous derivative of order ν on $[0, 1]$ then for any $x \in [0, 1]$ we have the relation*

$$f(x) = \int_0^1 f(t)\, dt + \sum_{k=1}^{\nu-1} \frac{B_k(x)}{k!} [f^{(k-1)}(1) - f^{(k-1)}(0)] -$$

$$- \frac{1}{\nu!} \int_0^1 f^{(\nu)}(t) [B_\nu^*(x - t) - B_\nu^*(x)]\, dt. \tag{1.4.1}$$

Proof. Consider the integral

$$\rho_\nu(x) = \rho_\nu = \frac{1}{\nu!} \int_0^1 f^{(\nu)}(t) B_\nu^*(x - t)\, dt.$$

Considering that $\nu > 1$, we integrate by parts. Because

$$\frac{d}{dt} B_\nu^*(x - t) = - \nu \, B_{\nu-1}^*(x - t)$$

$$B_\nu^*(x - 1) = B_\nu^*(x) = B(x)$$

then

$$\rho_\nu = \frac{B_\nu^*(x)}{\nu!} \, [f^{(\nu-1)}(1) - f^{(\nu-1)}(0)] +$$

$$+ \frac{1}{(\nu - 1)!} \int_0^1 f^{(\nu-1)}(t) \, B_{\nu-1}^*(x - t) \, dt =$$

$$= \frac{B_\nu(x)}{\nu!} \, [f^{(\nu-1)}(1) - f^{(\nu-1)}(0)] + \rho_{\nu-1}.$$

We carry out this operation $\nu - 1$ times:

$$\rho_\nu = \sum_{k=2}^{\nu} \frac{B_k(x)}{k!} \, [f^{(k-1)}(1) - f^{(k-1)}(0)] + \rho_1.$$

The function $B_1^*(x)$ has a jump of -1 at the integers; at all other points it has a derivative equal to $+1$,

$$B_1^*(+0) - B_1^*(-0) = -1, \qquad \frac{d}{dt} B_1^*(x - t) = -1.$$

In order to calculate ρ_1 we suppose at first $0 < x < 1$

$$\rho_1(x) = \int_0^x f'(t) \, B_1^*(x - t) \, dt + \int_x^1 f'(t) \, B_1^*(x - t) \, dt =$$

$$= B_1^*(+0) \, f(x) - B_1^*(x) \, f(0) + \int_0^x f(t) \, dt +$$

$$+ B_1^*(x - 1) \, f(1) - B_1^*(-0) \, f(x) + \int_x^1 f(t) \, dt =$$

$$= [B_1^*(+0) - B_1^*(-0)] \, f(x) + B_1(x) \, [f(1) - f(0)] + \int_0^1 f(t) \, dt.$$

For $\rho_\nu \; (\nu = 1, 2, \dots)$ we finally obtain

$$\rho_\nu = \frac{1}{\nu!} \int_0^1 f^{(\nu)}(t) \, B^*(x - t) \, dt$$

$$\rho_\nu = \sum_{k=1}^{\nu} \frac{B_k(x)}{k!} \left[f^{(k-1)}(1) - f^{(k-1)}(0) \right] - f(x) + \int_0^1 f(t)\, dt.$$

This result differs only in form from (1.4.1). The proof was carried out for the open segment $0 < x < 1$, but by continuity equation (1.4.1) is valid also for the closed segment $0 \le x \le 1$. This proves Theorem 1.

If f is defined on an arbitrary finite segment $[a, b]$ and has ν continuous derivatives there, then its expansion on $[a, b]$ in Bernoulli polynomials is obtained from (1.4.1) by means of a linear transformation of the variable

$$f(x) = \frac{1}{h} \int_a^b f(t)\, dt + \sum_{k=1}^{\nu-1} \frac{h^{k-1} B_k\left(\dfrac{x-a}{h}\right)}{k!} \left[f^{(k-1)}(b) - f^{(k-1)}(a) \right] -$$

$$- \frac{h^{\nu-1}}{\nu!} \int_a^b f^{(\nu)}(t) \left[B_\nu^*\left(\frac{x-t}{h}\right) - B_\nu^*\left(\frac{x-a}{h}\right) \right] dt, \qquad (1.4.2)$$

where $h = b - a$.

REFERENCES

A. O. Gel'fond, *Calculus of Finite Differences*, Gostekhizdat, Moscow, 1952, Chap. 4 (Russian).

G. H. Hardy, *Divergent Series*, Oxford, 1949, Chap. 13.

J. F. Steffensen, *Interpolation*, Baltimore, 1927, Chaps. 12, 13.

CHAPTER 2

Orthogonal Polynomials

2.1. GENERAL THEOREMS ABOUT ORTHOGONAL POLYNOMIALS

Much of this book is devoted to a study of integrals of the form

$$\int_a^b p(x) f(x) dx \tag{2.1.1}$$

where $p(x)$ is a given fixed function and $f(x)$ is an arbitrary function of some wide class. The theory of approximate evaluation of this type of integral is closely related to the theory of orthogonal polynomials.

The function $p(x)$ is called a weight function. We will usually restrict ourselves to nonnegative weight functions except in a few cases which will be specifically mentioned.

The theory of orthogonal polynomials for nonnegative weight functions has been developed to a high degree. We will discuss only the small portion of this theory which is necessary to construct certain special approximate integration formulas.

Let $[a, b]$ be any finite or infinite segment. For the present it suffices to assume that the weight function $p(x)$ satisfies the two conditions[1]

1. $p(x)$ is nonnegative, measurable, and not identically zero on the segment $[a, b]$,

2. the products $p(x)x^m$, for any nonnegative integer m, are summable on $[a, b]$.

The functions $f(x)$ and $g(x)$ are said to be *orthogonal* on the segment

---

[1] The reader who is not familiar with the Lebesgue integral can consider $p(x)$ to be a nonnegative function which has only a finite number of zeros on $[a, b]$ for which $\int_a^b p(x)|x|^m dx$ is finite for $m = 0, 1, 2, \ldots$.

$[a, b]$ with respect to the weight function $p(x)$ if the product $p(x)f(x) \times g(x)$ is summable and

$$\int_a^b p(x)f(x)g(x)dx = 0. \qquad (2.1.2)$$

The function $f(x)$ is said to be *normalized* on $[a, b]$ with respect to $p(x)$ if $p(x)f^2(x)$ is summable and

$$\int_a^b p(x)f^2(x)dx = 1. \qquad (2.1.3)$$

Hereafter, if it is clear which function is taken as the weight function, the phrase "with respect to the weight function $p(x)$" will be omitted.

We introduce the notation

$$c_m = \int_a^b p(x)x^m dx \qquad (m = 0, 1, 2, \ldots),$$

and let us consider the determinant

$$\Delta_n = \begin{vmatrix} c_0 & c_1 & \cdots & c_n \\ c_1 & c_2 & \cdots & c_{n+1} \\ \hdotsfor{4} \\ c_n & c_{n+1} & \cdots & c_{2n} \end{vmatrix}.$$

It is not difficult to see that Δ_n is different from zero. For this purpose we construct the homogeneous system of $n + 1$ equations in the $n + 1$ unknowns $a_0, a_1, \ldots, a_n$

$$\begin{aligned} a_0 c_0 + a_1 c_1 &+ \cdots + a_n c_n &= 0 \\ a_0 c_1 + a_1 c_2 &+ \cdots + a_n c_{n+1} &= 0 \\ &\hdotsfor{2} \\ a_0 c_n + a_1 c_{n+1} &+ \cdots + a_n c_{2n} &= 0 \end{aligned} \qquad (2.1.4)$$

If it were true that $\Delta_n = 0$, then this system would have nontrivial solutions, which we can show is impossible. Indeed, if we substitute in (2.1.4), the integrals which the c_m represent then the system becomes

$$\int_a^b p(x)[a_0 + a_1 x + \cdots + a_n x^n]dx = 0$$

$$\int_a^b p(x)x[a_0 + a_1 x + \cdots + a_n x^n]dx = 0$$

· ·

$$\int_a^b p(x) x^n [a_0 + a_1 x + \cdots + a_n x^n] dx = 0$$

Multiplying these equations respectively by $a_0, a_1, \ldots, a_n$ and adding we obtain

$$\int_a^b p(x) [a_0 + a_1 x + \cdots + a_n x^n]^2 dx = 0$$

which is possible only if the polynomial $a_0 + a_1 x + \cdots + a_n x^n$ is identically zero and consequently only if all of its coefficients $a_0, a_1, \ldots, a_n$ are zero. Therefore the system (2.1.4) can have only the trivial solution, and $\Delta_n \neq 0$.

Let n be any positive integer. In order to solve one of the problems in the theory of approximate integration it will be necessary to construct a polynomial of degree n:

$$P_n(x) = a_0 + a_1 x + \cdots + a_n x^n, \qquad a_n \neq 0, \qquad (2.1.5)$$

which will be orthogonal on $[a, b]$ to all polynomials of degree $< n$. This is the same as requiring $P_n(x)$ to satisfy the conditions

$$\int_a^b p(x) P_n(x) x^m dx = 0 \qquad (m = 0, 1, \ldots, n-1). \qquad (2.1.6)$$

The coefficients a_k are determined by the linear system of n equations in $n + 1$ unknowns:

$$
\begin{aligned}
a_0 c_0 \;\; + a_1 c_1 + \cdots + a_{n-1} c_{n-1} \;\; + a_n c_n \;\;\;\;\; &= 0 \\
a_0 c_1 \;\; + a_1 c_2 + \cdots + a_{n-1} c_n \;\;\;\;\; + a_n c_{n+1} &= 0 \\
\cdots\cdots\cdots\cdots\cdots\cdots\cdots\cdots\cdots\cdots\cdots\cdots& \\
a_0 c_{n-1} + a_1 c_n + \cdots + a_{n-1} c_{2n-2} + a_n c_{2n-1} &= 0.
\end{aligned}
\qquad (2.1.7)
$$

This is a homogeneous system, and since the number of equations is less than the number of unknowns it will have a nontrivial solution. This is true even if $p(x)$ changes sign on the interval of integration. However, without some additional assumption about $p(x)$ it is impossible to make any definite statement about the number of linearly independent solutions of the system or about the degree of the polynomial (2.1.5).

The determinant of the coefficients of $a_0, a_1, \ldots, a_{n-1}$ is Δ_{n-1}. If $p(x)$ is nonnegative then $\Delta_{n-1} \neq 0$. If a_n is fixed then the system will have a unique solution $a_0, a_1, \ldots, a_{n-1}$. The orthogonality conditions (2.1.6) determine $P_n(x)$ to within a constant factor; we will choose this factor so that

$$a_n > 0 \quad \text{and} \quad \int_a^b p(x) P_n^2(x)\, dx = 1.$$

We can prove the following theorem about the roots of $P_n(x)$.

Theorem 1. *If the polynomial $P_n(x)$ is orthogonal on the segment $[a, b]$ to all polynomials of degree less than n, with respect to the nonnegative weight function $p(x)$, then all the roots of $P_n(x)$ are real and distinct and lie inside $[a, b]$.*

Proof. Let us consider the roots of $P_n(x)$ which lie inside $[a, b]$ and which have odd multiplicities to be

$$\xi_1, \xi_2, \ldots, \xi_m.$$

To establish the theorem it suffices to prove that the number of such roots m is not less than n.

Let us assume the contrary: $m < n$. We can show that this is inconsistent with the orthogonality assumption. We construct the polynomial of degree m

$$Q_m(x) = (x - \xi_1)(x - \xi_2) \cdots (x - \xi_m).$$

$Q_m(x)$ changes sign at the same points inside $[a, b]$ as does $P_n(x)$. The product $P_n(x) Q_m(x)$ does not change sign inside $[a, b]$ and therefore the integral $\int_a^b p(x) P_n(x) Q_m(x)\, dx$ is different from zero. Because $Q_m(x)$ has degree $< n$ this contradicts the assumption that $P_n(x)$ is orthogonal to each polynomial of degree less than n. This proves the theorem.

The system of polynomials

$$P_0(x), P_1(x), \ldots, P_n(x), \ldots \tag{2.1.8}$$

is called an *orthogonal* and *normalized* system, or, for short, an *orthonormal* system, if it satisfies the requirements:

1. $P_n(x)$ is a polynomial of degree n.

2. $\displaystyle \int_a^b p(x) P_n(x) P_m(x)\, dx = \begin{cases} 0 & \text{for } m \neq n \\ 1 & \text{for } m = n. \end{cases}$

We will write the nth degree polynomial of an orthonormal system in the form

$$P_n(x) = a_n x^n + b_n x^{n-1} + \cdots \tag{2.1.9}$$

We now prove that three consecutive polynomials of an orthonormal system satisfy a recursion relation

$$xP_n(x) = \frac{a_n}{a_{n+1}} P_{n+1}(x) + \left(\frac{b_n}{a_n} - \frac{b_{n+1}}{a_{n+1}}\right) P_n(x) + \frac{a_{n-1}}{a_n} P_{n-1}(x). \qquad (2.1.10)$$

In fact, $xP_n(x)$ is a polynomial of degree $n + 1$ and can be represented in the form

$$xP_n(x) = \sum_{k=0}^{n+1} c_{n,k} P_k(x).$$

The coefficients $c_{n,k}$ are the Fourier coefficients:

$$c_{n,k} = \int_a^b p(x) xP_n(x) P_k(x) \, dx.$$

If $k < n - 1$ then $xP_k(x)$ is a polynomial of degree $k + 1 < n$ and $c_{n,k} = 0$ because $P_n(x)$ is orthogonal to each polynomial of degree less than n,

$$xP_n(x) = c_{n,n+1} P_{n+1}(x) + c_{n,n} P_n(x) + c_{n,n-1} P_{n-1}(x).$$

Let us substitute for $P_s(x)$ ($s = n - 1, n, n + 1$) its representation (2.1.9). Comparing the coefficients of the highest degree gives $c_{n,n+1} = \dfrac{a_n}{a_{n+1}}$. Since for any n and k we have the relation $c_{n,k} = c_{k,n}$ then we also have $c_{n,n-1} = \dfrac{a_{n-1}}{a_n}$. To obtain $c_{n,n}$ we can compare the coefficients of x^n; this gives

$$c_{n,n} = \frac{b_n}{a_n} - \frac{b_{n+1}}{a_{n+1}}.$$

This establishes (2.1.10) for $n = 1, 2, \ldots$, but that equation is also valid for $n = 0$ if we assume $a_{-1} = 0$ and $P_{-1}(x) \equiv 0$.

To calculate the coefficients in certain approximate integration formulas the Christoffel-Darboux relationship will be useful. To establish this relationship let us, at first, multiply the recursion relation (2.1.10) by $P_n(t)$.

$$xP_n(x) P_n(t) = \frac{a_n}{a_{n+1}} P_{n+1}(x) P_n(t) +$$

$$+ \left(\frac{b_n}{a_n} - \frac{b_{n+1}}{a_{n+1}}\right) P_n(x) P_n(t) + \frac{a_{n-1}}{a_n} P_{n-1}(x) P_n(t)$$

Let us form an equation similar to this by interchanging x and t in the above and then subtracting the resulting equation from the above. The middle terms will cancel and we will have

$$(x - t) P_n(x) P_n(t) = \frac{a_n}{a_{n+1}} [P_{n+1}(x) P_n(t) - P_n(x) P_{n+1}(t)] -$$

$$- \frac{a_{n-1}}{a_n} [P_n(x) P_{n-1}(t) - P_{n-1}(x) P_n(t)].$$

Let us write equations similar to the previous by replacing n in turn by $n - 1$, $n - 2, \ldots, 0$. If we add all of the resulting equations we obtain the Christoffel-Darboux identity

$$(x - t) \sum_{k=0}^{n} P_k(x) P_k(t) = \frac{a_n}{a_{n+1}} [P_{n+1}(x) P_n(t) - P_n(x) P_{n+1}(t)].$$

2.2. JACOBI AND LEGENDRE POLYNOMIALS

Jacobi polynomials are polynomials which form an orthogonal system on the segment $[-1, +1]$ with respect to the weight function $p(x) = (1 - x)^a (1 + x)^\beta$. They depend on two parameters a and β, and for any values of these parameters we can determine the function

$$P_n^{(a, \beta)}(x) =$$

$$= \frac{(-1)^n}{2^n n!} (1 - x)^{-a} (1 + x)^{-\beta} \frac{d^n}{dx^n} [(1 - x)^{a+n} (1 + x)^{\beta+n}]. \quad (2.2.1)$$

This equation is called the *Rodriguez formula for the Jacobi polynomial.* Usually we take those branches of this many-valued function for which

$$\arg(1 - x) = \arg(1 + x) = 0 \qquad \text{for } -1 < x < + 1.$$

Then (2.2.1) is a polynomial of degree not greater than n:

$$P_n^{(a, \beta)}(x) = A_n x^n + B_n x^{n-1} + \cdots.$$

This can be seen by differentiating $\dfrac{d^n}{dx^n} [(1 - x)^{a+n} (1 + x)^{\beta+n}]$ by the rule of Leibnitz and substituting the result in (2.2.1),

$$P_n^{(a, \beta)}(x) = \frac{(-1)^n}{2^n n!} \sum_{k=0}^{n} \frac{n!}{k! (n - k)!} \times$$

$$\times (a + n) \cdots (a + n - k + 1) (-1)^k (1 - x)^{n-k} \times$$

$$\times (\beta + n) \cdots (\beta + k + 1) (1 + x)^k.$$

The coefficient A_n, of the highest order term x^n, can be found if we take the highest order terms from the factors $(1 - x)^{n-k}$ and $(1 + x)^k$; these terms are respectively $(-1)^{n-k} x^{n-k}$ and x^k:

$$A_n x^n = \frac{1}{2^n n!} \sum_{k=0}^{n} \frac{n!}{k!(n-k)!} \times$$

$$\times (\alpha + n) \cdots (\alpha + n - k + 1)x^{n-k}(\beta + n) \cdots (\beta + k + 1)x^k.$$

The same result is obtained if we apply the rule of Leibnitz to calculate the derivative of order n in the function

$$\frac{1}{2^n n!} x^{-\alpha} x^{-\beta} \frac{d^n}{dx^n} (x^{\alpha+n} x^{\beta+n}) = \frac{1}{2^n n!} x^{-\alpha-\beta} \frac{d^n}{dx^n} (x^{\alpha+\beta+2n}).$$

Therefore

$$A_n = \frac{1}{2^n n!} (\alpha + \beta + 2n)(\alpha + \beta + 2n - 1) \cdots (\alpha + \beta + n + 1) =$$

$$= \frac{\Gamma(\alpha + \beta + 2n + 1)}{2^n n! \Gamma(\alpha + \beta + n + 1)}. \tag{2.2.2}$$

We will consider the parameters α, β to be real and α, $\beta > -1$[2] and show that the Jacobi polynomials $P_n^{(\alpha, \beta)}(x)$ $(n = 0, 1, 2, \ldots)$ form an orthogonal system on the segment $[-1, +1]$ with respect to the weight function $p(x) = (1-x)^{\alpha}(1+x)^{\beta}$:

$$I_{n,m} = \int_{-1}^{+1} (1-x)^{\alpha}(1+x)^{\beta} P_n^{(\alpha, \beta)}(x) P_m^{(\alpha,\beta)}(x)dx = 0 \tag{2.2.3}$$

For convenience we write

$$y_n = \frac{(-1)^n}{2^n n!} (1-x)^{\alpha+n}(1+x)^{\beta+n}.$$

Then

$$P_n^{(\alpha, \beta)}(x) = (1-x)^{-\alpha}(1+x)^{-\beta} y_n^{(n)}.$$

Let us assume $m \le n$ and substitute in $I_{n,m}$ the expression for $P_n^{(\alpha, \beta)}(x)$ in terms of y_n:

$$I_{n,m} = \int_{-1}^{+1} y_n^{(n)} P_m^{(\alpha, \beta)}(x)dx.$$

Integrating by parts gives

---

[2]To construct quadrature formulas for the integration of analytic functions of a complex variable it is necessary to take $\operatorname{Re}\alpha$, $\operatorname{Re}\beta > -1$.

$$I_{n,m} = y_n^{(n-1)} P_m^{(\alpha,\beta)}(x) \Big|_{-1}^{+1} - \int_{-1}^{+1} y_n^{(n-1)} [P_m^{(\alpha,\beta)}(x)]' dx =$$

$$= -\int_{-1}^{+1} y_n^{(n-1)} [P_m^{(\alpha,\beta)}(x)]' dx.$$

The term which does not involve the integral vanishes because $\alpha, \ \beta > -1$. Integrating by parts n times gives

$$I_{n,m} = (-1)^n \int_{-1}^{+1} y_n [P_m^{(\alpha,\beta)}(x)]^{(n)} dx. \qquad (2.2.4)$$

For $m < n$ we have $[P_m^{(\alpha,\beta)}(x)]^{(n)} \equiv 0$, and consequently $I_{n,m} = 0$, which proves orthogonality for two Jacobi polynomials of different degrees.

For $m = n$ equation $(2.2.3)$ gives

$$I_{n,n} = \int_{-1}^{+1} (1-x)^\alpha (1+x)^\beta [P_n^{(\alpha,\beta)}(x)]^2 dx =$$

$$= (-1)^n \int_{-1}^{+1} y_n \, n! A_n \, dx = \frac{n!}{2^n n!} A_n \int_{-1}^{+1} (1-x)^{\alpha+n}(1+x)^{\beta+n} dx.$$

The last integral reduces to the Euler integral of the first kind. Let us substitute $x = 2t - 1$:

$$\int_{-1}^{+1} (1-x)^{\alpha+n}(1+x)^{\beta+n} dx = 2^{\alpha+\beta+2n+1} \int_0^1 t^{\beta+n}(1-t)^{\alpha+n} dt =$$

$$= 2^{\alpha+\beta+2n+1} B(\alpha+n+1, \ \beta+n+1).$$

Since

$$B(p,q) = \frac{\Gamma(p)\Gamma(q)}{\Gamma(p+q)}$$

then

$$I_{n,n} = \frac{2^{\alpha+\beta+1} \Gamma(\alpha+n+1)\Gamma(\beta+n+1)}{(\alpha+\beta+2n+1)\, n! \, \Gamma(\alpha+\beta+n+1)}. \qquad (2.2.5)$$

If $n = 0$ and $\alpha + \beta + 1 = 0$, then

$$I_{0,0} = \Gamma(\alpha+1)\Gamma(\beta+1).$$

From $(2.2.4)$ and $(2.2.5)$ we see that an orthonormal system of polynomials on $[-1, +1]$ with respect to the weight function $(1-x)^\alpha (1+x)^\beta$ is given by

$$p_n^{(\alpha, \beta)}(x) = \frac{1}{\sqrt{I_{n,n}}} P_n^{(\alpha, \beta)}(x) \tag{2.2.6}$$

The leading coefficients of these are

$$a_n = \frac{1}{\sqrt{I_{n,n}}} A_n \tag{2.2.7}$$

Legendre polynomials are a special case of the Jacobi polynomials. They are the Jacobi polynomials for $\alpha = 0$, $\beta = 0$. The Rodriguez formula for Legendre polynomials is

$$P_n(x) = \frac{1}{2^n n!} \frac{d^n}{dx^n} (x^2 - 1)^n. \tag{2.2.8}$$

From this equation it is easy to find the expansion for $P_n(x)$ in powers of x

$$P_n(x) = \frac{(2n)!}{2^n (n!)^2} x^n - \frac{(2n-2)!}{2^n (n-1)!(n-2)!} x^{n-2} + \cdots.$$

The Legendre polynomials are orthogonal on $[-1, +1]$ with respect to the constant weight function $p(x) \equiv 1$. Equations $(2.2.4)$ and $(2.2.5)$ have the form:

$$\int_{-1}^{+1} P_n(x) P_m(x) dx = \begin{cases} 0 & \text{for } m \neq n \\ \dfrac{2}{2n+1} & \text{for } m = n. \end{cases} \tag{2.2.9}$$

An orthonormal system on $[-1, +1]$ with constant weight function is given by the polynomials

$$p_n(x) = \sqrt{\frac{2n+1}{2}} P_n(x). \tag{2.2.10}$$

The leading coefficient in $p_n(x)$ is

$$a_n = \sqrt{\frac{2n+1}{2}} \frac{(2n)!}{2^n (n!)^2}. \tag{2.2.11}$$

2.3. CHEBYSHEV POLYNOMIALS

The Chebyshev polynomials of the first kind can be defined by

$$T_n(x) = \cos(n \text{ arc cos } x) \qquad (n = 0, 1, 2, \ldots). \tag{2.3.1}$$

These polynomials are an orthogonal system on the segment $[-1, +1]$ with respect to the weight function $p(x) = \dfrac{1}{\sqrt{1 - x^2}}$. First of all, let us show that $T_n(x)$ is indeed a polynomial of degree n in x and that the coefficient of the highest degree term is 2^{n-1}:

$$T_n(x) = 2^{n-1}x^n + \cdots \tag{2.3.2}$$

We use the elementary trigonometric identity

$$\cos(n + 1)\theta + \cos(n - 1)\theta = 2\cos\theta\cos n\theta.$$

If we put $\theta = \arccos x$, we obtain the following recursion relation for $T_n(x)$:

$$T_{n+1}(x) = 2xT_n(x) - T_{n-1}(x).$$

It is evident that equation (2.3.2) is valid for $T_0(x) = 1$ and $T_1(x) = x$. By the recursion relation we can see that it is true for all n.

We will now establish the orthogonality property for the polynomials $T_n(x)$:

$$I_{n,m} = \int_{-1}^{1} \frac{T_n(x)T_m(x)}{\sqrt{1 - x^2}}\, dx = \begin{cases} 0 & \text{for } m \neq n \\ \pi/2 & \text{for } m = n. \end{cases} \tag{2.3.3}$$

This is equivalent to the statement that the polynomials $T_n(x)$ ($n = 0$, 1, 2, ...) form an orthogonal system on the segment $[-1, +1]$ with respect to the weight function $p(x) = \dfrac{1}{\sqrt{1 - x^2}}$.

Let us change the variable of integration in $I_{n,m}$ by substituting $x = \cos\theta$, $\theta = \arccos x$. As x varies from -1 to $+1$ we can take θ to vary from π to 0. Since $T_n(x) = \cos n\theta$, $T_m(x) = \cos m\theta$ and $dx = -\sin\theta\, d\theta$, then

$$I_{n,m} = \int_{0}^{\pi} \cos n\theta \cos m\theta\, d\theta = \begin{cases} 0 & \text{for } m \neq n \\ \pi/2 & \text{for } m = n, \end{cases}$$

which establishes (2.3.3).

The weight function $p(x) = (1 - x^2)^{-\frac{1}{2}}$ ($-1 \leq x \leq +1$) is a special case of the Jacobi weight function $(1 - x)^{\alpha}(1 + x)^{\beta}$ for $\alpha = \beta = -1/2$. For a given weight function the polynomials of the corresponding orthogonal system are defined to within a constant factor. Therefore the Jacobi polynomials $P_n^{\left(-\frac{1}{2}, -\frac{1}{2}\right)}(x)$ can differ from $T_n(x)$ by only a constant factor

$$P_n^{\left(-\frac{1}{2}, -\frac{1}{2}\right)}(x) = c_n T_n(x). \tag{2.3.4}$$

In order to find c_n it is sufficient to compare the leading coefficients

$$\frac{\Gamma(2n)}{2^n n! \Gamma(n)} = c_n 2^{n-1}$$

$$c_n = \frac{\Gamma(2n)}{2^{2n-1} \Gamma(n) \Gamma(n+1)}.$$

The polynomials $T_n(x)$ were introduced by P. L. Chebyshev in connection with the solution of the following problem:

Among all the polynomials of degree n which have leading coefficient equal to unity

$$P(x) = x^n + c_{n-1} x^{n-1} + \cdots$$

determine those which deviate least from zero in absolute value on the segment $[-1, +1]$. That is, determine the polynomials for which

$$\max_{-1 \leq x \leq 1} |P(x)|$$

has the least possible value.

We will show that the polynomials

$$T_n^*(x) = 2^{-n+1} T_n(x) = 2^{-n+1} \cos(n \text{ arc cos } x)$$

have this property. Indeed, $\max_{[-1, +1]} |T_n^*(x)| = 2^{-n+1}$ and we also have

$$T_n^*\left(\cos \frac{m\pi}{n}\right) = 2^{-n+1}(-1)^m \qquad (m = 0, 1, \ldots, n).$$

If there would be a polynomial $P(x)$ which would satisfy the condition $|P(x)| < 2^{-n+1}$ ($-1 \leq x \leq +1$), then the difference $R(x) = T_n^*(x) - P(x)$ would be a polynomial of degree less than n, for which $(-1)^m R\left(\cos \frac{m\pi}{n}\right) > 0$ ($m = 0, 1, \ldots, n$). The polynomial $R(x)$ would then have at least n roots in the interval $[-1, +1]$ which is impossible, because its degree is less than n.

A similar argument establishes the uniqueness of the polynomials of least deviation. Let $P(x)$ be an arbitrary polynomial of the indicated form for which $\max_{[-1, +1]} |P(x)| = \max_{[-1, +1]} |T_n^*(x)| = 2^{-n+1}$. The difference $S(x) = P(x) - T_n^*(x)$ will have degree less than n. At the points

$$x_m = \cos \frac{m\pi}{n}$$

$$S(x_m) = (-1)^m 2^{-m+1} - P(x_m)$$

and since $|P(x_m)| \leq 2^{-n+1}$,

$$(-1)^m S(x_m) \geq 0 \qquad (m = 0, 1, \ldots, n).$$

Hence it follows that $S(x)$ has no fewer than n zeros, either distinct or coincident. But because the degree of $S(x)$ is less than n then $S(x)$ is identically zero and $P(x) = T_n^*(x)$.

The Chebyshev polynomials of the second kind are defined as the polynomials

$$U_n(x) = \frac{\sin[(n+1) \text{ arc cos } x]}{\sqrt{1-x^2}} \qquad (n = 0, 1, 2, \ldots). \qquad (2.3.5)$$

It is possible to show that the functions $U_n(x)$ are indeed polynomials of degree n, having leading coefficient 2^n. To do this we use the trigonometric identity

$$\sin(n+2)\theta + \sin n\theta = 2\cos \theta \sin(n+1)\theta.$$

If we put $\cos \theta = x$, $\theta = $ arc cos x and divide both sides by $\sqrt{1-x^2}$, then we obtain the recursion formula for $U_n(x)$

$$U_{n+1}(x) = 2xU_n(x) - U_{n-1}(x). \qquad (2.3.6)$$

We note that $U_0(x) = 1$ and $U_1(x) = 2x$ have the indicated form. By means of induction it is easy to show from (2.3.6) that $U_n(x)$ is indeed a polynomial of the form $U_n(x) = 2^n x^n + \cdots$.

The polynomials $U_n(x)$ satisfy the relationship

$$I_{n,m} = \int_{-1}^{1} U_n(x)U_m(x)\sqrt{1-x^2}\, dx = \begin{cases} 0 & \text{for } m \neq n \\ \pi/2 & \text{for } m = n. \end{cases} \qquad (2.3.7)$$

In other words, the $U_n(x)$ ($n = 0, 1, 2, \ldots$) form an orthogonal system on the segment $[-1, +1]$ with respect to the weight function $p(x) = \sqrt{1-x^2}$. To prove this we change the variable of integration in the integral

$$I_{n,m} = \int_{-1}^{+1} \frac{\sin[(n+1) \text{ arc cos } x] \sin[(m+1) \text{ arc cos } x]}{\sqrt{1-x^2}}\, dx$$

by substituting $x = \cos \theta$; then it changes to the form

$$I_{n,m} = \int_0^{\pi} \sin(n+1)\theta \sin(m+1)\theta\, d\theta$$

and equation (2.3.7) is verified without difficulty.

The weight function $p(x) = \sqrt{1-x^2}$ is also a Jacobi weight function for $\alpha = \beta = 1/2$. Therefore the polynomials $U_n(x)$ can only differ by a constant factor from the Jacobi polynomials $P_n^{\left(\frac{1}{2}, \frac{1}{2}\right)}(x)$

$$P_n^{\left(\frac{1}{2}, \frac{1}{2}\right)}(x) = e_n U_n(x).$$

Comparison of the leading coefficients gives

$$e_n = \frac{(2n+1)!}{2^{2n} n! (n+1)!}.$$

The polynomials $U_n(x)$ possess the following minimal property:

Among all polynomials $P(x)$ of degree n with leading coefficient equal to unity, $2^{-n} U_n(x)$ minimizes the value of the integral

$$\int_{-1}^{1} |P(x)| \, dx. \tag{2.3.8}$$

In order to prove this it will be necessary to establish certain auxiliary results.

1. We will need a trignometric series for the function[3] $\sin x \,\text{sign}\,\sin px$, where p is an integer. In the theory of Fourier series the following expansion is known

$$\text{sign}\,\sin x = \frac{4}{\pi} \sum_{k=0}^{\infty} \frac{\sin(2k+1)x}{2k+1}.$$

Hence we see that

$$\text{sign}\,\sin px = \frac{4}{\pi} \sum_{k=0}^{\infty} \frac{\sin(2k+1)px}{2k+1}. \tag{2.3.9}$$

If this equation is multiplied by $\sin x$, then using the relation

$$2 \sin x \sin(2k+1)px = \cos[(2k+1)p - 1]x - \cos[(2k+1)p + 1]x$$

we immediately obtain the desired trigonometric series

$$\sin x \,\text{sign}\,\sin px =$$

$$= \frac{2}{\pi} \sum_{k=0}^{\infty} (2k+1)^{-1} \{\cos[(2k+1)p - 1]x - \cos[(2k+1)p + 1]x\}.$$

---

[3] The function $\text{sign}\,x$ is defined by

$$\text{sign}\,x = \begin{cases} -1 & \text{for } x < 0 \\ 0 & \text{for } x = 0 \\ +1 & \text{for } x > 0 \end{cases}$$

2. If n is a positive integer then for $r = 0, 1, \ldots, n - 1$ the following equation is satisfied:

$$\int_{-1}^{+1} x^r \operatorname{sign} U_n(x) \, dx = 0. \qquad (2.3.10)$$

If we substitute $x = \cos \theta$ in (2.3.10) we obtain

$$\int_0^\pi \cos^r \theta \, \sin \theta \, \operatorname{sign} \sin (n + 1) \, \theta \, d\theta = 0.$$

The powers $\cos^r \theta$ $(r = 0, 1, \ldots, n - 1)$ can be linearly expressed in terms of $\cos m\theta$ $(m = 0, 1, \ldots, n - 1)$ and conversely. Therefore the last equation is equivalent to

$$\int_0^\pi \cos m\theta \, \sin \theta \, \operatorname{sign} \sin (n + 1)\theta \, d\theta = 0 \quad (m = 0, 1, \ldots, n - 1).$$

Because the function under the integral sign is even, this is equivalent to

$$\int_{-\pi}^\pi \cos m\theta \, \sin \theta \, \operatorname{sign} \sin (n + 1)\theta \, d\theta = 0. \qquad (2.3.11)$$

The trigonometric series for $\sin \theta \, \operatorname{sign} \sin (n + 1)\theta$ is given by (2.3.9) for $p = n + 1$.

The smallest frequency in the terms of the series (2.3.9) is in the term corresponding to $k = 0$; this frequency is $(n + 1) - 1 = n$. Therefore, for $m = 0, 1, \ldots, n - 1$, equation (2.3.11) is known to be satisfied.

Using (2.3.10) it is easy to prove the above stated minimal property for $U_n(x)$. For simplicity we denote $2^{-n} U_n(x) = P(x)$ and let us take any polynomial $P^*(x)$ of degree n which has leading coefficient equal to unity:

$$\int_{-1}^{+1} |P(x)| \, dx = \int_{-1}^{+1} P(x) \operatorname{sign} U_n(x) \, dx =$$

$$= \int_{-1}^{+1} P^*(x) \operatorname{sign} U_n(x) \, dx +$$

$$+ \int_{-1}^{+1} [P(x) - P^*(x)] \operatorname{sign} U_n(x) \, dx.$$

The last of these integrals is equal to zero by (2.3.10) and by the fact that the difference $P(x) - P^*(x)$ is a polynomial of degree less than n.

Also

$$\int_{-1}^{+1} P^*(x) \operatorname{sign} U_n(x)\, dx \le \int_{-1}^{+1} P^*(x) \operatorname{sign} P^*(x)\, dx = \int_{-1}^{+1} |P^*(x)|\, dx.$$

Consequently

$$\int_{-1}^{+1} |P(x)|\, dx \le \int_{-1}^{+1} |P^*(x)|\, dx. \qquad (2.3.12)$$

This proves the assertion. We make two more remarks. From the above argument we see that equality is possible in (2.3.12) only when

$$\operatorname{sign} P^*(x) = \operatorname{sign} U_n(x) \qquad \text{for } -1 < x < 1.$$

The polynomials

$$U_n(x) = \frac{\sin\left[(n+1)\,\arccos x\right]}{\sqrt{1-x^2}} = \frac{\sin(n+1)\,\theta}{\sin\theta}$$

have n roots $x_k = \cos\dfrac{k\pi}{n+1}$ $(k = 1,\, 2,\, \ldots,\, n)$ in the interval $-1 < x < 1$.

If $\operatorname{sign} P^*(x) = \operatorname{sign} U_n(x)$, $-1 < x < +1$, then the points x_k must also be roots of $P^*(x)$. The polynomial $P^*(x)$ has degree n and therefore the x_k are roots of multiplicity one and $P^*(x)$ has no other roots. Since $P^*(x)$ and $P(x) = 2^{-n} U_n(x)$ have identical leading coefficients we must have

$$P^*(x) = P(x).$$

Equality in (2.3.12) is possible only when $P^*(x) = P(x) = 2^{-n} U_n(x)$.

Let us now calculate the minimal value of the integral (2.3.8):

$$2^{-n} \int_{-1}^{+1} |U_n(x)|\, dx = 2^{-n} \int_0^{\pi} |\sin(n+1)\,\theta|\, d\theta =$$

$$= 2^{-n}(n+1) \int_0^{\frac{\pi}{n+1}} \sin(n+1)\,\theta\, d\theta =$$

$$= -2^{-n}(n+1) \left[\frac{\cos(n+1)\,\theta}{n+1}\right]_0^{\frac{\pi}{n+1}} = 2^{-n+1}.$$

Thus we have proven the theorem:

Theorem 2. *For any polynomial of degree n which has leading coefficient equal to unity*

$$P(x) = x^n + c_{n-1}x^{n-1} + \cdots + c_0$$

the following inequality is valid:

$$\int_{-1}^{+1} |P(x)|\, dx \geq 2^{-n+1}.$$

Equality is possible only when

$$P(x) = 2^{-n}U_n(x).$$

2.4. CHEBYSHEV-HERMITE POLYNOMIALS

The Chebyshev-Hermite polynomials are orthogonal on the entire line $-\infty < x < \infty$ with respect to the weight function $p(x) = e^{-x^2}$. These polynomials can be defined by the formula[4]

$$H_n(x) = (-1)^n e^{x^2} \frac{d^n}{dx^n} e^{-x^2}. \tag{2.4.1}$$

Let us write $\phi = e^{-x^2}$. Then $\phi^{(n)} = (-1)^n e^{-x^2} H_n(x)$. Differentiating gives

$$\phi^{(n+1)} = (-1)^n [-2xH_n(x) + H_n'(x)]\, e^{-x^2},$$

and since $\phi^{(n+1)} = (-1)^{n+1} e^{-x^2} H_{n+1}(x)$, then

$$H_{n+1}(x) = 2xH_n(x) - H_n'(x). \tag{2.4.2}$$

Hence, from $H_0(x) = 1$, it is easy to obtain, by induction, that $H_n(x)$ is a polynomial of degree n of the form

$$H_n(x) = 2^n x^n + \cdots.$$

The polynomials $H_n(x)$ satisfy the following relationship:

$$\int_{-\infty}^{\infty} e^{-x^2} H_n(x) H_m(x)\, dx = \begin{cases} 0 & \text{for } m \neq n \\ 2^n \sqrt{\pi}\, n! & \text{for } m = n. \end{cases}$$

In other words the $H_n(x)$ $(n = 0, 1, 2, \ldots)$ form an orthogonal system on the entire line $(-\infty, +\infty)$ with respect to the weight function e^{-x^2}. To prove this let us suppose $m \leq n$:

---

[4]Sometimes other Chebyshev-Hermite polynomials are used:

$$H_n^*(x) = (-1)^n e^{\frac{1}{2}x^2} \frac{d^n}{dx^n} e^{-\frac{1}{2}x^2}.$$

These are related to the polynomials (2.4.1) by $H_n^*(x) = 2^{-\frac{1}{2}n} H_n\left(\frac{x}{\sqrt{2}}\right)$.

$$I = \int_{-\infty}^{\infty} e^{-x^2} H_n(x) H_m(x)\, dx = (-1)^n \int_{-\infty}^{\infty} \phi^{(n)} H_m(x)\, dx =$$

$$= (-1)^n\, \phi^{(n-1)} H_m(x) \Big|_{-\infty}^{\infty} + (-1)^{n-1} \int_{-\infty}^{\infty} \phi^{(n-1)} H_m'(x)\, dx =$$

$$= (-1)^{n-1} \int_{-\infty}^{\infty} \phi^{(n-1)} H_m'(x)\, dx = \cdots = \int_{-\infty}^{\infty} \phi H_m^{(n)}(x)\, dx.$$

For $m < n$, $H_m^{(n)}(x) \equiv 0$ and thus $I = 0$. If $m = n$ then

$$I = 2^n n! \int_{-\infty}^{\infty} \phi\, dx = 2^n n! \int_{-\infty}^{\infty} e^{-x^2}\, dx = 2^n n! \sqrt{\pi}.$$

An orthonormal system is formed by the polynomials

$$h_n(x) = \frac{H_n(x)}{2^{\frac{1}{2}n} (n!)^{\frac{1}{2}} \pi^{\frac{1}{4}}}. \tag{2.4.3}$$

The leading coefficients of these are

$$a_n = 2^{\frac{1}{2}n} (n!)^{-\frac{1}{2}} \pi^{-\frac{1}{4}}. \tag{2.4.4}$$

2.5. CHEBYSHEV-LAGUERRE POLYNOMIALS

The Chebyshev-Laguerre polynomials are orthogonal on the half-line $0 \le x < \infty$ with respect to the weight function $p(x) = x^\alpha e^{-x}$. Let α be any number. We choose the branch of the many-valued function x^α defined by the condition $\arg x = 0$, for $x > 0$. We can define the Chebyshev-Laguerre polynomials by the formula

$$L_n^{(\alpha)}(x) = (-1)^n x^{-\alpha} e^x \frac{d^n}{dx^n} (x^{\alpha+n} e^{-x}). \tag{2.5.1}$$

Differentiating by the rule of Leibnitz we find the expansion of $L_n^{(\alpha)}$ in powers of x to be

$$L_n^{(\alpha)}(x) = x^n - \frac{n}{1!}(n + \alpha) x^{n-1} +$$

$$+ \frac{n(n-1)}{2!}(n + \alpha)(n + \alpha - 1) x^{n-2} - \cdots. \tag{2.5.2}$$

We will consider α to be a real number $\alpha > -1$. We can show that $L_n^{(\alpha)}(x)$ possesses the following property:

$$I = \int_0^\infty x^a e^{-x} L_n^{(a)}(x) L_m^{(a)}(x)\, dx = \begin{cases} 0 & \text{for } m \neq n \\ n!\,\Gamma(n + a + 1) & \text{for } m = n. \end{cases} \quad (2.5.3)$$

Let us denote, for simplicity, $x^{a+n} e^{-x} = \phi_n$. Then

$$L_n^{(a)}(x) = (-1)^n x^{-a} e^x \phi_n^{(n)}.$$

Consider $m \leq n$ and substitute, in I, for the polynomial $L_n^{(a)}(x)$ its expression in terms of ϕ_n:

$$I = (-1)^n \int_0^\infty \phi_n^{(n)} L_m^{(a)}(x)\, dx =$$

$$= (-1)^n \phi_n^{(n-1)} L_m^{(a)}(x) \Big|_0^\infty + (-1)^{n-1} \int_0^\infty \phi_n^{(n-1)} [L_m^{(a)}(x)]'\, dx =$$

$$= (-1)^{n-1} \int_0^\infty \phi_n^{(n-1)} [L_m^{(a)}(x)]'\, dx.$$

The term which does not involve the integral vanishes because $a > -1$. Carrying out the integration by parts n times we obtain

$$I = \int_0^\infty \phi_n [L_m^{(a)}(x)]^{(n)}\, dx.$$

For $m < n$, we have $[L_m^{(a)}(x)]^{(n)} \equiv 0$ and therefore $I = 0$. When $m = n$,

$$I = n! \int_0^\infty \phi_n\, dx = n! \int_0^\infty x^{a+n} e^{-x}\, dx = n!\,\Gamma(a + n + 1).$$

The orthonormal Chebyshev-Laguerre polynomials are

$$l_n^{(a)}(x) = \frac{L_n^{(a)}(x)}{[n!\,\Gamma(a + n + 1)]^{\frac{1}{2}}}. \quad (2.5.4)$$

The coefficients of x^n in these are

$$a_n = [n!\,\Gamma(a + n + 1)]^{-\frac{1}{2}}. \quad (2.5.5)$$

REFERENCES

V. L. Goncharov, *Theory of Interpolation and Approximation of Functions*, Gostekhizdat, Moscow, 1954, Chap. 3, 4 (Russian).

D. Jackson, *Fourier Series and Orthogonal Polynomials*, Carus Monograph No. 6, Math. Assoc. of Amer., 1941, Chap. 2, 7–10.

A. N. Korkin and E. I. Zolotarev, "Sur un certain minimum." Collected
    works of E. I. Zolotarev, Vol. 1, pp. 138–153.

I. P. Natanson, *Constructive Theory of Functions*, Gostekhizdat, Mos-
    cow, 1949 (Russian). (There is a German translation of this book:
    Konstruktive Funktionentheorie, Berlin, 1955.)

G. Szego, *Orthogonal Polynomials*, Amer. Math. Soc. Colloquium Publ.,
    Vol. 23, 1959.

CHAPTER 3

Interpolation of Functions

3.1. FINITE DIFFERENCES AND DIVIDED DIFFERENCES

The theory of approximate integration uses in many ways results from the theory of interpolation which in turn makes wide use of finite differences. Here we develop only the simplest results from the theory of differences.

Suppose that we know the values of $f(x)$ at the following equally spaced points of interval h:

$$x_k = x_0 + kh \quad (k = 0, 1, 2, \ldots)$$

$$f_0 = f(x_0), \quad f_1 = f(x_0 + h), \ldots, \quad f_k = f(x_0 + kh), \ldots.$$

We call the quantities

$$\Delta f_0 = f_1 - f_0, \quad \Delta f_1 = f_2 - f_1, \quad \ldots, \quad \Delta f_n = f_{n+1} - f_n, \ldots$$

finite differences of the first order, and the quantities

$$\Delta^2 f_0 = \Delta f_1 - \Delta f_0, \quad \Delta^2 f_1 = \Delta f_2 - \Delta f_1, \ldots, \quad \Delta^2 f_n = \Delta f_{n+1} - \Delta f_n, \ldots$$

are called *differences of the second order*, and so forth.

Differences of order n are defined from differences of the preceding order by

$$\Delta^n f_0 = \Delta^{n-1} f_1 - \Delta^{n-1} f_0, \quad \Delta^n f_1 = \Delta^{n-1} f_2 - \Delta^{n-1} f_1, \ldots$$

This provides a recursive definition of finite differences of all orders. We can find an expression for differences of any order in terms of the values f_k of the function

$$\Delta^n f = f_n - \frac{n}{1!} f_{n-1} + \frac{n(n-1)}{2!} f_{n-2} -$$

$$- \frac{n(n-1)(n-2)}{3!} f_{n-3} + \cdots + (-1)^n f_0. \tag{3.1.1}$$

This equation is obviously true for $n = 1$, and it can easily be proved for any n by induction. If we introduce the operator which increases the argument by step h

$$Ef(x) = f(x + h) \quad \text{or} \quad Ef_k = f_{k+1}$$

then (3.1.1) can be written in the symbolic form

$$\Delta^n f_0 = (E - 1)^n f_0. \tag{3.1.2}$$

It is also useful to note that any value of the function f_n can be expressed in terms of f_0 and the differences Δf_0, $\Delta^2 f_0, \ldots$ by the relationship

$$f_n = f_0 + \frac{n}{1!} \Delta f_0 + \frac{n(n-1)}{2!} \Delta^2 f_0 + \cdots + \Delta^n f_0. \tag{3.1.3}$$

This equation is true for $n = 1$ since $f_1 - f_0 = \Delta f_0$, and it can be established for any n by induction. In symbolic form (3.1.3) is

$$f_n = (1 + \Delta)^n f_0. \tag{3.1.4}$$

In interpolation problems it is not always possible to use equally spaced values of the function. For example, one can not always obtain astronomical observations at equally spaced intervals of time.

For unequally spaced values of the argument finite differences are replaced by quantities which are usually called *divided differences* or *difference ratios*.

Let $x_0, x_1, x_2, \ldots, x_n, \ldots$ be arbitrary values of the argument. Divided differences of the first order are defined

$$f(x_0, x_1) = \frac{f(x_1) - f(x_0)}{x_1 - x_0}, \quad f(x_1, x_2) = \frac{f(x_2) - f(x_1)}{x_2 - x_1}, \ldots.$$

The quantities

$$f(x_0, x_1, x_2) = \frac{f(x_1, x_2) - f(x_0, x_1)}{x_2 - x_0}$$

$$f(x_1, x_2, x_3) = \frac{f(x_2, x_3) - f(x_1, x_2)}{x_3 - x_1}$$

are divided differences of the second order; and

$$f(x_0, x_1, x_2, x_3) = \frac{f(x_1, x_2, x_3) - f(x_0, x_1, x_2)}{x_3 - x_0}$$

is a divided difference of third order, and so forth.

The function $f(x_0, x_1, \ldots, x_n)$ is a linear function of $f(x_0), \ldots, f(x_n)$ and it can be shown that

$$f(x_0, \ldots, x_n) = \tag{3.1.5}$$
$$= \sum_{\nu=0}^{n} \frac{f(x_\nu)}{(x_\nu - x_0) \cdots (x_\nu - x_{\nu-1})(x_\nu - x_{\nu+1}) \cdots (x_\nu - x_n)}.$$

This equation is true for $n = 1$ since

$$f(x_0, x_1) = \frac{f(x_1) - f(x_0)}{x_1 - x_0} = \frac{f(x_0)}{x_0 - x_1} + \frac{f(x_1)}{x_1 - x_0}.$$

Assuming that (3.1.5) is true for divided differences of order n we can verify it for order $n + 1$:

$$f(x_0, x_1, \ldots, x_{n+1}) =$$
$$= (x_{n+1} - x_0)^{-1}[f(x_1, x_2, \ldots, x_{n+1}) - f(x_0, x_1, \ldots, x_n)] =$$
$$= (x_{n+1} - x_0)^{-1}\left[\sum_{\nu=1}^{n+1} \frac{f(x_\nu)}{(x_\nu - x_1) \cdots (x_\nu - x_{n+1})} - \right.$$
$$\left. - \sum_{\nu=0}^{n} \frac{f(x_\nu)}{(x_\nu - x_0) \cdots (x_\nu - x_n)} \right] =$$
$$= \sum_{\nu=0}^{n+1} \frac{f(x_\nu)}{(x_\nu - x_0) \cdots (x_\nu - x_{n+1})}.$$

Equation (3.1.5) can be written in a shorter form by introducing the polynomial

$$\omega(x) = (x - x_0)(x - x_1) \cdots (x - x_n).$$

Then

$$f(x_0, x_1, \ldots x_n) = \sum_{\nu=0}^{n} \frac{f(x_\nu)}{\omega'(x_\nu)}. \tag{3.1.6}$$

A permutation of $x_0, x_1, \ldots, x_n$ in the right side of (3.1.5) only changes the order of the summands, and therefore $f(x_0, x_1, \ldots, x_n)$ is a symmetric function of its arguments $x_0, x_1, \ldots, x_n$.

By means of induction we can also verify the following formula which expresses any value of the function $f(x_n)$ in terms of $f(x_0)$ and the divided differences $f(x_0, x_1, \ldots, x_k)$ ($k = 1, 2, \ldots, n$):

$$f(x_n) = f(x_0) + (x_n - x_0) f(x_0, x_1) +$$
$$+ (x_n - x_0)(x_n - x_1) f(x_0, x_1, x_2) + \cdots \tag{3.1.7}$$
$$\cdots + (x_n - x_0)(x_n - x_1) \cdots (x_n - x_{n-1}) f(x_0, x_1, \ldots, x_n).$$

When the values of the argument are equally spaced the divided differences can be simply expressed in terms of finite differences:

$$f(x_0, x_0 + h) = \frac{f(x_0 + h) - f(x_0)}{h} = \frac{\Delta f_0}{1! h}$$

$$f(x_0, x_0 + h, x_0 + 2h) = \frac{f(x_0 + h, x_0 + 2h) - f(x_0, x_0 + h)}{2h} = \frac{\Delta^2 f_0}{2! h^2} \tag{3.1.8}$$

$$f(x_0, x_0 + h, \ldots, x_0 + nh) = \frac{\Delta^n f_0}{n! h^n}.$$

It is often useful in applications to be able to relate finite differences and divided differences to derivatives. We assume that the points $x_0, x_1, \ldots, x_n$ lie in the segment $[a, b]$.

Theorem 1. *If $f(x)$ has a continuous derivative of order n on $[a, b]$ then the following equation is valid:*

$$f(x_0, \ldots, x_n) = \int_0^1 \int_0^{t_1} \cdots \int_0^{t_{n-1}} f^{(n)} \left(x_0 + \sum_{\nu=1}^n t_\nu (x_\nu - x_{\nu-1}) \right) \times \tag{3.1.9}$$
$$\times \, dt_n \cdots dt_2 dt_1$$

Proof: This equation is easily verified for $n = 1$:

$$\int_0^1 f'(x_0 + t_1(x_1 - x_0)) \, dt_1 = \frac{f(x_1) - f(x_0)}{x_1 - x_0} = f(x_0, x_1).$$

Assuming that (3.1.9) is true for divided differences of order $n - 1$ we can show that it is true for differences of order n. Denoting the integral on the right side of (3.1.9) by $I(x_0, x_1, \ldots, x_n)$, and carrying out the integration with respect to t_n gives:

$$I(x_0, x_1, \ldots, x_n) =$$
$$= \int_0^1 \int_0^{t_1} \cdots \int_0^{t_{n-2}} (x_n - x_{n-1})^{-1} \{ f^{(n-1)}(x_0 + t_1(x_1 - x_0) +$$
$$+ \cdots + t_{n-1}(x_n - x_{n-2})) - f^{(n-1)}(x_0 + t_1(x_1 - x_0) +$$
$$+ \cdots + t_{n-1}(x_{n-1} - x_{n-2})) \} \, dt_{n-1} \cdots dt_2 dt_1 =$$

$$= (x_n - x_{n-1})^{-1} \left[f(x_0, \ldots, x_{n-2}, x_n) - f(x_0, \ldots, x_{n-2}, x_{n-1}) \right] =$$

$$= f(x_{n-1}, x_0, x_1, \ldots, x_{n-2}, x_n) =$$

$$= f(x_0, x_1, \ldots, x_{n-2}, x_{n-1}, x_n).$$

This proves the theorem.

As a corollary to (3.1.9) we can obtain a simpler relationship between $f(x_0, x_1, \ldots, x_n)$ and $f^{(n)}(x)$. The region of integration in (3.1.9) is a simplex in the n-dimensional space $(t_1, t_2, \ldots, t_n)$. This is the simplex defined by the inequalities

$$0 \le t_n \le t_{n-1} \le \cdots \le t_1 \le 1. \tag{3.1.10}$$

The volume of this simplex is

$$\int_0^1 \int_0^{t_1} \cdots \int_0^{t_{n-1}} dt_n \cdots dt_2 dt_1 = \frac{1}{n!}.$$

Consider the quantity

$$\xi = x_0 + \sum_{\nu=1}^n t_\nu (x_\nu - x_{\nu-1}) =$$

$$= (1 - t_1) x_0 + (t_1 - t_2) x_1 + \cdots + (t_{n-1} - t_n) x_{n-1} + t_n x_n.$$

From (3.1.10) we see that the multipliers of all the x_k are nonnegative. Since the sum of these multipliers is unity, ξ is a weighted average of the abscissas x_k $(k = 0, 1, \ldots, n)$ and therefore ξ will certainly lie in the segment $[a, b]$. A point in the interior of the simplex (3.1.10) thus corresponds to an interior point ξ of $[a, b]$.

Applying the mean value theorem to the integral (3.1.9) gives:

Theorem 2. *If $f(x)$ has a continuous derivative of order n on $[a, b]$ then there exists an interior point ξ of $[a, b]$ for which*

$$f(x_0, x_1, \ldots, x_n) = \frac{f^{(n)}(\xi)}{n!}. \tag{3.1.11}$$

The relationship between finite differences and derivatives then follows from (3.1.8), (3.1.9), and (3.1.11):

$$\Delta^n f_0 = n! \, h^n \int_0^1 \int_0^{t_1} \cdots \int_0^{t_{n-1}} f^{(n)} \left(x + h \sum_{\nu=1}^n t_\nu \right) \times$$

$$\times \, dt_n \cdots dt_2 dt_1 = h^n f^{(n)}(\xi) \quad x_0 < \xi < x_0 + nh. \tag{3.1.12}$$

Thus if we divide the interval size h by λ, then the finite difference $\Delta^n f_0$ will be divided by about λ^n.

3.2. THE INTERPOLATING POLYNOMIAL AND ITS REMAINDER

Suppose that for $n + 1$ arbitrary points $x_0, x_1, \ldots, x_n$, which we will call the nodes (or points) of interpolation, we are given the values of the function $f(x_k)$. We wish to construct a polynomial of degree $\leq n$

$$P_n(x) = a_0 x^n + a_1 x^{n-1} + \cdots + a_n = \sum_{\nu=1}^{n} a_\nu x^{n-\nu} \qquad (3.2.1)$$

which has the same value as $f(x)$ at the nodes x_k:

$$P_n(x_k) = f(x_k) \quad (k = 0, 1, \ldots, n). \qquad (3.2.2)$$

To find the coefficients a_ν of this polynomial we must solve the system of $n + 1$ linear equations

$$\sum_{\nu=0}^{n} a_\nu x_k^{n-\nu} = f(x_k) \quad (k = 0, 1, \ldots, n).$$

The determinant of this system is the Vandermonde determinant

$$W(x_0, \ldots, x_n) = \begin{vmatrix} x_0^n & x_0^{n-1} & \cdots & 1 \\ x_1^n & x_1^{n-1} & \cdots & 1 \\ \cdots\cdots\cdots\cdots\cdots\cdots \\ x_n^n & x_n^{n-1} & \cdots & 1 \end{vmatrix}$$

which is different from zero since no two of the nodes x_k coincide. Therefore for any values $f(x_k)$ we can construct one and only one polynomial $P_n(x)$ which satisfies (3.2.2).

The polynomial $P_n(x)$ can be represented in different forms; the most convenient form depends on how it is to be used. Below we derive two of the most useful representations for $P_n(x)$.

From the nodes x_k we construct the auxiliary polynomial $\omega_k(x)$ defined by

$$\omega_k(x_i) = \begin{cases} 0 & \text{for } i \neq k \\ 1 & \text{for } i = k. \end{cases} \qquad (3.2.3)$$

It is easy to see that this polynomial can be written in the form

$$\omega_k(x) = \frac{(x - x_0) \cdots (x - x_{k-1})(x - x_{k+1}) \cdots (x - x_n)}{(x_k - x_0) \cdots (x_k - x_{k-1})(x_k - x_{k+1}) \cdots (x_k - x_n)}.$$

or in terms of the polynomial $\omega(x) = (x - x_0)(x - x_1) \cdots (x - x_n)$

$$\omega_k(x) = \frac{\omega(x)}{(x - x_k)\omega'(x_k)}. \qquad (3.2.4)$$

The polynomial $\omega_k(x)$ is called the *Lagrangian coefficient* corresponding to the node x_k.

The interpolating polynomial $P_n(x)$ can now be written in the form

$$P_n(x) = \sum_{k=0}^{n} \frac{\omega(x)}{(x - x_k)\omega'(x_k)} f(x_k) \qquad (3.2.5)$$

which is due to Lagrange.

Since the $\omega_k(x)$ are polynomials of degree n, then (3.2.5) is a polynomial of degree not greater than n. From (3.2.3) it is easy to see that (3.2.5) satisfies conditions (3.2.2).

In some cases the Lagrangian representation for $P_n(x)$ is inconvenient. It is often impossible to say beforehand how many nodes x_k will be necessary to achieve the desired precision in the interpolation. Suppose that for the number of nodes first chosen the required precision is not achieved. Then we must use one or more additional nodes. Introducing one more node completely changes all the terms in (3.2.5). It is desirable then to have a representation for $P_n(x)$ for which the previous calculations do not have to be repeated with the addition of one more node but for which it is only necessary to add one new term.

Newton's representation of the interpolating polynomial has this property:

$$P_n(x) = f(x_0) + (x - x_0)f(x_0, x_1) + (x - x_0)(x - x_1)f(x_0, x_1, x_2) + \cdots$$
$$\cdots + (x - x_0)(x - x_1) \cdots (x - x_{n-1})f(x_0, x_1, \ldots, x_n). \qquad (3.2.6)$$

The right side of (3.2.6) is a polynomial of degree not greater than n. From (3.1.7) we can see that it indeed satisfies conditions (3.2.2) since for $x = x_0$ the right side of (3.2.6) reduces to $f(x_0)$ and for $x = x_1$ it reduces to $f(x_0) + (x_1 - x_0)f(x_0, x_1) = f(x_1)$, and so forth.

We call the difference between $f(x)$ and the interpolating polynomial $P_n(x)$ the *remainder of the interpolation*:

$$R_n(x) = f(x) - P_n(x) = f(x) - \sum_{k=0}^{n} \frac{\omega(x)f(x_k)}{(x - x_k)\omega'(x_k)}. \qquad (3.2.7)$$

The remainder $R_n(x)$ depends on the properties of the function $f(x)$ and on the location of x and the nodes x_k. It can be expected that $R_n(x)$ will be smaller for functions $f(x)$ which are smoother, that is for functions with higher order derivatives. When $f(x)$ is analytic it can be expected that the further the singular points of $f(x)$ lie from x and the x_k the smaller $R_n(x)$ will be.

From the representation (3.2.7) for $R_n(x)$ it is difficult to see how the properties of $f(x)$ influence the remainder and it will be useful to have other representations from which we can more easily estimate $R_n(x)$ for different classes of functions. Many representations for $R_n(x)$ have been constructed[1]. Here we derive only two of the simplest.

1. Let the point x and the nodes x_k ($k = 0, 1, \ldots, n$) belong to the segment $[a, b]$.

Theorem 3. *If $f(x)$ has a continuous derivative of order $n + 1$ on $[a, b]$ then the remainder $R_n(x)$ of the interpolation can be written:*

$$R_n(x) = \omega(x) \int_0^1 \int_0^{t_1} \cdots \int_0^{t_n} \times$$

$$\times f^{(n+1)} \left(x + \sum_{\nu=0}^{n} t_{\nu+1}(x_\nu - x_{\nu-1}) \right) dt_{n+1} \cdots dt_2 dt_1 \tag{3.2.8}$$

where $x_{-1} = x$.

Proof. This theorem can be obtained as a corollary to Theorem 1. Consider the values $f(x_0), f(x_1), \ldots, f(x_n), f(x)$ of the function. From (3.1.7) we see that

$$f(x) = f(x_0) + (x - x_0)f(x_0, x_1) + \cdots +$$
$$+ (x - x_0)(x - x_1) \cdots (x - x_{n-1})f(x_0, x_1, \ldots, x_n) +$$
$$+ (x - x_0)(x - x_1) \cdots (x - x_n)f(x_0, x_1, \ldots, x_n, x).$$

The terms on the right side of this equation with the last term omitted is Newton's form for $P_n(x)$. Therefore the last term must be the remainder of the interpolation:

$$R_n(x) = \omega(x)f(x_0, x_1, \ldots, x_n, x) = \omega(x)f(x, x_0, x_1, \ldots, x_n). \tag{3.2.9}$$

This result combined with Theorem 1 gives (3.2.8).

If we apply the mean value theorem to the integral in (3.2.8) we obtain a simpler representation for $R_n(x)$.

Theorem 4. *If $f(x)$ has a continuous derivative of order $n + 1$ on $[a, b]$ then there exists an interior point ξ of $[a, b]$ for which*

$$R_n(x) = \frac{\omega(x)}{(n+1)!} f^{(n+1)}(\xi). \tag{3.2.10}$$

This is often called the Lagrange form for $R_n(x)$.

---

[1]See the references at the end of this chapter.

2. Suppose that $f(z)$ is an analytic function of the complex variable z and is holomorphic in a domain B which has in its interior the points x and x_k $(k = 0, 1, \ldots, n)$. For simplicity we assume that B is simply connected.

Theorem 5. *The remainder of the interpolation for $f(z)$ at the point x can be represented as the contour integral*

$$R_n(x) = \frac{\omega(x)}{2\pi i} \int_l \frac{f(z)}{\omega(z)(z-x)} \, dz \qquad (3.2.11)$$

where l is any simple closed curve inside B which encloses x and x_k $(k = 0, 1, \ldots, n)$.

This theorem is easily proved by verifying (3.2.7). The function $\dfrac{f(z)}{\omega(z)(z-x)}$ has simple poles at the points $z = x$, $z = x_k$ $(k = 0, 1, \ldots, n)$. Thus calculating the integral in (3.2.11) by residues we at once obtain (3.2.7).

3.3. INTERPOLATION WITH MULTIPLE NODES

We assume that we are given m distinct nodes $x_1, x_2, \ldots, x_m$ and that at the first node x_1 we are given the value of the function $f(x)$ and its derivatives up to order $a_1 - 1$

$$f(x_1), f'(x_1), \ldots, f^{(a_1-1)}(x_1).$$

At the second node x_2 we assume that we are given the value of $f(x)$ and its derivatives up to order $a_2 - 1$

$$f(x_2), f'(x_2), \ldots, f^{(a_2-1)}(x_2),$$

and so forth. The numbers $a_1, a_2, \ldots, a_m$ are called the multiplicities of the nodes $x_1, x_2, \ldots, x_m$.

Let $n + 1$ denote the number of conditions given about $f(x)$:

$$a_1 + a_2 + \cdots + a_m = n + 1.$$

We wish to construct a polynomial $P_n(f; x)$ of degree not greater than n which will satisfy the conditions [2]

$$P_n^{(i)}(f; x_k) = f^{(i)}(x_k), \ i = 0, 1, \ldots, a_k - 1; \quad k = 1, 2, \ldots, m. \quad (3.3.1)$$

That $P_n(f; x)$ will be unique can be proved from well-known theorems of algebra. Suppose that there exists two polynomials $P_n(f; x)$ which satisfy

---

[2] We use the convention $f^{(0)}(x) \equiv f(x)$.

conditions (3.3.1) and let $Q(x)$ be their difference. Then $Q(x)$ is a polynomial of degree not greater than n which satisfies the conditions

$$Q^{(i)}(x_k) = 0, \quad i = 0, 1, \ldots, a_k - 1; \; k = 1, 2, \ldots, m.$$

Thus each node x_k is a zero of $Q(x)$ of multiplicity not less than a_k. The sum of the multiplicities of these zeros will be not less than $a_1 + a_2 + \cdots + a_m = n + 1$. But it is known that the sum of the multiplicities of the zeros can exceed the degree of the polynomial only when $Q(x)$ is identically zero. This proves that $P_n(f; x)$ will be unique.

It is clear that the interpolating polynomial $P_n(f; x)$ can be written in the form

$$P_n(f; x) = \sum_{k=1}^{m} \sum_{i=0}^{a_k-1} L_{k,i}(x) f^{(i)}(x_k) \tag{3.3.2}$$

where the $L_{k,i}(x)$ are polynomials of degree $\leq n$. To construct these polynomials we assume at first that $f(x)$ is an analytic function.

Assume that $f(z)$ is a function of the complex variable z which is holomorphic in a certain domain B which contains in its interior the points x and x_k ($k = 1, \ldots, m$). As above, we again assume that B is simply connected. We take any simple closed curve l contained in B which encloses x and the x_k. Everywhere inside l the function $f(z)$ can be represented as a Cauchy integral

$$f(x) = \frac{1}{2\pi i} \int_l \frac{f(z)}{z - x} \, dz. \tag{3.3.3}$$

This equation permits us to investigate $f(z)$ by a study of the elementary function $\dfrac{1}{z - x}$ which is often called the Cauchy kernel.

Instead of studying the interpolating polynomial for $\dfrac{1}{z - x}$ it will be more convenient to study the remainder of the interpolation:

$$R_n\left(\frac{1}{z - x}; x\right) = \frac{1}{z - x} - P_n\left(\frac{1}{z - x}; x\right) =$$

$$= \frac{1}{z - x} - \sum_{k=1}^{m} \sum_{i=0}^{a_k-1} L_{k,i}(x) \frac{i!}{(z - x_k)^{i+1}}. \tag{3.3.4}$$

We consider (3.3.4) to be a function of the parameter z. This is a proper rational fraction for which (3.3.4) is the expansion in sums of simple fractions. We note that the point $z = x$ is a simple pole of $R_n\left(\dfrac{1}{z - x}; x\right)$ with residue equal to unity.

We now reduce the fraction on the right of (3.3.4) to a common denominator.

Setting

$$A(z) = \prod_{k=1}^{m} (z - x_k)^{\alpha_k}$$

we obtain a fractional representation for R_n of the form

$$R_n\left(\frac{1}{z-x}; x\right) = \frac{B(z, x)}{A(z)(z-x)}. \tag{3.3.5}$$

Since the fraction (3.3.5) is proper the numerator $B(z, x)$ is a polynomial in z of degree not greater than $n+1$. We can show that $B(z, x)$ is independent of z and equals $A(x)$. To do this we will find an expansion of (3.3.5) for values of z with large modulus. If $|z|$ is large then

$$\frac{1}{z-x} = \sum_{\nu=0}^{\infty} \frac{x}{z^{\nu+1}}.$$

Because the remainder operator R_n is linear we can see that

$$R_n\left(\frac{1}{z-x}; x\right) = \sum_{\nu=0}^{\infty} \frac{1}{z^{\nu+1}} R_n(x^\nu; x).$$

Here $R_n(x^\nu; x)$ is the remainder of the interpolation for x^ν. But for a polynomial of degree not greater than n the interpolation is exact and therefore

$$R_n(x^\nu; x) = 0, \quad \nu = 0, 1, \ldots, n$$

$$R_n\left(\frac{1}{z-x}; x\right) = \sum_{\nu=n+1}^{\infty} \frac{1}{z^{\nu+1}} R_n(x^\nu; x).$$

The highest degree of $1/z$ in the expansion (3.3.5) for large $|z|$ must be $\frac{1}{z^{n+2}}$. This means that the degree of the numerator $B(z, x)$ with respect to z must be $n+2$ lower than the degree of the denominator and therefore must not depend on z: $B(z, x) = B(x)$. At the pole $z = x$ the residue of (3.3.5) is unity and therefore $B(x) = A(x)$ and

$$R_n\left(\frac{1}{z-x}; x\right) = \frac{A(x)}{A(z)(z-x)}. \tag{3.3.6}$$

Now we multiply (3.3.4) by $\frac{f(z)}{2\pi i}$ and integrate around l. Using (3.3.3)

and (3.3.6) we obtain an expression for the remainder of the interpolation
for $f(z)$ at the point x of the form

$$R_n(f; x) = f(x) - \sum_{k=1}^{m} \sum_{i=1}^{a_k-1} L_{k,i}(x) f^{(i)}(x_k) =$$

$$= \frac{A(x)}{2\pi i} \int_l \frac{f(z)}{A(z)(z-x)} dz. \quad (3.3.7)$$

Evaluating the integral in (3.3.7) by residues we can find the interpolating polynomial (3.3.4):

$$P_n(f; x) = f(x) - R_n(f; x).$$

At the pole $z = x$ the residue of $\dfrac{A(x)f(z)}{A(z)(z-x)}$ is $f(x)$. Let us now find
the residue of this function at the pole $z = x_k$. For z close to x_k we
have the following expansions in powers of $z - x_k$.

$$f(z) = \sum_{s=0}^{\infty} \frac{f^{(s)}(x_k)}{s!} (z - x_k)^s$$

$$\frac{1}{z-x} = \frac{1}{(z-x_k)-(x-x_k)} = -\sum_{s=0}^{\infty} \frac{(z-x)^s}{(x-x_k)^{s+1}}$$

$$\frac{(z-x_k)^{a_k}}{A(z)} = \sum_{s=0}^{\infty} c_s^{(k)} (z-x_k)^s.$$

The residue of the function

$$\frac{f(z)}{A(z)(z-x)} = \frac{1}{(z-x_k)^{a_k}} \frac{(z-x_k)^{a_k}}{A(z)} \frac{1}{z-x} f(z)$$

is obtained by multiplying the above three series together and determining
the coefficient of $(x - x_k)^{a_k-1}$. A simple calculation shows that this coefficient is

$$-\sum_{i=0}^{a_k-1} f^{(i)}(x_k) \frac{1}{i!} \sum_{s=0}^{a_k-1-i} c_s^{(k)} (x-x_k)^{-a_k+i+s}.$$

The residue of $\dfrac{A(x)f(z)}{A(z)(z-x)}$ is this expression multiplied by $A(x)$.

Thus we have obtained the following expression for $P_n(f; x)$ which is
due to Hermite:

$$P_n(f; x) = \sum_{k=1}^{m} \sum_{i=0}^{a_k-1} f^{(i)}(x_k) \frac{1}{i!} \frac{A(x)}{(x-x_k)^{a_k}} \times$$

$$\times \sum_{s=0}^{a_k-1-i} c_s^{(k)}(x-x_k)^{i+s}. \qquad (3.3.8)$$

If $f(z)$ is defined on the real line and is not an analytic function then the representation (3.3.8) for its interpolating polynomial remains valid, but the representation of the remainder for $P_n(f; x)$ as a contour integral is no longer valid.

For a nonanalytic function $f(x)$ we give another representation for R_n for functions with sufficiently high order derivatives.

Let the points x and x_k $(k = 1, 2, \ldots, m)$ belong to a certain segment $[a, b]$.

Theorem 6. *If $f(x)$ has a continuous derivative of order $n + 1$ on $[a, b]$ then there exists an interior point ξ of $[a, b]$ for which*

$$R_n(f; x) = \frac{A(x)}{(n+1)!} f^{(n+1)}(\xi). \qquad (3.3.9)$$

The proof of this theorem follows from an application of the following variation of Rolle's theorem to the function

$$F(z) = f(z) - P_n(f; z) - \frac{A(z)}{A(x)}[f(x) - P_n(f; x)].$$

Let $a_1 < a_2 < \cdots < a_m$ and let $f(x)$ satisfy the conditions

$$f^{(i)}(a_k) = 0, \quad (i = 0, 1, \ldots, a_k - 1; k = 1, 2, \ldots, m).$$

Then if $f(x)$ has a continuous derivative of order $r = a_1 + \cdots + a_m$ then between a_1 and a_m there exists a point ξ for which $f^{(r)}(\xi) = 0$.

REFERENCES

V. L. Goncharov, *The Theory of Interpolation and Approximation of Functions*, Moscow, 1954, Chap. I and V (Russian).

C. Hermite, "Sur la formule d'interpolation de Lagrange," *J. Reine Angew. Math.*, Vol. 84, 1878, pp. 70–79.

E. Ia. Remez, "On the remainder term in certain formulas of approximate analysis," *Dokl. Akad. Nauk SSSR*, Vol. 26, 1940, pp. 130–134 (Russian).

E. T. Whittaker and G. Robinson, *The Calculus of Observations*, Glasgow, 1940.

CHAPTER 4

Linear Normed Spaces.
Linear Operators

4.1. LINEAR NORMED SPACES

Functional analysis provides a useful method for studying certain questions related to quadrature formulas. Using concepts from this branch of mathematics we can study many different questions related to quadrature formulas from a single point of view. In this chapter we develop only a few of the simple concepts and results from functional analysis which will be needed in the remainder of this book.

Let $X = \{x\}$ be a set of certain "elements" x. The nature of these elements is arbitrary: they may be points, lines, functions or any other quantities.

The set X is called *linear* if the following two operations are defined on the elements of X: addition $x + y$, and multiplication λx by a (real or complex) number λ, in such a way that the result of these operations produces a new element of the set. To be more specific these operations are required to satisfy:

1. associativity of addition $(x + y) + z = x + (y + z)$;
2. commutivity of addition $x + y = y + x$;
3. the existence of a *zero* element θ, which for every $x \in X$ satisfies

$$x + \theta = x;$$

4. for each x of X there exists an *inverse* element $-x$, for which

$$x + (-x) = \theta;$$

5. associativity of multiplication

$$\lambda(\mu x) = (\lambda\mu)x;$$

6. the distributive laws

$$(\lambda + \mu)x = \lambda x + \mu x, \qquad \lambda(x + y) = \lambda x + \lambda y;$$

7. $1 \cdot x = x$;
8. $0 \cdot x = \theta$;
9. if $\lambda x = \theta$ and $x \neq \theta$, then $\lambda = 0$.

A linear set X is called a *linear normed* or *vector* space, if for each element $x \in X$ there is defined a norm $\|x\|$, that is a real number possessing the properties of the length of a vector:

1. $\|x\| \geq 0$, and $\|x\| = 0$ if and only if $x = \theta$,
2. $\|x + y\| \leq \|x\| + \|y\|$,
3. $\|\lambda x\| = |\lambda| \cdot \|x\|$.

By means of the norm we can define convergence of a sequence of elements: we say that $x_n \longrightarrow x$, or $\lim_{n \to \infty} x_n = x$, if $\|x_n - x\| \longrightarrow 0$ as $n \longrightarrow \infty$.

Closely related to the concept of convergence is the concept of completeness of the space. If the sequence x_n $(n = 1, 2, \ldots)$ converges to a certain element x, then such a sequence satisfies the Bolzano-Cauchy criterion: for each $\epsilon > 0$ there exists an integer $N(\epsilon)$ such that for $n > N(\epsilon)$ and any $m > 0$

$$\|x_{n+m} - x_n\| < \epsilon.$$

The converse may be false: if a sequence x_n $(n = 1, 2, \ldots)$ satisfies the Bolzano-Cauchy criterion then it is still possible that there does not exist an element x in X to which the sequence x_n converges as $n \longrightarrow \infty$.

The space X is called *complete* if for every sequence x_n, which satisfies the Bolzano-Cauchy criterion, there exists an element x in X to which the sequence x_n converges as $n \longrightarrow \infty$. A complete, normed, linear space is called a *Banach* space.

Let us give some examples of Banach spaces.

1. The space *C*.

Let $[a, b]$ be any finite segment. The elements of C are all continuous functions on $[a, b]$. Addition of the elements and multiplication of them by a number is the usual addition of functions and multiplication of functions by a number. For the norm of the function $x = x(t)$ we take

$$\|x\| = \max_{t \in [a, b]} |x(t)|. \tag{4.1.1}$$

Convergence of elements of C corresponds to uniform convergence of sequences of functions.

The space C is complete. From

$$\|x_{n+m} - x_n\| = \max_{t \in [a,\, b]} |x_{n+m}(t) - x_n(t)| < \epsilon$$

there follows the convergence of the sequence of functions $x_n(t)$ $(n = 1, 2, \ldots)$ for every t: $\lim\limits_{n \to \infty} x_n(t) = x(t)$ and because the limit of a uniformly convergent sequence of continuous functions is also a continuous function then $x(t) \in C$.

2. The space L_p $(p \geq 1)$.

This is the space of measurable functions on $[a, b]$ which are p^{th} power summable. Addition and multiplication by a number is also the usual addition of functions and multiplication of them by a number. The norm is defined by

$$\|x\| = \left(\int_a^b |x(t)|^p \, dx \right)^{\frac{1}{p}}. \tag{4.1.2}$$

Functions which differ only on a set of points of measure zero are considered equivalent.

The conditions which must be satisfied by the norm (4.1.2) are easily verified. Conditions 1 and 3 are obviously fulfilled. Condition 2 is the well known Minkowski inequality for integrals[1]:

$$\left(\int_a^b |x(t) + y(t)|^p \, dt \right)^{\frac{1}{p}} \leq \left(\int_a^b |x(t)|^p \, dt \right)^{\frac{1}{p}} + \left(\int_a^b |y(t)|^p \, dt \right)^{\frac{1}{p}}.$$

The space L_p is complete.[2]

3. The space L_2.

These are the functions that are square summable and the special case of L_p for $p = 2$. The norm in L_2 is

$$\|x\| = \left(\int_a^b x^2(t) \, dt \right)^{\frac{1}{2}}. \tag{4.1.3}$$

---

[1]See, for example, L. A. Lyusternik and V. I. Sobolev, *Elements of Functional Analysis*, Gostekhizdat, Moscow, 1951 (Russian; or the German translation of this book, *Elemente der Funktionalanalysis*, Berlin, 1955, pp. 244-6).

[2]See, for example, L. A. Lyusternik and V. I. Sobolev, *ibid.*, pp. 35-7 (or in the German translation, pp. 18-19).

Convergence of elements here means convergence of functions in the sense of mean square deviation.

4. The space L of summable functions on $[a, b]$.

It is also a particular case of L_p for $p = 1$. The norm in L is defined as:

$$\|x\| = \int_a^b |x(t)| dt \qquad (4.1.4)$$

and has the geometric meaning of the area between the t axis and the graph of the function $x(t)$ over the interval a to b.

5. The space V.

For functions of bounded variation on $[a, b]$, for which $x(a) = 0$, to obtain the norm in V we take the total variation of $x(t)$ on $[a, b]$

$$\|x\| = \underset{[a, b]}{\mathrm{Var}}\ x(t). \qquad (4.1.5)$$

It is clear that conditions 1 and 3 for the norm are fulfilled. The fulfillment of the second condition follows from the inequality

$$\underset{[a, b]}{\mathrm{Var}}\ [x(t) + y(t)] \leq \underset{[a, b]}{\mathrm{Var}}\ x(t) + \underset{[a, b]}{\mathrm{Var}}\ y(t).$$

The space V is complete. Indeed, let the Bolzano-Cauchy criterion be satisfied for the sequence of elements x_n $(n = 1, 2, \ldots)$:

$$|x_{n+m}(t) - x_n(t)| = \left| \int_a^t d[x_{n+m}(t) - x_n(t)] \right| \leq$$

$$\leq \underset{[a, b]}{\mathrm{Var}}\ [x_{n+m}(t) - x_n(t)] = \|x_{n+m} - x_n\| < \epsilon$$

for $n > N(\epsilon)$. Hence we see that the sequence of functions $x_n(t)$ converges for every $t \in [a, b]$

$$\lim_{n \to \infty} x(t) = x(t).$$

From the Bolzano-Cauchy criterion it must follow that the norms $\|x_n\|$ $(n = 1, 2, \ldots)$ are bounded from above by a certain number[3]

---

[3]Take $\epsilon > 0$ and choose N so that for $n, m > N$ we will have $\|x_m - x_n\| < \epsilon$. Fix any value of $m > N$. Thus $\|x_n\| \leq \|x_m\| + \|x_m - x_n\| < \|x_m\| + \epsilon$. Let M be the greatest of the numbers $\|x_1\|, \ldots, \|x_N\|, \|x_m\| + \epsilon$. Then for every n we will have $\|x_n\| \leq M$.

$$\|x_n\| \leq M \qquad (n = 1, 2, \ldots).$$

Divide $[a, b]$ into parts by the points $a = t_0 < t_1 < \cdots < t_k = b$. The following inequality holds for $x_n(t)$:

$$\sum_{i=0}^{k-1} |x_n(t_{i+1}) - x_n(t_i)| \leq \operatorname*{Var}_{[a, b]} x_n(t) = \|x_n\| \leq M.$$

If we now pass to the limit as $n \longrightarrow \infty$ we obtain

$$\sum_{i=0}^{k-1} |x(t_{i+1}) - x(t_i)| \leq M$$

which is equivalent to $\operatorname*{Var}_{[a, b]} x(t) \leq M$, and consequently $x(t)$ is an element of the space V.

Let $\epsilon > 0$ and choose the number N so that for $n, m > N$ we will have $\|x_m - x_n\| \leq \epsilon$. If in the inequality

$$\sum_{i=0}^{k-1} |[x_m(t_{i+1}) - x_n(t_{i+1})] - [x_m(t_i) - x_n(t_i)]| \leq$$

$$\leq \operatorname*{Var}_{[a, b]} [x_m(t) - x_n(t)] = \|x_m - x_n\| \leq \epsilon$$

we pass to the limit as $m \longrightarrow \infty$, then we easily obtain

$$\sum_{i=0}^{k-1} |[x(t_{i+1}) - x_n(t_{i+1})] - [x(t_i) - x_n(t_i)]| \leq \epsilon.$$

Because this is valid for any choice of points t_i $(i = 0, 1, \ldots, k)$ then there follows

$$\operatorname*{Var}_{[a, b]} [x(t) - x_n(t)] = \|x - x_n\|$$

and consequently the functions $x_n(t)$ converge to $x(t)$ with respect to the norm (4.1.5).

4.2. LINEAR OPERATORS

Let $X = \{x\}$ and $Y = \{y\}$ be two arbitrary sets of elements x and y. If for each element x there corresponds by some rule a certain element y: $y = H(x)$, then we will say that we are given an operator H. The set X is the domain on which H is defined and the domain of values of H is the set Y.

In the particular case when Y is the set of real or complex numbers so that to each element x there corresponds a certain number, the operator H is called a *functional*.

The concept of an operator is a direct and far-reaching generalization of the concept of a function.

If in the sets X and Y there is a rule for passing to the limit then we can define a continuous operator. The operator H is called *continuous* if from $x_n \longrightarrow x$ (in the set X) it follows that $H(x_n) \longrightarrow H(x)$ (in the set Y).

Below we will always assume that X and Y are linear normed spaces. The operator H is called *additive* if for any two elements x_1 and x_2 of X we have:[4]

$$H(x_1 + x_2) = Hx_1 + Hx_2.$$

The operator is called *linear* if it is additive and continuous.

If there exists a number M so that for every x there is satisfied the inequality

$$\|Hx\| \le M \|x\|$$

then H is called a *bounded* operator. Let us prove the assertion:

In order that an additive operator be continuous it is necessary and sufficient that it be bounded.

Proof of necessity. Let us suppose that the linear operator H is unbounded and show that this leads to a contradiction. Thus we could find a sequence of elements x_n for which

$$\|Hx_n\| \ge n \|x_n\|.$$

Consider the elements

$$x'_n = \frac{x_n}{n \|x_n\|}.$$

It is evident that $x'_n \longrightarrow \theta$ as $n \longrightarrow \infty$.

On the other hand

$$Hx'_n = \frac{1}{n \|x_n\|} Hx_n$$

and

$$\|Hx'_n\| = \frac{1}{n \|x_n\|} \|Hx_n\| \ge 1;$$

---

[4]We will often omit the parentheses around the argument of an operator.

$\|Hx'_n\| \not\to 0$ as $n \longrightarrow \infty$ and the operator H is not continuous at the zero element θ.

Proof of sufficiency. We will assume the operator H to be additive and bounded and take any element x. If $x_n \longrightarrow x$, that is $\|x_n - x\| \longrightarrow 0$, then

$$\|Hx_n - Hx\| = \|H(x_n - x)\| \leq M\|x_n - x\| \longrightarrow 0,$$

as $n \longrightarrow \infty$. $Hx_n \longrightarrow Hx$ and the operator H is continuous.

If H is a linear operator then the smallest of the constants M which satisfy the inequality

$$\|Hx\| \leq M\|x\|,$$

is called the *norm of the operator H* and is designated by $\|H\|$:

$$\|H\| = \min M.$$

In certain cases the following easily proved relation can be useful to find the norm

$$\|H\| = \sup_{\|x\| \leq 1} \|Hx\|. \tag{4.2.1}$$

Indeed, for $\|x\| \leq 1$, $\|Hx\| \leq \|H\| \|x\| \leq \|H\|$. Therefore

$$\sup_{\|x\| \leq 1} \|Hx\| \leq \|H\|. \tag{4.2.2}$$

By the definition of the norm, for each $\epsilon > 0$ there exists an element x' for which

$$\|Hx'\| > (\|H\| - \epsilon)\|x'\|.$$

Let us put

$$x = \frac{x'}{\|x'\|},$$

$$\|Hx\| = \frac{1}{\|x'\|} \|Hx'\| > \frac{1}{\|x'\|} (\|H\| - \epsilon)\|x'\| = \|H\| - \epsilon.$$

Because $\|x\| = 1$, then

$$\sup_{\|x\| \leq 1} \|Hx\| > \|H\| - \epsilon.$$

By the arbitrariness of ϵ and by (4.2.2) we obtain (4.2.1).

Let us find the norms of certain linear functionals which we will encounter later.

1. X **is the space** $C[a, b]$.

Consider the functional

$$Fx = \int_a^b f(t)\,x(t)\,dt, \qquad (4.2.3)$$

where $f(t)$ is a measurable and summable function on $[a, b]$. We have

$$\|F\| = \sup_{\|x\| \le 1} \left| \int_a^b f(t)\,x(t)\,dt \right| =$$

$$= \sup_{|x(t)| \le 1} \int_a^b |f(t)|\,|x(t)|\,dt \le \int_a^b |f(t)|\,dt.$$

Consider the function sign $f(t)$. It is measurable and $|\operatorname{sign} f(t)| \le 1$. For it

$$\int_a^b f(t)\,\operatorname{sign} f(t)\,dt = \int_a^b |f(t)|\,dt.$$

Because sign $f(t)$ is a measurable function there certainly exists a continuous function $x^*(t)$ for which $|x^*(t)| \le 1$ and which differs from sign $f(t)$ only on a set of arbitrarily small measure. Such a function can always be found so that $\int_a^b fx^*\,dt$ differs from $\int_a^b f \operatorname{sign} f\,dt$ by as little as we please. Therefore

$$\sup_{|x(t)| \le 1} \int_a^b f(t)\,x(t)\,dt \ge \int_a^b |f(t)|\,dt.$$

and consequently

$$\|F\| = \int_a^b |f(t)|\,dt. \qquad (4.2.4)$$

2. X **is** $L[a, b]$.

$$Fx = \int_a^b f(t)\,x(t)\,dt. \qquad (4.2.5)$$

Here f is a continuous function on $[a, b]$. The norm in L is defined by equation (4.1.4). We have

$$|Fx| = \left| \int_a^b fx\,dt \right| \leq \max_t |f(t)| \int_a^b |x(t)|\,dt = \max_t |f(t)| \cdot \|x\|$$

Hence we see that $\|F\| \leq \max_t |f(t)|$.

We can convince ourselves that in this estimate for $\|F\|$ the righthand side can not be decreased. Let ϵ be any small positive number.

Let us put $M = \max_t |f(t)|$ and let us suppose that this maximum is achieved at the point ξ. In order to be definite let us consider that $f(\xi)$ is a positive number: $f(\xi) = M > 0$. By the continuity of $f(t)$, close to ξ there exists a segment $\alpha < t < \beta$ in which

$$f(t) > M - \epsilon.$$

We define the function $x(t)$ by the relationships

$$x(t) = \begin{cases} \dfrac{1}{\beta - \alpha} & \text{for } t \in (\alpha,\, \beta) \\[2ex] 0 & \text{for } t \,\overline{\in}\, (\alpha,\, \beta). \end{cases}$$

Clearly

$$\|x\| = \int_a^b |x(t)|\,dt = 1,$$

$$|Fx| = \left| \int_a^b f(t)\,x(t)\,dt \right| =$$

$$= \frac{1}{\beta - \alpha} \int_\alpha^\beta f(t)\,dt > \frac{1}{\beta - \alpha}\,(M - \epsilon)\,(\beta - \alpha) = M - \epsilon.$$

Therefore

$$\|F\| = \sup_{\|x\| \leq 1} |Fx| > M - \epsilon,$$

and because ϵ is an arbitrary number, from this and from the previously obtained upper bound for $\|F\|$ we have

$$\|F\| = M = \max_t |f(t)|. \tag{4.2.6}$$

3. Let X be the space V.

Consider the functional

$$Fx = \int_a^b f(t)\,dx(t) \tag{4.2.7}$$

where $f(t)$ is a continuous function on $[a, b]$; we show that

$$\|F\| = \max_t |f(t)|, \qquad (4.2.8)$$

$$|Fx| = \left| \int_a^b f dx \right| \le \max_t |f(t)| \operatorname*{Var}_{[a, b]} x(t) = \max_t |f(t)| \cdot \|x\|.$$

Let us suppose that $\max |f(t)|$ is achieved at the point t_0 and suppose that t_0 lies inside $[a, b]$. Taking

$$x(t) = \begin{cases} 0 & \text{for } t < t_0 \\ \frac{1}{2} & \text{for } t = t_0 \\ 1 & \text{for } t > t_0 \end{cases}$$

we can see that the upper estimate which we have obtained for $|Fx|$ is achieved. This proves (4.2.8).

4.3. CONVERGENCE OF A SEQUENCE OF LINEAR OPERATORS

Let X and Y be Banach spaces. Consider the sequence of linear operators H_n $(n = 1, 2, \dots)$ defined on X and taking values in Y. The sequence H_n will be called convergent if for each $x \in X$ there is a convergent sequence of elements $y_n = H_n x$ (in the space Y). Let us denote $\lim H_n x = y = H x$. The operator H is additive. In fact if in the equation $H_n(x_1 + x_2) = H_n x_1 + H_n x_2$ we pass to the limit as $n \longrightarrow \infty$ then we obtain

$$H(x_1 + x_2) = H x_1 + H x_2.$$

We can also show that the operator H is linear. We prove a preliminary lemma.

Lemma. *If the sequence of operators H_n $(n = 1, 2, \dots)$ converges then their norms $\|H_n\|$ $(n = 1, 2, \dots)$ have a common bound:*

$$\|H_n\| \le M. \qquad (4.3.1)$$

Proof. Let us suppose the contrary. The set of elements x which satisfy the condition $\|x - x_0\| \le \epsilon$ will be called a *closed sphere* of radius ϵ with center x_0 and denoted by $S(x_0, \epsilon)$. We show that $\|H_n x\|$ can not have a common bound in that closed sphere. In fact let

$$\|H_n x\| \le K \qquad (4.3.2)$$

for $n = 1, 2, \dots$, and for every x in the sphere $S(x_0, \epsilon)$. For any x in X the element

$$x' = \frac{\epsilon}{\|x\|} x + x_0$$

belongs to $S(x_0, \epsilon)$. Therefore

$$\|H_n x'\| = \left\| \frac{\epsilon}{\|x\|} H_n x + H_n x_0 \right\| \leq K$$

and

$$\frac{\epsilon}{\|x\|} \|H_n x\| - \|H_n x_0\| \leq K.$$

Hence

$$\|H_n x\| \leq \frac{K + \|H_n x_0\|}{\epsilon} \|x\|.$$

The sequence of elements $H_n x_0$ converges and their norms $\|H_n x_0\|$ have a common bound. There must, then, exist a number K_1 independent of n and x for which

$$\|H_n x\| \leq K_1 \|x\|.$$

Therefore

$$\|H_n\| = \sup_{\|x\| \leq 1} \|H_n x\| \leq K_1$$

and this contradicts the assumption and inequality (4.3.2) can not be valid.

Let us take an arbitrary closed sphere $S_0(x_0, \epsilon)$. In this sphere the sequence $\|H_n x\|$ is unbounded. We can find, then, an operator H_{n_1} and an element $x_1 \in S_0$ for which

$$\|H_{n_1} x_1\| > 1.$$

By the continuity of the operator H_{n_1} this inequality will be satisfied in a certain closed sphere $S_1(x_1, \epsilon_1)$ contained in S_0. By an analogous argument we can find an operator H_{n_2} and an element $x_2 \in S_1$ for which

$$\|H_{n_2} x_2\| > 2$$

and so forth. We can assume that $\epsilon_n \longrightarrow 0$ as $n \longrightarrow \infty$. The constructed sequence of elements $x_1, x_2, \ldots$ will satisfy the Bolzano-Cauchy criterion. The space X is complete and the sequence will converge to a certain element $x^* \in X$

$$x_n \longrightarrow x^*, \qquad \text{as } n \longrightarrow \infty;$$

x^* belongs to all the spheres S_k $(k = 1, 2, \ldots)$.

Thus for the element x^*

$$\|H_{n_k} x^*\| > k.$$

This then contradicts the assumption that the sequence $H_n x$ converges for arbitrary $x \in X$.

The linearity of the limit operator H is now easily proved by using the lemma. Passing to the limit as $n \longrightarrow \infty$ in the inequality

$$\|H_n x\| \leq M \|x\|$$

we obtain

$$\|H x\| \leq M \|x\|.$$

The operator H is bounded and, in view of its addivity, is continuous and linear.

The conditions which must be satisfied by the sequence of operators H_n $(n = 1, 2, \dots)$ in order that they be convergent is expressed in the following theorem of Banach.

Theorem 1. *In order that the sequence of linear operators H_n $(n = 1, 2, \dots)$ be convergent it is necessary and sufficient that they satisfy the two conditions*:

1. The norms of the operators $\|H_n\|$ have a common bound.

2. $H_n x$ is convergent for each x in a set E which is everywhere dense in X[5].

Proof. The necessity of the second condition is obvious. The necessity of the first follows from the lemma.

The sufficiency of the conditions can be verified in the following way. Let $\|H_n\| \leq M$. Let us take an arbitrary $x \in X$ and select an element $x' \in E$ for which $\|x - x'\| < \dfrac{\epsilon}{3M}$. The sequence $H_n x'$ converges by condition 2 and for large n we will have

$$\left\|H_{n+m} x' - H_n x'\right\| < \frac{\epsilon}{3}.$$

Then

$$\left\|H_{n+m} x - H_n x\right\| \leq \left\|H_{n+m} x - H_{n+m} x'\right\| + \left\|H_{n+m} x' - H_n x'\right\| +$$

$$+ \left\|H_n x' - H_n x\right\| \leq 2M \|x - x'\| + \frac{\epsilon}{3} < \frac{2\epsilon}{3} + \frac{\epsilon}{3} = \epsilon.$$

Therefore the sequence $H_n x$ satisfies the Bolzano-Cauchy condition and in view of the completeness of Y there exists for each $x \in X$ a limit element $y = Hx = \lim\limits_{n \to \infty} H_n x$.

As shown above the limit operator H is linear.

---

[5]A set E is called *everywhere dense* in X if every element $x \in X$ is arbitrarily close, with respect to the norm, to an element of E.

REFERENCES

L. V. Kantorovich, "Functional analysis and applied mathematics," *Uspehi Matem. Nauk.* (*N.S.*), Vol. 3, no. 6, 1948, pp. 89–185 (Russian).

L. A. Lyusternik, "Basic concepts of functional analysis," *Uspehi Matem. Nauk.*, Vol. 1, 1936, pp. 77–148 (Russian).

L. A. Lyusternik and V. I. Sobolev, *Elements of Functional Analysis*, Gostekhizdat, Moscow, 1951 (Russian). (There is a German translation of this book, *Elemente der Funktionalanalysis*, Berlin, 1955.)

Part Two

APPROXIMATE CALCULATION OF DEFINITE INTEGRALS

CHAPTER 5

Quadrature Sums and Problems Related to Them. The Remainder in Approximate Quadrature

5.1. QUADRATURE SUMS

The problem of finding the numerical value of the integral of a function of one variable, because of its geometrical meaning, is often for simplicity called quadrature. In this book we study methods of quadrature which are used to approximately evaluate the integral by means of a finite number of values of the integrand and derivatives of the integrand. These methods are universal and they can be applied where other methods for calculating integrals fail. In many cases these methods also require less work than other methods.

Let us consider an integral of the form

$$\int_a^b p(x) f(x) \, dx$$

where $[a, b]$ is any finite or infinite segment of the real line, and $f(x)$ is an arbitrary function of a certain class. To simplify the discussion we assume in the beginning of this chapter that all functions $f(x)$ are continuous. We assume that $p(x)$ is a certain fixed function, which is measurable on $[a, b]$ and is not the identically zero function, and that the product $p(x) f(x)$ is summable on $[a, b]$. At first we will not make any additional assumptions about $p(x)$.

The most widely applied quadrature formulas are those which approximate the integral by a linear combination of values of the function

$$\int_a^b p(x) f(x) \, dx \approx \sum_{k=1}^n A_k f(x_k). \tag{5.1.1}$$

The sum $\sum_{k=1}^n A_k f(x_k)$ we will call a *quadrature sum*. Equations of the form (5.1.1) have received the name of mechanical quadrature formulas[1]. They contain the following $2n + 1$ parameters which can be selected in the construction: n abscissa or "nodes" x_k, n coefficients A_k and the number n. It is necessary to choose all of these parameters so that formula (5.1.1) will give a "sufficiently small error" for all functions f of a certain wide class. For the following discussion of ideas related to the construction of quadrature sums we will not precisely define the words "small error" and how wide the class of functions must be. The precise meaning of these words will be made clear later.

It is immediately clear, by counting the choices of the x_k and A_k, that the larger the value of n the more precise can (5.1.1) be made. Therefore for the construction of approximate quadrature formulas n is considered an arbitrary but fixed natural number.

In applying (5.1.1) the greatest amount of difficulty is usually in finding the values $f(x_k)$ $(k = 1, 2, \ldots, n)$. After the $f(x_k)$ have been found the construction of the quadrature sum $\sum_{k=1}^n A_k f(x_k)$, if n is not very large, is carried out comparatively easily. Therefore it is natural to try to achieve the necessary precision in the calculation with as small a number of nodes x_k as possible. For the construction of quadrature sums

---

[1] It is easy to attach a mechanical meaning to (5.1.1). Let us introduce the quantity $P = \int_a^b p \, dx$ and write (5.1.1) in the form $P^{-1} \int_a^b pf \, dx \approx \sum_{k=1}^n B_k f(x_k)$. Here the coefficients B_k will be abstract numbers. Let us agree to interpret them as "weights" belonging to the corresponding values $f(x_k)$. If we require that the equation be exact whenever f is a constant function then the B_k must satisfy $\sum_{k=1}^n B_k = 1$. The sum $\sum_{k=1}^n B_k f(x_k)$ then will have the meaning of an average of the values $f(x_k)$. The problem of construction of the equation reduces to finding weights B_k so that the average weighted value of $f(x_k)$ will approximately equal the mean integral value of f on the segment $[a, b]$: $P^{-1} \int_a^b pf \, dx$.

this is equivalent then to choosing the x_k and A_k to increase the precision of formula (5.1.1) for a given n. We now discuss the principle ways that have been investigated to achieve this.

1. Let us suppose we have been given a certain class F of functions f. In relation to this class we consider the system of functions

$$\omega_m(x) \quad (m = 1, 2, \ldots) \tag{5.1.2}$$

for which the products $p(x)\omega_m(x)$ are summable on $[a, b]$. Let us form a linear combination

$$s_n(x) = \sum_{k=1}^{n} a_k \omega_k(x).$$

For the evaluation of the integral $\displaystyle\int_a^b p(x)f(x)\,dx$ we take as the "distance" between f and s_n the value

$$\rho(f, s_n) = \int_a^b |p(f - s_n)|\,dx. \tag{5.1.3}$$

We will consider the system (5.1.2) to be complete in the class F, that is for each function $f \epsilon F$ and any $\epsilon > 0$ there exists a linear combination s_n for which $\rho(f, s_n) < \epsilon$.

In view of the inequality

$$\left| \int_a^b pf\,dx - \int_a^b ps_n\,dx \right| \leq \int_a^b |p(f - s_n)|\,dx = \rho(f, s_n)$$

it follows that the integral $\displaystyle\int_a^b pf\,dx$ can be calculated to as high a degree of accuracy as desired, if the integrand f is replaced by an appropriate linear combination s_n.

Thus it is evident that we can achieve a high degree of precision in the calculation by taking a large number of functions ω_k in the formation of s_n.

We can expect that if we choose the nodes x_k and coefficients A_k in the formula (5.1.1) to give good precision in integrating the functions ω_m, then the formula must also give good precision in the calculation of the integral for each function $f \epsilon F$. These simple considerations serve, of course, only for motivation and the error of the constructed formulas must be subjected to precise analysis and estimation. But it is useful to indicate a simple principle for the selection of the x_k and A_k: we will attempt to choose the x_k and A_k so that formula (5.1.1) gives an exact result for as many of the functions $\omega_m(x)$ as possible.

We say that equation (5.1.1) has degree of precision m with respect to the functions (5.1.2) if it is exact for $\omega_1, \omega_2, \ldots, \omega_m$:

$$\int_a^b p\,\omega_i\,dx = \sum_{k=1}^n A_k \omega_i(x_k) \quad (i = 1, 2, \ldots, m)$$

and it is not exact for ω_{m+1}. This way for choosing the x_k and A_k is a way to increase the degree of precision of equation (5.1.1). Of special interest are formulas of approximate quadrature which possess the highest possible degree of precision. Some formulas of this type will be discussed in Chapter 7.

If the class F is given, then for the construction of equation (5.1.1) for the integration of the function f, the choice of the system of functions ω_n $(n = 1, 2, \ldots)$ is still arbitrary. The requirement of completeness, which must be satisfied by the system, does not fully define it and there are still many ways to select the ω_n.

Approximate quadrature formulas which we will now consider take into account the properties of the functions ω_n. If we want the formulas to give good precision then the ω_n must necessarily be chosen so that the properties of ω_n will agree with the properties of f and we can expect that the error in (5.1.1) will be smaller the more closely the linear combination s_n approximates the function f for a fixed n.

We mention now some examples of the choice of ω_n. Let $[a, b]$ be any finite segment. It is known that for any continuous function f on $[a, b]$ and for any $\epsilon > 0$ there exists a polynomial $P(x)$ which differs from $f(x)$, for any $x \,\epsilon\, [a, b]$, by less than ϵ:

$$\left| f(x) - P(x) \right| < \epsilon.$$

This is the property of completeness of algebraic polynomials in the space of continuous functions C. From this then there follows, at once, the completeness of the system of polynomials in the sense of the metric (5.1.3).

We take the system of powers of x: $1, x, x^2, \ldots$ as the functions ω_n and we will say that equation (5.1.1) has algebraic degree of precision m, if it is exact for all possible polynomials of degree m and not exact for all polynomials of degree $m + 1$. This is equivalent to the equation

$$\int_a^b p x^i dx = \sum_{k=1}^n A_k x_k^i$$

being fulfilled for $i = 0, 1, \ldots, m$ and not fulfilled for $i = m + 1$.

We can expect that (5.1.1) will have a smaller error for more continuous functions on $[a, b]$ for the higher the algebraic degree of precision.

The system of powers x^n $(n = 0, 1, \ldots)$ are a sufficiently convenient basis for the construction of quadrature formulas of the highest degree of precision for any finite segment $[a, b]$.

Let us suppose now that the segment of integration is infinite, for example, let it be the segment $0 \le x < \infty$. We will take some subset F of continuous functions f on $[0, \infty)$. On each finite segment $0 \le x \le b < \infty$ we can construct a polynomial $P(x)$ which approximates f uniformly with any preassigned degree of precision. But $P(x)$ can not give a uniform approximation to f on the entire half-line and the difference $f - P$, for large x, can have a large value. In spite of this, providing the weight function $p(x)$ decreases sufficiently fast as $x \longrightarrow \infty$, it can happen that for any $f \epsilon F$ the integral $\int_0^\infty |p(f - P)| \, dx$ can be made as small as we desire and the system of powers x^n will then be complete in the class F with respect to the metric (5.1.3). In this case the quadrature formulas of the highest algebraic degree of precision also can be applied for the approximate calculation of integrals of the form $\int_0^\infty pfdx$. These formulas will be discussed in Chapter 7.

In this connection we wish to give an example to clarify how to choose the functions ω_n so that their properties closely agree with the class F of functions to be integrated. Let us suppose f is continuous and has an asymptotic representation, on the segment $0 \le x < \infty$, of the form

$$f(x) \sim c_0 + \frac{c_1}{x} + \frac{c_2}{x^2} + \cdots.$$

Each polynomial $P(x)$ of degree greater than zero grows without bound as $x \longrightarrow \infty$, and the order of growth is higher for the polynomials of higher degree. The behavior of the polynomials on the half-line $[0, \infty)$ naturally differs from the behavior of bounded functions and polynomials can not be successful for the approximation of bounded functions on $[0, \infty)$. For certain weight functions $p(x)$ it can happen that approximate quadratures of the highest degree of accuracy, with basis functions the system of powers x^n, will give slow convergence, as $n \longrightarrow \infty$, to the value of the integral and to obtain the necessary precision a large number of nodes will be required.

For the approximation of functions of the type mentioned above it is more suitable to use rational functions of some special form, for example, the functions $(x + 1)^{-k}$ $(k = 0, 1, 2, \ldots)$. If we take these for the basis functions and construct the corresponding quadrature formulas of the highest degree of precision[2] then they might be expected to give better precision for the same number of nodes than formulas based on $\omega_k(x) = x^k$.

We mention now another example for the choice of $\omega_n(x)$. Let f be a periodic function of period 2π and suppose that we want to evaluate the integral

$$\int_0^{2\pi} f(x)\,dx.$$

For the functions ω_n it is then natural to choose the trigonometric functions $\cos kx$, $\sin kx$ $(k = 0,1,2,\ldots)$. As it turns out in this case the formulas of the highest degree of accuracy are elementary and their construction is quite simple. Because of their simplicity we will not postpone their construction to a later chapter. However, in order that we do not interrupt the discussion related to the choice of the nodes and coefficients we delay the study of these formulas to the following section.

2. Let us suppose that we are given a class F of functions f. We endeavor to construct a quadrature formula (5.1.1) which will be in some sense, which we will clarify below, "best" for a given class. For each function f the error in the formula (5.1.1) has the value

$$R(f) = \int_a^b p(x)f(x)\,dx - \sum_{k=1}^n A_k f(x_k).$$

---

[2]Trans. note. A quadrature formula for the segment $0 \leq x < \infty$ which is exact for $(1 + x)^{-k}$ $(k = 2, 3, \ldots, m + 2)$:

$$\int_0^\infty (1 + x)^{-k}dx = \sum_{i=1}^n A_i (1 + x_i)^{-k} \quad (k = 2, 3, \ldots, m + 2)$$

may be obtained by a transformation of a formula for the segment $0 \leq y \leq 1$ which is exact for y^k $(k = 0, 1, \ldots, m)$:

$$\int_0^1 y^k dy = \sum_{i=1}^n B_i y_i^k \quad (k = 0, 1, \ldots, m).$$

Using the transformation $y = \dfrac{1}{1 + x}$, $dy = \dfrac{-dx}{(1 + x)^2}$ we see that

$$\int_0^1 y^k dy = \int_0^\infty (1 + x)^{-k-2}dx \quad (k = 0, 1, 2, \ldots)$$

and it is then not difficult to see that the nodes x_i and coefficients A_i given by

$$x_i = \frac{1 - y_i}{y_i} \qquad A_i = \frac{B_i}{y_i^2} \qquad (i = 1, 2, \ldots, n)$$

give the desired formula for $[0, \infty)$.

As a quantity which characterizes the precision of the quadrature formula for all functions f, we can take the upper bound of $|R(f)|$.

$$R = \sup_f |R(f)|.$$

Here R depends on the x_k and A_k. Desiring to achieve possibly better accuracy for all functions $f \in F$ we can choose the x_k and A_k so that R will have the least possible value. Such formulas we will call formulas with least estimate of the remainder in the class F.

3. We have now indicated two possibilities with regard to the choice of the nodes and coefficients. There are still other methods for constructing quadrature formulas by subjecting the nodes and coefficients to meet other demands. We indicate one problem of this type. First of all we note that to make formula (5.1.1) exact for functions having constant value on $[a, b]$ we have only the choice of the coefficients A_k at our disposal. If it is required that (5.1.1) be exact for $f \equiv 1$, then we obtain the following condition:

$$\sum_{k=1}^{n} A_k = \int_a^b p(x)\,dx. \tag{5.1.4}$$

Let us assume that the values $f(x_k)$, of the function f, entering into the quadrature sum are to be found from measurements and contain accidental errors. Let us suppose in addition that all of the $f(x_k)$ have been obtained as the result of measurements of equivalent precision.

The values of the quadrature sums will also contain accidental errors. We can state the problem thus: in what manner shall we choose the coefficients A_k, which fulfill condition (5.1.4), so that the quadrature sum $\sum_{k=1}^{n} A_k f(x_k)$ will have the least square error. It is known that if the arguments $z_1, \ldots, z_n$ of a linear function $y = a_1 z_1 + \cdots + a_n z_n$ are random quantities subjected to the law of normal distribution with one and the same standard deviation and if the coefficients of the linear function are subjected to the condition $\sum_{k=1}^{n} a_k = 1$, then the average squared error of the sum will be the least when all the coefficients are equal[3].

---

[3]If the random variables $z_1, \ldots, z_n$ are normally distributed with standard deviations $\sigma_1, \ldots, \sigma_n$ and if y is a linear function of them: $y = a_1 z_1 + \cdots + a_n z_n$, then y is also normally distributed with standard deviation $S = (a_1^2 \sigma_1^2 + \cdots + a_n^2 \sigma_n^2)^{\frac{1}{2}}$ (see, for example, S. N. Bernstein, *Theory of Probability* (in Russian). Gostekhizdat, Moscow, 1946, pp. 269–72; or H. D. Brunk, *An Introduction to Mathematical Statistics*, Ginn, 1960, pp. 88–9). For $\sigma_1 = \sigma_2 = \cdots = \sigma_n = \sigma$ we will have $S = \sigma(a_1^2 + \cdots + a_n^2)^{\frac{1}{2}}$ and for the condition $a_1 + \cdots + a_n = 1$, S will have a minimum in the case when all of the a_k are equal.

Therefore a quadrature formula with equal coefficients

$$\int_a^b p(x) f(x)\, dx \approx C\, [f(x_1) + \cdots + f(x_n)] \tag{5.1.5}$$

will have the least square error. At the same time such formulas are especially convenient for graphical calculation because the sum of the ordinates can be removed from the drawing with the help of the simplest graphical equipment.

We mention now one more property of quadrature sums which will have great value in the remainder of the book. For calculations, almost always, it is necessary to know the approximate values $f(x_k)$ exact to a certain number of decimal places.

Let all values $f(x_k)$ be known with error not exceeding in absolute value the number ϵ. Calculating, from the approximate values $f(x_k)$, the quadrature sum $\sum_{k=1}^n A_k f(x_k)$ we obtain its value with error which must be estimated by the quantity

$$\epsilon \sum_{k=1}^n |A_k|.$$

Such an estimate is exact and can not be decreased. If the sum $\sum_{k=1}^n |A_k|$ is very large, then even a small error in the values $f(x_k)$ can lead to a large error in the approximate value of the integral. For the construction of quadrature formulas therefore we should always strive so that the sum of the absolute values of the coefficients will have the smallest possible value.

In one important special case it is easy to indicate the condition for which $\sum_{k=1}^n |A_k|$ will have the smallest possible value. We will consider the weight function $p(x)$ to be nonnegative

$$p(x) \geq 0 \quad \text{for} \quad x \,\epsilon\, [a,\, b].$$

In addition we suppose that the quadrature formula is exact for $f(x) \equiv 1$, which is equivalent to equation (5.1.4) for the coefficients A_k. Then, evidently, $\sum_{k=1}^n |A_k|$ will have the least value when all the coefficients A_k are positive: $A_k > 0$. This fact is one of the reasons why quadrature formulas with positive coefficients are especially important for applications.

5.2. REMARKS ON THE APPROXIMATE INTEGRATION OF PERIODIC FUNCTIONS

Let the segment of integration $[a, b]$ be finite. It is always possible by a linear transformation to transform this segment to the segment $[0, 2\pi]$. Let us consider integrals of the form

$$\int_0^{2\pi} f(x)\, dx \qquad (5.2.1)$$

where $f(x)$ is a function with period 2π. As above we will study approximate quadrature formulas of the form

$$\int_0^{2\pi} f(x)\, dx \approx \sum_{k=1}^{n} A_k f(x_k). \qquad (5.2.2)$$

Here x_k belongs to the segment $0 \le x \le 2\pi$.

For the obvious reason, for the approximation of a periodic function we take not algebraic, but trigonometric polynomials. We recall that trigonometric polynomials of degree m are functions of the form

$$T_m(x) = a_0 + \sum_{k=1}^{m} (a_k \cos kx + b_k \sin kx) \qquad (5.2.3)$$

where a_0, a_k, b_k $(k = 1, \ldots, m)$ are certain constants.

We will say that formula (5.2.2) has trigonometric degree of precision m if it is exact for all possible trigonometric polynomials up to degree m inclusive and there exists a polynomial of degree $m + 1$ for which it is not exact.

It is easy to verify that no matter how we choose the nodes x_k and coefficients A_k formula (5.2.2) can not be exact for all trigonometric polynomials of degree n.

Let us construct the function

$$T(x) = \prod_{k=1}^{n} \sin^2 \frac{x - x_k}{2}.$$

Because $\sin^2 \dfrac{x - x_k}{2} = \frac{1}{2}[1 - \cos(x - x_k)]$ it is clear that $T(x)$ is a polynomial of degree n. Then the quadrature formula (5.2.2) can not be exact for it because $\displaystyle\int_0^{2\pi} T(x)\, dx > 0$, but $\displaystyle\sum_{k=1}^{n} A_k T(x_k) = 0$ because all of the nodes x_k are roots of the polynomial $T(x)$.

The trigonometric degree of precision of (5.2.2) is therefore always less than n and the A_k and x_k can be taken to make the degree, at the most, equal to $n-1$.

It turns out that the highest degree of precision $n-1$ is achieved by the simplest quadrature formula with equal coefficients:

$$A_k = \frac{2\pi}{n} \quad (k = 1, 2, \ldots, n)$$

and equally spaced nodes.

Let us consider any set of equally-spaced points on the real axis with interval $h = \frac{2\pi}{n}$. Let a be the point of the set nearest to the origin from the right or coinciding with the origin. The points $a + kh$ $(k = 0, 1, \ldots, n-1)$ lie in the segment $0 \leq x < 2\pi$. Let us take these as the nodes x_k and construct a quadrature formula

$$\int_0^{2\pi} f(x)\, dx \approx \frac{2\pi}{n} \sum_{k=1}^{n} f\left(a + (k-1)\frac{2\pi}{n}\right). \tag{5.2.4}$$

We can show that it is exact for all trigonometric polynomials up to degree $n-1$ inclusive. To do this it is sufficient to show that equation (5.2.4) will be exact for the functions e^{imx} $(m = 0, 1, \ldots, n-1)$. For $m = 0$ the assertion is evidently true. For $1 \leq m \leq n-1$

$$\int_0^{2\pi} e^{imx} dx = \frac{1}{im}\left(e^{im2\pi} - 1\right) = 0$$

and

$$\sum_{k=1}^{n} e^{im[a+(k-1)h]} = e^{ima} \sum_{k=1}^{n} e^{i(k-1)mh} = e^{ima} \frac{e^{imnh} - 1}{e^{imh} - 1} =$$

$$= e^{ima} \frac{e^{im2\pi} - 1}{e^{imh} - 1} = 0$$

which then proves the assertion.

5.3. THE REMAINDER IN APPROXIMATE QUADRATURE AND ITS REPRESENTATION

The value of the remainder

$$R(f) = \int_a^b p(x) f(x)\, dx - \sum_{k=1}^{n} A_k f(x_k) \tag{5.3.1}$$

of the quadrature depends on the choice of the quadrature formula, that is on the choice of the x_k and A_k, and also on the properties of the integrand function f. Formula (5.3.1) is one of the possible representations of the remainder, but to determine from it the influence of the structural properties[4] of $f(x)$ on $R(f)$ is very difficult. The expression (5.3.1) is defined for a very wide class of functions. It is valid for any function f for which the integral $\int_a^b pf dx$ exists and which has finite value at each node x_k. Because of its generality it does not take into account other properties of f.

In order to simplify the study of $R(f)$ we will construct another representation for it by which we can easily study the influence on $R(f)$ of such properties of the function f as its order of differentiability, the value of $\max_x |f(x)|$ and so forth. The representation which we will derive will be especially useful to determine how the structure of the class influences $R(f)$.

We will consider that we are given a set F of integrand functions f. The remainder $R(f)$ is a functional defined on the set F. In functional analysis there are theorems concerning the general forms of linear functionals defined on certain concrete linear spaces. These theorems can be used, in many cases, to construct a representation of the remainder term $R(f)$ for the set F.[5]

For this problem of finding the desired representation of the remainder we will make use of some simple methods of classical analysis.

If we consider a class F of functions which possess some structural property then it is often possible to give a formula which will represent each function of the class F and only functions of this class. Such a formula is called the *characteristic representation of the class F* or its *structural formula*.

If the structural formula of the class F is known then from it we can in principle obtain all of the necessary information about the class F and in particular we can construct the representation of the remainder in the quadrature for functions of the class F. Such a representation will be constructed each time that it is required in the presentation.

We will now give one example to illustrate the above remarks.

We say that the function f belongs to the class $C_r[a, b]$ if it has r continuous derivatives on $[a, b]$. The characteristic representation of a function of this class is furnished by its Taylor series. If $f \in C_r[a, b]$ and if α is any point of the segment $[a, b]$ then

---

[4]By "structural properties" of the function we mean such properties as bounded variation, absolute continuity, satisfaction of a Lipshitz condition, belonging to a certain class of differentiability and so forth.

[5]See the references at the end of this chapter.

$$f(x) = \sum_{i=0}^{r-1} \frac{f^{(i)}(a)}{i!}(x-a)^i + \int_a^x f^{(r)}(t)\frac{(x-t)^{r-1}}{(r-1)!}dt. \qquad (5.3.2)$$

It will be convenient to replace the integral having a variable limit by a definite integral over the segment $[a, b]$. This can be done by introducing the "jump" function which annihilates the superfluous section of integration. We define $E(x)$ by

$$E(x) = \begin{cases} 1 & \text{for} \quad x > 0 \\ \frac{1}{2} & \text{for} \quad x = 0 \\ 0 & \text{for} \quad x < 0. \end{cases}$$

It is easy to verify that equation (5.3.2) can be written in the form

$$f(x) = \sum_{i=0}^{r-1} \frac{f^{(i)}(a)}{i!}(x-a)^i + \int_a^b f^{(r)}(t)[E(x-t) -$$

$$\qquad (5.3.3)$$

$$- E(a-t)]\frac{(x-t)^{r-1}}{(r-1)!}dt.$$

On the righthand side of (5.3.3) there are r numerical parameters $f^{(i)}(a)$ $(i = 0, 1, \ldots, r-1)$ and the functional parameter $f^{(r)}(t)$ which is a continuous function on $[a, b]$.

Each function f of $C_r[a, b]$ has a representation of the form (5.3.3). Conversely, for any numerical parameters $f^{(i)}(a)$ $(i = 0, 1, \ldots, r-1)$ and any function $f^{(r)}(t)$ continuous on $[a, b]$, the function $f(x)$ defined by equation (5.3.3) belongs to $C_r[a, b]$.

If the interval of integration is not the entire real axis then, in order not to introduce an additional parameter, a is often taken as one of the end points of $[a, b]$. For example, if we take a as the left end point a, then formula (5.3.3) has the simplified form:

$$f(x) = \sum_{i=0}^{r-1} \frac{f^{(i)}(a)}{i!}(x-a)^i + \int_a^b f^{(r)}(t)E(x-t)\frac{(x-t)^{r-1}}{(r-1)!}dt \qquad (5.3.4)$$

where $r \geq 1$.

Where there is no ambiguity in the designation of the class of functions $C_r[a, b]$ the symbol $[a, b]$ will be omitted.

Let the integrand f belong to the class C_r. We will attempt to determine how the r-fold differentiability of f affects the remainder and the convergence of the quadrature process. To do this we obtain a representation for $R(f)$ which is characteristic of the class C_r. This can be found if we replace in (5.3.1) the expression (5.3.3) for f:

$$R(f) = \sum_{i=0}^{r-1} \frac{f^{(i)}(a)}{i!} R[(x-a)^i] +$$

$$+ R \int_a^b f^{(r)}(t) [E(x-t) - E(a-t)] \frac{(x-t)^{r-1}}{(r-1)!} dt \; . \tag{5.3.5}$$

In the double integral

$$\int_a^b p(x) \int_a^b f^{(r)}(t)[E(x-t) - E(a-t)] \frac{(x-t)^{r-1}}{(r-1)!} dt dx$$

which occurs in the last term on the righthand side of (5.3.5), we assume that we can change the order of integration. By the assumptions that we have made about the weight function $p(x)$ this is certainly possible if $[a, b]$ is a finite segment. Then (5.3.5) can be written as

$$R(f) = \sum_{i=0}^{r-1} \frac{f^{(i)}(a)}{i!} R[(x-a)^i] + \int_a^b f^{(r)}(t) K(t) dt \tag{5.3.6}$$

where the kernel $K(t)$ has the form

$$K(t) = \int_a^b p(x) \; [E(x-t) - E(a-t)] \frac{(x-t)^{r-1}}{(r-1)!} \; dx -$$

$$- \sum_{k=1}^n A_k [E(x_k-t) - E(a-t)] \frac{(x_k-t)^{r-1}}{(r-1)!} . \tag{5.3.7}$$

If $t \neq a$ and $t \neq x_k$ $(k = 1, \ldots, n)$ then we easily obtain the following equations for $K(t)$:

$$t < a, \;\; K(t) = -\int_a^t p(x) \frac{(x-t)^{r-1}}{(r-1)!} \; dx + \sum_{x_k < t} A_k \frac{(x_k-t)^{r-1}}{(r-1)!}$$

$$t > a, \;\; K(t) = \int_t^b p(x) \frac{(x-t)^{r-1}}{(r-1)!} dx - \sum_{x_k > t} A_k \frac{(x_k-t)^{r-1}}{(r-1)!} . \tag{5.3.8}$$

Analogously, we can construct representations for the remainder for other classes of functions when we know a characteristic representation for them, for example, for analytic functions.

In Chapters 8 and 12 we will see that the specialized representation of the remainder which we discussed above permits a sufficiently simple solution of the problems of finding a precise estimate for $R(f)$ and of convergence of the quadrature process for certain classes of functions.

REFERENCES

A. Kneschke, "Theorie der genaherten Quadratur," *J. Reine Angew. Math.*, Vol. 187, 1949, pp. 115–28.

J. Radon, "Restausdrucke bei Interpolations und Quadraturformeln durch bestimmte Integrale," *Monatsh. Math. Phys.*, Vol. 42, 1935, pp. 389–96.

E. Ia. Remez, "On certain classes of linear functionals in the spaces C_r and on the remainder term in formulas of approximate analysis I, II," *Trudy In-Ta. Mat. Akad. Nauk Ukrain. SSR*, Vol. 3, 1939, pp. 21–62; Vol. 4, 1940, pp. 47–82 (Ukrainian).

E. Ia. Remez, "On the remainder terms in certain formulas of approximate analysis," *Dokl. Akad. Nauk SSSR*, Vol. 26, 1940, pp. 130–4 (Russian).

A. Sard, "Integral representations of remainders," *Duke Math. J.*, Vol. 15, 1948, pp. 333–45.

CHAPTER 6

Interpolatory Quadratures

6.1. INTERPOLATORY QUADRATURE FORMULAS AND THEIR REMAINDER TERMS

Quadrature formulas are often constructed from interpolating polynomials. In this way we can, in many cases, obtain quadrature formulas which are convenient to use and which will give sufficiently accurate results.

Let us choose n arbitrary points $x_1, x_2, \ldots, x_n$ in the segment $[a, b]$ and, using these points,[1] construct the interpolating polynomial for $f(x)$:

$$f(x) = P(x) + r(x) \tag{6.1.1}$$

$$P(x) = \sum_{k=1}^{n} \frac{\omega(x)}{(x - x_k)\omega'(x_k)} f(x_k),$$

$$\omega(x) = (x - x_1) \cdots (x - x_k). \tag{6.1.2}$$

Here $r(x)$ is the remainder of the interpolation.

The exact value of the integral $\displaystyle\int_a^b p(x) f(x) dx$ is

$$\int_a^b p(x) f(x) dx = \int_a^b p(x) P(x) dx + \int_a^b p(x) r(x) dx.$$

---

[1] If we assume that $f(x)$ is defined only on the segment $[a, b]$ then we must choose the x_k to belong to $[a, b]$. If $f(x)$ is also defined outside the segment of integration then it is not necessary that all the x_k belong to $[a, b]$. Quadrature formulas which contain nodes lying outside $[a, b]$ can be used for the integration of analytic functions. It is usually desirable, however, to have the points belong to the segment of integration.

If the interpolation (6.1.1) is sufficiently precise so that the remainder $r(x)$ is small throughout the segment $[a, b]$ then the second term in this last equation can be neglected. Thus we obtain the approximate equation

$$\int_a^b p(x)f(x)dx \approx \sum_{k=1}^n A_k f(x_k) \qquad (6.1.3)$$

where

$$A_k = \int_a^b p(x) \frac{\omega(x)}{(x - x_k)\omega'(x_k)} dx. \qquad (6.1.4)$$

Quadrature formulas of the form (6.1.3), in which the coefficients have the form (6.1.4), are called *interpolatory* quadrature formulas. Interpolatory quadrature formulas can be characterized by the following theorem:

Theorem 1. *In order that the quadrature formula* (6.1.3) *be interpolatory it is necessary and sufficient that it be exact for all possible polynomials of degree less than or equal to* $n - 1$.

Proof. Each polynomial $P(x)$ of degree $\leq n - 1$ can be represented in the form $P(x) = \displaystyle\sum_{k=1}^n \frac{\omega(x)}{(x - x_k)\omega'(x_k)} P(x_k)$. If we take the coefficients A_k to have the values (6.1.4) then the quadrature formula (6.1.3) will be exact for $P(x)$.

In the previous paragraph the values $P(x_k)$, in the representation for $P(x)$, may be any real, finite numbers. The requirement that (6.1.3) be exact for all polynomials of degree $\leq n - 1$ is equivalent to requiring that the equation

$$\int_a^b p(x) \sum_{k=1}^n \frac{\omega(x)}{(x - x_k)\omega'(x_k)} P(x_k)dx = \sum_{k=1}^n A_k P(x_k)$$

be valid for every set of $P(x_k)$. But then the coefficients A_k must have the values (6.1.4) and formula (6.1.3) will be interpolatory. This completes the proof.

This theorem shows that specifying the n nodes x_k will completely define the quadrature formula—that is, the coefficients A_k will also be completely determined—if we require that the formula be exact for each polynomial of degree $\leq n - 1$. The nodes, however, may still be chosen in any manner we desire in order to make the quadrature formula meet some special demand.

Everything that was said in Section 5.3 holds true for the remainder of an interpolatory quadrature formula (6.1.3). In addition we can obtain, for this type of quadrature formula, a few deeper results.

The remainder of the quadrature (6.1.3) is the integral of the remainder $r(x)$ of the interpolation,

$$R(f) = \int_a^b p(x)r(x)dx = \int_a^b p(x)\omega(x)f(x, x_1, \ldots, x_n)dx. \qquad (6.1.5)$$

To study $R(f)$ we can now use theorems concerning the remainder $r(x)$. For example, if $f(x)$ has n continuous derivatives on $[a, b]$ then $r(x)$ can be represented in the form (3.2.8). Using the notation of Chapter 3 we obtain the following expression for the remainder $R(f)$:

$$R(f) = \int_a^b \int_0^1 \int_0^{t_1} \cdots \int_0^{t_{n-1}} p(x)\omega(x) \times$$

$$\times f^{(n)}\left(x + \sum_{\nu=1}^n t_\nu(x_\nu - x_{\nu-1})\right)dt_n \cdots dt_2\, dt_1\, dx \qquad (6.1.6)$$

where $x = x_0$. It is often preferable to use the simpler expression for $R(f)$ which is obtained from the Lagrangian form of $r(x)$:

$$r(x) = \frac{1}{n!}\omega(x)f^{(n)}(\xi), \qquad a < \xi < b$$

$$R(f) = \frac{1}{n!}\int_a^b p(x)\omega(x)f^{(n)}(\xi)\, dx. \qquad (6.1.7)$$

It is difficult to find an exact estimate for $R(f)$ from (6.1.7) because we cannot determine how ξ depends on x.

If the nth derivative of $f(x)$ is bounded in absolute value on $[a, b]$ by the number M_n:

$$|f^{(n)}(x)| \le M_n, \qquad x \in [a, b] \qquad (6.1.8)$$

then from (6.1.7) we obtain the estimate

$$|R(f)| \le \frac{M_n}{n!}\int_a^b |p(x)\omega(x)|dx. \qquad (6.1.9)$$

If $p(x)\omega(x)$ does not change sign on $[a, b]$ then the estimate (6.1.9) cannot be improved. For an arbitrary $p(x)$ and an arbitrary set of n nodes x_k we can obtain a precise estimate for $R(f)$ for any function satisfying (6.1.8). To do this we use (5.3.6). If in that equation we put $r = n$ and use the fact that the remainder in the quadrature formula is zero for every polynomial of degree $< n$, then we obtain

$$R(f) = \int_a^b f^{(n)}(t) K(t) dt \qquad (6.1.10)$$

where $K(t)$ has the form (5.3.7). For a function which satisfies (6.1.8) we obtain from (6.1.10) the precise estimate[2]

$$|R(f)| \leq M_n \int_a^b |K(t)| dt. \qquad (6.1.11)$$

6.2. NEWTON-COTES FORMULAS

The earliest known quadrature formulas are those which are now known as the Newton-Cotes formulas. Some of these are still widely used because of their simplicity. They are formulas for a constant weight function and a finite interval of integration.

Let us consider the integral

$$\int_a^b f(x) dx. \qquad (6.2.1)$$

Let us divide the segment $[a, b]$ into n equal subsegments, of length $h = \dfrac{b-a}{h}$, with endpoints a, $a + h$, $a + 2h$, ..., $a + nh = b$. We will construct an interpolatory quadrature formula using these points as the nodes. To find the values of the coefficients A_k in a form which is independent of the segment $[a, b]$ let us write (6.1.3) in the form

$$\int_a^b f(x) dx = (b - a) \sum_{k=0}^n B_k^n f(a + kh). \qquad (6.2.2)$$

The coefficients $B_k^n = (b - a)^{-1} A_k$ are given by:

$$B_k^n = (b - a)^{-1} \int_a^b \frac{\omega(x)}{(x - a - kh)\omega'(a + kh)} dx,$$

where $\omega(x) = (x - a)(x - a - h) \cdots (x - a - nh)$. If we introduce a new variable t, by substituting $x = a + th$, then we will have

$$x - a - kh = h(t - k),$$

$$\omega(x) = h^{n+1} t(t - 1)(t - 2) \cdots (t - n),$$

---

[2]Trans. note: For a better discussion of this result, and some examples, see the book by S. M. Nikol'skii listed in the references at the end of this chapter.

$$\omega'(a + kh) = (-1)^{n-k} h^n k!(n - k)!.$$

This gives

$$B_k^n = \frac{(-1)^{n-k}}{nk!(n - k)!} \int_0^n t(t - 1) \cdots (t - k + 1) \times$$

$$\times (t - k - 1) \cdots (t - n)\, dt. \qquad (6.2.3)$$

Here we give[3] the values of the coefficients B_k^n for $n = 1$ to 10. Since $B_k^n = B_{n-k}^n$ we have tabulated only those coefficients for $k \leq \frac{1}{2}n$.

n	B_0^n	B_1^n	B_2^n	B_3^n	B_4^n	B_5^n
1	$\dfrac{1}{2}$					
2	$\dfrac{1}{6}$	$\dfrac{4}{6}$				
3	$\dfrac{1}{8}$	$\dfrac{3}{8}$				
4	$\dfrac{7}{90}$	$\dfrac{32}{90}$	$\dfrac{12}{90}$			
5	$\dfrac{19}{288}$	$\dfrac{75}{288}$	$\dfrac{50}{288}$			
6	$\dfrac{41}{840}$	$\dfrac{216}{840}$	$\dfrac{27}{840}$	$\dfrac{272}{840}$		
7	$\dfrac{751}{17280}$	$\dfrac{3577}{17280}$	$\dfrac{1323}{17280}$	$\dfrac{2989}{17280}$		
8	$\dfrac{989}{28350}$	$\dfrac{5888}{28350}$	$\dfrac{-928}{28350}$	$\dfrac{10496}{28350}$	$\dfrac{-4540}{28350}$	
9	$\dfrac{2857}{89600}$	$\dfrac{15741}{89600}$	$\dfrac{1080}{89600}$	$\dfrac{19344}{89600}$	$\dfrac{5778}{89600}$	
10	$\dfrac{16067}{598752}$	$\dfrac{106300}{598752}$	$\dfrac{-48525}{598752}$	$\dfrac{272400}{598752}$	$\dfrac{-260550}{598752}$	$\dfrac{427368}{598752}$

---

[3]Trans. note: The values of B_k^n for $n = 1$ to 20 are given in the paper by W. W. Johnson and in the book by Z. Kopal.

Even this short table of the B_k^n shows the irregularity of these coefficients. In order to appraise the Newton-Cotes formulas for a large number of nodes we will derive[4] asymptotic representations for the B_k^n for large n. To do this let us transform the integral

$$I = \int_0^n \frac{x(x-1)\cdots(x-n)}{x-k}\,dx$$

occurring in (6.2.3). Using the relationships

$$x(x-1)\cdots(x-n) = \frac{\Gamma(x+1)}{\Gamma(x-n)}$$

$$\frac{1}{\Gamma(z)} = \frac{\Gamma(1-z)\sin\pi z}{\pi}$$

we obtain

$$x(x-1)\cdots(x-n) = \frac{(-1)^n}{\pi}\Gamma(x+1)\Gamma(n+1-x)\sin\pi x$$

$$I = (-1)^n \int_0^n \frac{\Gamma(x+1)\Gamma(n+1-x)\sin\pi x}{\pi(x-k)}\,dx.$$

Let us divide this integral into 3 parts:

$$\int_0^n = \int_0^3 + \int_3^{n-3} + \int_{n-3}^n = \alpha + \beta + \gamma.$$

We will first obtain an estimate for the integral β. From the theory of the function $\Gamma(z)$ it is known[5] that $\dfrac{\Gamma'(z)}{\Gamma(z)}$ is a monotonically increasing function for $z > 0$. Thus $\dfrac{\Gamma'(x+1)}{\Gamma(x+1)} - \dfrac{\Gamma'(n+1-x)}{\Gamma(n+1-x)}$, for $-1 < x < \dfrac{n}{2}$, will be negative and, for $\dfrac{n}{2} < x < n+1$, will be positive. From this it

---

[4]See the paper by R. O. Kuz'min.

[5]This can be seen from the expansion

$$\frac{\Gamma'(z)}{\Gamma(z)} = -\frac{1}{z} - C + \sum_{k=1}^{\infty}\left(\frac{1}{k} - \frac{1}{k+z}\right).$$

See, for example, V. I. Smirnov, *Course of Higher Mathematics*, Gostekhizdat, Moscow, 1949, Vol. 3, sec. 73 (Russian); or E. D. Rainville, *Special Functions*, Macmillan, New York, 1960, p. 10.

follows that $ln \, \Gamma(x + 1)\Gamma(n + 1 - x)$ and also $\Gamma(x + 1)\Gamma(n + 1 - x)$ will, for $3 \leq x \leq n - 3$, have its largest value on the end of this segment:

$$0 < \Gamma(x + 1)\Gamma(n + 1 - x) \leq \Gamma(4)\Gamma(n - 2) = 6\Gamma(n - 2).$$

For every x we have $\left| \dfrac{\sin \pi x}{\pi(x - k)} \right| \leq 1$ and thus

$$|\beta| \leq 6n\Gamma(n - 2) = \frac{6\Gamma(n + 1)}{(n - 2)(n - 1)} = O\left(\frac{\Gamma(n + 1)}{n^2}\right).$$

We will, at first, study the integrals α and γ for $1 \leq k \leq n - 1$. It will be sufficient to consider the integral α. Using Taylor's formula and the fact that the derivative of the function $\dfrac{\Gamma'(z)}{\Gamma(z)}$ is, for large z, of the order $\dfrac{1}{z}$ we obtain:

$$ln \, \Gamma(n + 1 - x) = ln \, \Gamma(n + 1) - \frac{x\Gamma'(n + 1)}{\Gamma(n + 1)} + O\left(\frac{1}{n}\right).$$

Thus using the fact that for large z the approximation $\dfrac{\Gamma'(z)}{\Gamma(z)} = ln \, z + O\left(\dfrac{1}{z}\right)$ is valid,[6] we obtain

$$\Gamma(n + 1 - x) = \Gamma(n + 1) \, e^{-x \, ln \, n} \left[1 + O\left(\frac{1}{n}\right)\right].$$

For $0 \leq x \leq 3$ it is then evident that we have

$$\Gamma(x + 1) \, \frac{\sin \pi x}{\pi(x - k)} = -\frac{x}{k} + O\left(\frac{x^2}{k}\right),$$

$$\alpha = \int_0^3 \Gamma(n + 1) \, e^{-x \, ln \, n} \left[1 + O\left(\frac{1}{n}\right)\right] \left[-\frac{x}{k} + O\left(\frac{x^2}{k}\right)\right] dx.$$

Because

$$\int_0^3 e^{-x \, ln \, n} \, x \, dx = \frac{1}{ln^2 \, n} - \frac{1}{n^3}\left[\frac{3}{ln \, n} + \frac{1}{ln^2 \, n}\right]$$

---

[6]See, for example, E. Jahnke and F. Emde, *Tables of Functions*, 4th ed., New York, 1945, p. 19; or N. Nielsen, *Handbuch der Theorie der Gammafunktion*, Leipzig, 1906, p. 15.

and

$$\int_0^3 e^{-x \, ln \, n} x^2 \, dx = \frac{2}{ln^3 \, n} - \frac{1}{n^3}\left[\frac{9}{ln \, n} + \frac{6}{ln^2 \, n} + \frac{2}{ln^3 \, n}\right]$$

then

$$a = -\frac{\Gamma(n+1)}{k \, ln^2 \, n}\left[1 + O\left(\frac{1}{ln \, n}\right)\right]$$

for $1 \le k \le n-1$. In a similar way we can obtain the following estimate for the integral γ:

$$\gamma = (-1)^{n-1}\frac{\Gamma(n+1)}{(n-k) \, ln^2 \, n}\left[1 + O\left(\frac{1}{ln \, n}\right)\right]$$

for $1 \le k \le n-1$.

If we use the estimate which we obtained above for the integral β, then we obtain the following estimate for the integral I:

$$I = \frac{(-1)^{n-1}\Gamma(n+1)}{ln^2 n}\left[\frac{1}{k} + \frac{(-1)^n}{n-k}\right]\left[1 + O\left(\frac{1}{ln \, n}\right)\right].$$

This leads to the asymptotic representation of the Newton-Cotes coefficients for $1 \le k \le n-1$:

$$B_k^n = \frac{(-1)^{k-1}n!}{k! \, (n-k)! \, n \, ln^2 n}\left[\frac{1}{k} + \frac{(-1)^n}{(n-k)}\right]\left[1 + O\left(\frac{1}{ln \, n}\right)\right]. \quad (6.2.4)$$

A similar calculation for B_0^n and B_n^n gives

$$B_0^n = B_n^n = \frac{1}{n \, ln \, n}\left[1 + O\left(\frac{1}{ln \, n}\right)\right]. \quad (6.2.5)$$

From these expressions for B_k^n we see that for large n the Newton-Cotes formulas will have both positive and negative coefficients which exceed in absolute value any arbitrary large number. Thus, for large n, a small discrepancy in the values of the function $f(a + kh)$ can lead to a large error in the quadrature sum. Therefore the Newton-Cotes formulas with large numbers of nodes are of little use for practical calculations.

The expression (6.1.5) for the remainder $R(f)$ of the Newton-Cotes formulas is:

$$R(f) = \int_a^b \omega(x)f(x, a, a+h, \dots, a+nh) \, dx. \quad (6.2.6)$$

This equation can be reduced to a very simple form which is much more convenient for application.

6.2. Newton-Cotes Formulas 87

Let us consider, at first, the case when n is an even number; in this case the Newton-Cotes formulas have an odd number of nodes. The polynomial $\omega(x) = (x - a)(x - a - h) \cdots (x - a - nh)$ possesses the property $\omega(a + z) = -\omega(a + nh - z)$ and the graph of it will be symmetric with respect to the midpoint $\dfrac{a + b}{2}$ of the segment $[a, b]$. An example of the form of the graph is illustrated by Figure 2.

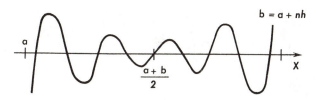

Figure 2.

Let us consider the function $\Omega(x) = \displaystyle\int_a^x \omega(t)\,dt$. We note, first of all, that $\Omega(a) = 0$ and $\Omega(a + nh) = \Omega(b) = 0$. This last equation follows from the symmetry properties of the function $\omega(x)$. We show now that $\Omega(x)$ is not zero anywhere inside $[a, b]$. To do this consider the integrals $I_\nu = \displaystyle\int_{a+\nu h}^{a+(\nu+1)h} \omega(x)\,dx$. The assertion will be proved if we establish that the sequence of numbers $I_0, I_1, \ldots, I_{\frac{n}{2}-1}$ decrease in absolute value.

If in the integral $I_\nu = \displaystyle\int_{a+\nu h}^{a+(\nu+1)h} (x - a)(x - a - h) \cdots (x - a - nh)\,dx$

we set $x = y + h$, then the integral is transformed to the form

$$I_\nu = \int_{a+(\nu-1)h}^{a+\nu h} (y - a + h) \cdots (y - a) \cdots (y - a - (n - 1)h)\,dy =$$

$$= \int_{a+(\nu-1)h}^{a+\nu h} \frac{y - a + h}{y - a - nh}\,\omega(y)\,dy = \frac{\eta - a + h}{\eta - a - nh}\,I_{\nu-1},$$

where $a + (\nu - 1)h < \eta < a + \nu h$. In order that $|I_\nu| < |I_{\nu-1}|$ the inequality $\eta - a + h < nh - \eta + a$ or $\eta - a < \frac{1}{2}(n - 1)h$ must be satisfied. But this last inequality is indeed true because

$$\eta < a + \nu h, \qquad \eta - a < \nu h \leq \left(\frac{n}{2} - 1\right) h.$$

Let us integrate (6.2.6) by parts and apply the mean value theorem:

$$R(f) = \Omega(x) f(x, a, \ldots, a + nh) \bigg|_a^b - \int_a^b f_x'(x, a, \ldots, a + nh) \, \Omega(x) \, dx =$$

$$= -f_x'(\eta, a, \ldots, a + nh) \int_a^b \Omega(x) \, dx, \qquad a < \eta < b.$$

From

$$f(x, a, \ldots, a + nh) =$$

$$= \int_0^1 \cdots \int_0^{t_n} f^{(n+1)} \left(x + t_1(a - x) + \cdots + h \sum_{\nu=2}^{n+1} t_\nu \right) dt_{n+1} \cdots dt_1$$

it follows that

$$f_x'(x, a, \ldots, a + nh) = \int_0^1 \cdots \int_0^{t_n} (1 - t_1) \times$$

$$\times f^{(n+2)} \left(x + t_1(a - x) + \cdots + h \sum_{\nu=2}^{n+1} t_\nu \right) dt_{n+1} \cdots dt_1$$

and applying the mean value theorem to this last integral gives

$$f_x'(\eta, a, \ldots, a + nh) = \frac{f^{(n+2)}(\xi)}{(n + 2)!}, \qquad a < \xi < b.$$

Finally

$$\int_a^b \Omega(x) \, dx = x \Omega(x) \bigg|_a^b - \int_a^b x \Omega'(x) \, dx = -\int_a^b x \omega(x) \, dx.$$

This proves that the remainder term of the Newton-Cotes quadrature formula, for an odd number of nodes, can be expressed as

$$R(f) = \frac{f^{(n+2)}(\xi)}{(n + 2)!} \int_a^b x \omega(x) \, dx. \qquad (6.2.7)$$

We will now find the sign of the coefficient of $f^{(n+2)}(\xi)$. The function $\Omega(x) = \int_a^x \omega(t) \, dt$ does not change sign on the segment $[a, b]$ and therefore it is sufficient to calculate its sign at one point; let us use the point $x = a + h$:

$$\Omega(a + h) = \int_a^{a+h} \omega(t) \, dt.$$

For $a < t < a + h$ the first factor in the product $\omega(t) = (t - a)(t - a - h) \cdots (t - a - nh)$ is positive and all the remaining factors are negative. Thus the sign of $\Omega(t)$ is $(-1)^n$ for $t \in (a, b)$. Because

$$\int_a^b x\omega(x)\,dx = -\int_a^b \Omega(x)\,dx$$

it follows that the sign of $\int_a^b x\omega(x)\,dx$ is $(-1)^{n+1} = -1$ since n is even.

Thus we have established the theorem:

Theorem 2. *If the number of nodes, which is $n + 1$, in the Newton-Cotes formula (6.2.2) is odd and if the function $f(x)$ has a continuous derivative of order $n + 2$ on $[a, b]$, then the expression for $R(f)$ is given by (6.2.7) where ξ is a point inside $[a, b]$. The coefficient of $f^{(n+2)}(\xi)$ is negative.*

We indicate two consequences of this theorem.

1. If the number of nodes in formula (6.2.2) is odd then the algebraic degree of precision of this formula is $n + 1$.

From the representation (6.2.7) for the error, formula (6.2.2) will be exact whenever $f(x)$ is a polynomial of degree $\leq n + 1$. If $f(x)$ is a polynomial of degree $n + 2$ then $f^{(n+2)}(x)$ will be different from zero and $R(f) \neq 0$.

2. Let us assume that $f^{(n+2)}(x)$ exists and is continuous on $[a, b]$. We will construct the representation (5.3.6) for the error. We have $r = n + 2$ and for simplicity we take $\alpha = a$. Because the degree of precision of (6.2.2) is $n + 1$, the terms under the summation sign in (5.3.6) will be zero and we will have the following expression for the error:

$$R(f) = \int_a^b f^{(n+2)}(t) K(t)\,dt. \tag{6.2.8}$$

Using the fact that $p(x) \equiv 1$, the kernel $K(t)$ is calculated to be

$$K(t) = \frac{(b - t)^{n+2}}{(n + 2)!} - \sum_{k=1}^n A_k E(a + kh - t)\frac{(a + kh - t)^{n+1}}{(n + 1)!}.$$

We can show that the function $K(t)$ is nonpositive on $[a, b]$.

From (6.2.7) we see that if $f^{(n+2)}(x)$ does not become zero at any point of $[a, b]$ then $R(f) \neq 0$. If the function $K(t)$ would change sign on $[a, b]$ then there would exist a function $f^{(n+2)}(x)$, which is different from zero throughout $[a, b]$, for which $\int_a^b f^{(n+2)}(t) K(t)\,dt = 0$. From the derivative

$f^{(n+2)}(x)$ we could reconstruct the function $f(x)$ in the usual way. For such a function $R(f) = 0$ which contradicts our assumption. Because the coefficient of $f^{(n+2)}(\xi)$ in (6.2.7) is negative the kernel $K(t)$ must be a nonpositive function on $[a, b]$:

$$K(t) \leq 0.$$

Let us now consider the case when n is an odd number; in this case there are an even number of nodes in formula (6.2.2). The polynomial $\omega(x)$ takes on equal values at the points $a + t$ and $b - t$, $t < b - a$. This means that the graph of $\omega(x)$ is symmetric with respect to the line $x = \dfrac{a+b}{2}$.

In order to simplify the expression (6.2.6) for the remainder let us split the segment $[a, b]$ into two parts $[a, a + (n-1)h]$ and $[a + (n-1)h, b]$. The polynomial $\omega(x)$ does not change sign on the second part of the segment and we can apply the mean value theorem to the integral over this segment:

$$R(f) = \int_a^{a+(n-1)h} \omega(x) f(x, a, \ldots, a + nh)\, dx +$$

$$+ \frac{f^{(n+1)}(\xi_1)}{(n+1)!} \int_{a+(n-1)h}^b \omega(x)\, dx = I + II.$$

Let us now look at the integral over the first part of the segment. From $\omega(x)$ we separate the factor $x - a - nh$ and write $\omega(x) = (x - a - nh)\omega_1(x)$. By the definition of the divided difference

$$f(a + nh, \ldots, a, x) = \frac{f[a + (n-1)h, \ldots, a, x] - f[a + nh, \ldots, a]}{x - a - nh}$$

and thus

$$I = \int_a^{a+(n-1)h} \omega_1(x) f(x, a, \ldots, a + (n-1)h)\, dx -$$

$$- f(a, \ldots, a + nh) \int_a^{a+(n-1)h} \omega_1(x)\, dx.$$

Because n is an even number $\displaystyle\int_a^{a+(n-1)h} \omega_1(x)\, dx = 0$, and the second

term in the above expression for I vanishes. The first term is an integral of the form (6.2.6) for an odd number of nodes and it can be expressed as:

$$I = \frac{f^{(n+1)}(\xi_2)}{(n+1)!} \int_a^{a+(n-1)h} x\omega_1(x)\,dx,$$

where we recall that the coefficient of $f^{(n+1)}(\xi_2)$ is a negative number.

Since $\int_a^{a+(n-1)h} \omega_1(x)\,dx = 0$, we can replace $x\omega_1(x)$ by $(x-a-nh) \times$

$\times\, \omega_1(x) = \omega(x)$ in the integral I. Thus we obtain the expression

$$R(f) = \frac{f^{(n+1)}(\xi_2)}{(n+1)!} \int_a^{a+(n-1)h} \omega(x)\,dx + \frac{f^{(n+1)}(\xi_1)}{(n+1)!} \int_{a+(n-1)h}^b \omega(x)\,dx.$$

For $a + (n-1)h < x < b$ the last factor in

$$\omega(x) = (x-a)(x-a-h)\cdots(x-a-nh)$$

is negative and the other factors are positive. This means that

$$\int_{a+(n-1)h}^b \omega(x)\,dx < 0.$$

Since the coefficients of both $f^{(n+1)}(\xi_2)$ and $f^{(n+1)}(\xi_1)$, in the last expression for $R(f)$, are different from zero and of the same sign and since $f^{(n+1)}(x)$ is a continuous function, then between ξ_1 and ξ_2 there exists a point ξ for which

$$R(f) = \frac{f^{(n+1)}(\xi)}{(n+1)!} \int_a^b \omega(x)\,dx. \qquad (6.2.9)$$

This proves:

Theorem 3. *If the number of nodes in the Newton-Cotes formula* (6.2.2) *is even and if $f(x)$ has a continuous derivative of order $n+1$ on $[a, b]$ then the remainder $R(f)$ is given by* (6.2.9) *where ξ is a point inside the segment. The coefficient of $f^{(n+1)}(\xi)$ in this expression is negative.*

As in the case for an odd number of nodes there are two immediate consequences of this theorem.

1. If formula (6.2.2) has an even number of nodes $n+1$, then its algebraic degree of precision is $n+1$.

2. If (6.2.2) has an even number of nodes and if $f(x)$ has a continuous derivative of order $n+1$ on $[a, b]$ then the remainder in this formula can be represented in the form

$$R(f) = \int_a^b f^{(n+1)}(t)K(t)\,dt \qquad (6.2.10)$$

where the kernel $K(t)$ is nonpositive on $[a, b]$ and is given by

$$K(t) = \frac{(b-t)^{n+1}}{(n+1)!} - \sum_{k=1}^{n} A_k E(a + kh - t) \frac{(a+kh-t)^n}{n!}. \quad (6.2.11)$$

6.3. CERTAIN OF THE SIMPLEST NEWTON-COTES FORMULAS

Newton-Cotes formulas with a large number of nodes are seldom applied in practical calculations for the reasons pointed out in the previous section. It is preferable to use a formula with a small number of nodes and to increase its accuracy split up the segment $[a, b]$ into many subintervals and apply the formula to each of these smaller intervals.

Let us consider, at first, the case $n = 1$. Here we interpolate $f(x)$ using its values at the endpoints a, b of the segment of integration. Equation (6.2.2) then becomes the well known formula:

$$\int_a^b f(x)\,dx \approx (b-a)\left[\frac{1}{2}f(a) + \frac{1}{2}f(b)\right] \quad (6.3.1)$$

which is called the trapezoidal formula. In this case we have $\omega(x) = (x-a)(x-b)$ and (6.2.9) gives

$$R(f) = -\frac{(b-a)^3}{12} f''(\xi), \qquad a < \xi < b. \quad (6.3.2)$$

To study the error in the formula (6.3.1) when the interval of integration is split up into subsegments we will obtain a representation for the remainder which is different from (6.2.10).

Let us assume that $f(x)$ has a continuous second derivative. We will expand it in Bernoulli polynomials using equation (1.4.2) with $\nu = 2$:

$$f(x) = (b-a)^{-1} \int_a^b f(t)\,dt + B_1\left(\frac{x-a}{b-a}\right)[f(b) - f(a)] -$$

$$- \frac{(b-a)}{2} \int_a^b f''(t)\left[B_2^*\left(\frac{x-t}{b-a}\right) - B_2^*\left(\frac{x-a}{b-a}\right)\right]dt.$$

Because (6.3.1) is exact for linear functions, $R(f)$ reduces to the remainder of the quadrature for the last term on the right hand side of this equation. The remainder of this term is the same if we replace $B_2^*(z)$ by $y_2^*(z) = B_2^*(z) - B_2$:

$$R(f) = -\frac{(b-a)}{2} \int_a^b f''(t)\, R_x\left[y_2^*\left(\frac{x-t}{b-a}\right) - y_2^*\left(\frac{x-a}{b-a}\right)\right]dt.$$

The symbol R_x denotes the remainder of the quadrature with respect to the variable x.

In the following calculations we use the rule for integration of Bernoulli polynomials; the fact that y_2^* and B_2^* have period one; and the relations $y_2(1) = y_2(0) = 0$

$$R_x\left[y_2^*\left(\frac{x-t}{b-a}\right) - y_2^*\left(\frac{x-a}{b-a}\right)\right] = \int_a^b\left[y_2^*\left(\frac{x-t}{b-a}\right) - y_2^*\left(\frac{x-a}{b-a}\right)\right]dt -$$

$$-\frac{(b-a)}{2}\left\{\left[y_2^*\left(\frac{a-t}{b-a}\right) - y_2^*(0)\right] + \left[y_2^*\left(\frac{b-t}{b-a}\right) - y_2^*(1)\right]\right\} =$$

$$= -(b-a)\,y_2^*\left(\frac{b-t}{b-a}\right),$$

$$R(f) = \frac{(b-a)^2}{2!}\int_a^b f''(t)\,y_2^*\left(\frac{b-t}{b-a}\right)\,dt. \tag{6.3.3}$$

Now we split up the segment $[a, b]$ into n equal subsegments of length $h = \frac{b-a}{n}$. Consider the segment $[a + kh, a + (k+1)h]$ and let us apply formula (6.3.1)

$$\int_{a+kh}^{a+(k+1)h} f(x)\,dx = \frac{h}{2}\left[f(a+kh) + f(a+(k+1)h)\right] +$$

$$+ \frac{h^2}{2!}\int_{a+kh}^{a+(k+1)h} f''(t)\,y_2^*\left(\frac{a+kh-t}{h}\right)\,dt.$$

Since $y_2^*(x)$ has period one we have $y_2^*\left(\frac{a+kh-t}{h}\right) = y_2^*\left(\frac{a-t}{h}\right)$. We carry out this calculation for each subsegment and sum the results to obtain the repeated trapezoidal formula with remainder in the form of a definite integral

$$\int_a^b f(x)\,dx = h\left[\frac{1}{2}\,f_0 + f_1 + \cdots + f_{n-1} + \frac{1}{2}\,f_n\right] +$$

$$+ \frac{h^2}{2!}\int_a^b f''(t)\,y_2^*\left(\frac{a-t}{h}\right)\,dt \tag{6.3.4}$$

where we have written $f_k = f(a+kh)$. The kernel $y_2^*\left(\frac{a-t}{h}\right)$ of the remainder does not change sign and the mean value theorem can be applied to the integral in (6.3.4) to give

$$R(f) = -\frac{(b-a)^3}{12n^2} f''(\xi).$$

We go now to the case $n = 2$. Here we interpolate $f(x)$ using its values at the three points a, $\dfrac{a+b}{2}$, b.

Quadrature formula (6.2.2) then becomes

$$\int_a^b f(x)\,dx \approx (b-a)\left[\frac{1}{6}f(a) + \frac{4}{6}f\left(\frac{a+b}{2}\right) + \frac{1}{6}f(b)\right] \qquad (6.3.5)$$

which is known as Simpson's formula. The remainder is found by (6.2.7) to be

$$R(f) = \frac{f^{(4)}(\xi)}{4!} \int_a^b x(x-a)\left(x - \frac{a+b}{2}\right)(x-b)\,dx =$$

$$= -\frac{1}{90}\left(\frac{b-a}{2}\right)^5 f^{(4)}(\xi). \qquad (6.3.6)$$

Assuming that $f(x)$ has four continuous derivatives on $[a, b]$ we expand it in Bernoulli polynomials as follows:

$$f(x) = (b-a)^{-1}\int_a^b f(t)\,dt +$$

$$+ \sum_{k=1}^{3} \frac{(b-a)^{k-1}}{k!} B_k^*\left(\frac{x-a}{b-a}\right)[f^{(k-1)}(b) - f^{(k-1)}(a)] -$$

$$- \frac{(b-a)^3}{4!}\int_a^b f^{(4)}(t)\left[y_4^*\left(\frac{x-t}{b-a}\right) - y_4^*\left(\frac{x-a}{b-a}\right)\right]dt. \quad (6.3.7)$$

Equation (6.3.5) is exact for all polynomials of third degree. Therefore $R(f)$ will be the remainder when the quadrature is applied to the last term on the right hand side of this equation:

$$R(f) = -\frac{(b-a)^3}{4!}\int_a^b f^{(4)}(t)\,R_x\left[y_4^*\left(\frac{x-t}{b-a}\right) - y_4^*\left(\frac{x-a}{b-a}\right)\right]dt. \quad (6.3.8)$$

$$R_x\left[y_4^*\left(\frac{x-t}{b-a}\right) - y_4^*\left(\frac{x-a}{b-a}\right)\right] = \int_a^b \left[y_4^*\left(\frac{x-t}{b-a}\right) - y_4^*\left(\frac{x-a}{b-a}\right)\right]dx -$$

$$-(b-a)\left\{\frac{1}{6}\left[y_4^*\left(\frac{a-t}{b-a}\right)-y_4^*(0)\right]+\right.$$

$$+\frac{4}{6}\left[y_4^*\left(\frac{\frac{1}{2}(a+b)-t}{b-a}\right)-y_4^*\left(\frac{1}{2}\right)\right]+$$

$$\left.+\frac{1}{6}\left[y_4^*\left(\frac{b-t}{b-a}\right)-y_4^*(1)\right]\right\}=$$

$$=-(b-a)\left\{\frac{1}{3}y_4^*\left(\frac{a-t}{b-a}\right)+\frac{2}{3}y_4^*\left(\frac{\frac{1}{2}(a+b)-t}{b-a}\right)-\frac{1}{24}\right\}.$$

In these calculations we have made use of the values $B_n\left(\frac{1}{2}\right)$ given in (1.2.14):

$$y_4^*\left(\frac{1}{2}\right)=B_4\left(\frac{1}{2}\right)-B_4=-(2-2^{-3})B_4=\frac{1}{16},$$

$$R(f)=\frac{(b-a)^4}{4!}\int_a^b f^{(4)}(t)\left\{\frac{1}{3}y_4^*\left(\frac{a-t}{b-a}\right)+\right.$$

$$\left.+\frac{2}{3}y_4^*\left(\frac{\frac{1}{2}(a+b)-t}{b-a}\right)-\frac{1}{24}\right\}dt. \tag{6.3.9}$$

Let us divide $[a, b]$ into n equal subsegments of length $h = \dfrac{b-a}{n}$ where n is an even integer. Let us apply formula (6.3.5) with remainder (6.3.9) to the segment $[a+(k-1)h, a+(k+1)h]$ which consists of an adjacent pair of subsegments:

$$\int_{a+(k-1)h}^{a+(k+1)h}f(x)dx=2h\left[\frac{1}{6}f_{k-1}+\frac{4}{6}f_k+\frac{1}{6}f_{k+1}\right]+$$

$$+\frac{2}{9}h^4\int_{a+(k-1)h}^{a+(k+1)h}f^{(4)}(t)\left\{y_4^*\left(\frac{a+(k-1)h-t}{2h}\right)+\right.$$

$$\left.+2y_4^*\left(\frac{a+kh-t}{2h}\right)-\frac{1}{8}\right\}dt.$$

Carrying out this last calculation for the segments

$$[a, a + 2h], \quad [a + 2h, a + 4h], \ldots, \quad [a + (n - 2)h, a + nh]$$

and summing the results we obtain the repeated Simpson's rule:

$$\int_a^b f(x)dx = \frac{h}{3}[f_0 + f_n + 2(f_2 + f_4 + \cdots + f_{n-2}) +$$

$$+ 4(f_1 + f_3 + \cdots + f_{n-1})] + \frac{2}{9}h^4 \int_a^b f^{(4)}(t) \times$$

$$\times \left\{ \gamma_4^* \left(\frac{a - t}{2h} \right) + 2\gamma_4^* \left(\frac{a + h - t}{2h} \right) - \frac{1}{8} \right\} dt. \quad (6.3.10)$$

The remainder term in (6.3.10) differs only in notation from (6.2.10) and the kernel of the remainder is therefore a function which does not change sign on the interval $[a, b]$. Applying the mean value theorem to this integral permits us to write the remainder term of (6.3.10) in the form

$$R(f) = -\frac{(b - a)^5}{180n^4} f^{(4)}(\xi). \quad (6.3.11)$$

For $n = 3$ we obtain a formula which is sometimes called the "three-eighths rule,"

$$\int_a^b f(x)dx \approx H\left[\frac{1}{8} f(a) + \frac{3}{8} f\left(a + \frac{1}{3}H \right) +$$

$$+ \frac{3}{8} f\left(a + \frac{2}{3}H \right) + \frac{1}{8} f(a + H) \right], \quad (6.3.12)$$

$$\omega(x) = (x - a)\left(x - a - \frac{1}{3}H \right)\left(x - a - \frac{2}{3}H \right)(x - a - H)$$

$$R(f) = \frac{f^{(4)}(\xi)}{4!} \int_a^b \omega(x)dx = -\frac{(b - a)^5}{6480} f^{(4)}(\xi) \quad (6.3.13)$$

$$H = b - a.$$

In order to obtain the integral representation for the remainder $R(f)$ in the repeated three-eighths rule we expand $f(x)$ in Bernoulli polynomials in the form (6.3.7). Equation (6.3.12) is exact for all polynomials of degree ≤ 3 and $R(f)$ has the form (6.3.8), but with other values of the integrand. In the present case[7]

$$R_x\left[y_4^*\left(\frac{x-t}{H}\right) - y_4^*\left(\frac{x-a}{H}\right)\right] = \int_a^b\left[y_4^*\left(\frac{x-t}{H}\right) - y_4^*\left(\frac{x-a}{H}\right)\right]dx -$$

$$- H\left\{\frac{1}{8}\left[y_4^*\left(\frac{a-t}{H}\right) - y_4^*(0)\right] + \frac{3}{8}\left[y_4^*\left(\frac{a-t}{H} + \frac{1}{3}\right) - y_4^*\left(\frac{1}{3}\right)\right] +\right.$$

$$+ \frac{3}{8}\left[y_4^*\left(\frac{a-t}{H} + \frac{2}{3}\right) - y_4^*\left(\frac{2}{3}\right)\right] + \frac{1}{8}\left[y_4^*\left(\frac{a-t}{H} + 1\right) - y_4^*(1)\right]\right\} =$$

$$= -\frac{H}{8}\left\{2y_4^*\left(\frac{a-t}{H}\right) + 3y_4^*\left(\frac{a-t}{H} + \frac{1}{3}\right) + 3y_4^*\left(\frac{a-t}{H} + \frac{2}{3}\right) - \frac{8}{27}\right\}.$$

Thus we obtain

$$R(f) = \frac{H^4}{4!8}\int_a^b f^{(4)}(t)\left\{2y^*\left(\frac{a-t}{H}\right) + 3y^*\left(\frac{a-t}{H} + \frac{1}{3}\right) +\right.$$

$$\left. + 3y_4^*\left(\frac{a-t}{H} + \frac{2}{3}\right) - \frac{8}{27}\right\}dt. \qquad (6.3.14)$$

Let n be a multiple of three. We divide $[a, b]$ into n equal parts of length $h = \frac{b-a}{n}$. Let us take the segment $[a + kh, a + (k + 3)h]$ and apply to it the three-eighths rule with remainder in the form (6.3.14)

$$\int_{a+kh}^{a+(k+3)h} f(x)dx = \frac{3h}{8}\{f[a + kh] + 3f[a + (k + 1)h] +$$

$$+ 3f[a + (k + 2)h] + f[a + (k + 3)h]\} +$$

$$+ \frac{27h^4}{64}\int_{a+kh}^{a+(k+3)h} f^{(4)}(t)\left\{2y_4^*\left(\frac{a + kh - t}{3h}\right) +\right.$$

---

[7] Here we make use of the following relationships:

a. y_4^* has period one, that is $y_4^*(z + 1) = y_4^*(z)$

b. If we put $n = 4$, $x = 1/3$, $m = 3$ in (1.2.8) we find

$$B_4 = 3^3\left[B_4\left(\frac{1}{3}\right) + B_4\left(\frac{2}{3}\right) + B_4(1)\right] = 3^3\left[2B_4\left(\frac{1}{3}\right) + B_4\right]$$

$$B_4\left(\frac{1}{3}\right) = B_4\left(\frac{2}{3}\right) = -\frac{13}{27}B_4 = \frac{13}{810}$$

$$y_4^*\left(\frac{1}{3}\right) = y_4^*\left(\frac{2}{3}\right) = B_4\left(\frac{1}{3}\right) - B_4 = \frac{7}{405}.$$

$$+ 3y_4^* \left(\frac{a + kh - t}{3h} + \frac{1}{3} \right) +$$

$$+ 3y_4^* \left(\frac{a + kh - t}{3h} + \frac{2}{3} \right) - \frac{8}{27} \Bigg\} \, dt.$$

Writing equations like this for the segments

$$[a, a + 3h], \quad [a + 3h, a + 6h], \ldots, \quad [a + (n - 3)h, a + nh]$$

and summing the results we obtain the repeated three-eighths rule:

$$\int_a^b f(x)dx = \frac{3h}{8} \{ f_0 + f_n + 2(f_3 + f_6 + \cdots + f_{n-3}) +$$

$$+ 3(f_1 + f_2 + f_4 + f_5 + \cdots + f_{n-2} + f_{n-1})\} +$$

$$+ \frac{27h^4}{64} \int_a^b f^{(4)}(t) \left\{ 2y_4^* \left(\frac{a - t}{3h} \right) + 3y_4^* \left(\frac{a - t}{3h} + \frac{1}{3} \right) + \right.$$

$$\left. + 3y_4^* \left(\frac{a - t}{3h} + \frac{2}{3} \right) - \frac{8}{27} \right\} dt. \tag{6.3.15}$$

We can also apply the mean value theorem to the integral representation for the remainder in the last expression. The remainder of the three-eighths rule can thus be written

$$R(f) = -\frac{(b - a)^5}{80n^4} f^{(4)}(\xi), \quad a < \xi < b. \tag{6.3.16}$$

When the number of segments n is a multiple of both 2 and 3 we can approximate the integral by both Simpson's rule and the three-eighths rule. Both of these formulas have the same algebraic degree of precision and are almost equally simple to use. The choice between these formulas must be based on the error of the final results. Comparison of the remainder terms (6.3.11) and (6.3.16) shows that use of the three-eighths rule may lead to an error which is more than twice as great as the error obtained by use of Simpson's rule. Thus we are forced to prefer Simpson's rule over the three-eighths rule.

REFERENCES

W. W. Johnson, "On Cotesian numbers: their history, computation and values to $n = 20$," *Quart. J. Pure Appl. Math.*, Vol. 46, 1915, pp. 52–65.

Z. Kopal, *Numerical Analysis*, Wiley, London and New York, 1955.

R. O. Kuz'min, "On the theory of mechanical quadrature," *Izv. Leningrad. Polytehn. In.-Ta. Otd. Tehn. Estest. Mat.*, Vol. 32, 1931 (Russian).

S. M. Nikol'skii, *Quadrature Formulas*, Fizmatgiz, Moscow, 1958 (Russian).

J. F. Steffensen, *Interpolation*, Williams and Wilkins, Baltimore, 1927.

E. T. Whittaker and G. Robinson, *The Calculus of Observations*, Blackie and Son, Glasgow, 1940.

CHAPTER 7

Quadratures of the Highest Algebraic Degree of Precision

7.1. GENERAL THEOREMS

In the beginning of this section we make the same assumptions about the weight function $p(x)$ as we made in Chapter 5.

The quadrature formula

$$\int_a^b p(x)f(x)dx \approx \sum_{k=1}^n A_k f(x_k), \qquad (7.1.1)$$

for a fixed n, contains the $2n$ parameters A_k and x_k $(k = 1, 2, \ldots, n)$. The problem is to select these parameters so that formula (7.1.1) will be exact for all polynomials of the highest possible degree (that is, for all polynomials of degree $\leq k$, where k is as large as possible).

In Section 6.1 we showed, by counting the choices of the coefficients A_k, that for any arrangement of x_k we can find an equation (7.1.1) which is exact for all polynomials of degree $\leq n - 1$. This requirement completely defines the coefficients A_k: formula (7.1.1) must be interpolatory and its coefficients must be given by (6.1.4).

In order to increase the precision of (7.1.1) the choice of the nodes x_k is still at our disposal. We might hope that for some choice of these nodes the degree of precision can be increased by n and that the formula can be made exact for all polynomials of degree $\leq 2n - 1$. Under what circumstances this can be achieved will be seen below.

We will now establish the conditions which must be satisfied by the

A_k and x_k in order that the degree of precision of formula (7.1.1) will be not less than $2n - 1$.

We prefer to consider the polynomial $\omega(x) = (x - x_1)(x - x_2) \cdots (x - x_n)$ instead of the nodes x_k themselves. If we know the x_k, then we can easily construct the polynomial $\omega(x)$. Conversely, if we know the polynomial $\omega(x) = x^n + a_1 x^{n-1} + \ldots$, then determining the roots of $\omega(x)$ will give us the x_k.

We must remember that if we determine $\omega(x)$ instead of the x_k directly then we must be careful that the roots of $\omega(x)$ will be real, distinct and located in the segment $[a, b]$.

Theorem 1. *If formula* (7.1.1) *is to be exact for all polynomials of degree* $\leq 2n - 1$, *then it is necessary and sufficient that* (7.1.1) *be interpolatory and that the polynomial* $\omega(x)$ *be orthogonal with respect to* $p(x)$ *to all polynomials* $Q(x)$ *of degree* $< n$:

$$\int_a^b p(x)\,\omega(x)\,Q(x)\,dx = 0.$$

Proof. First we establish the necessity. If (7.1.1) is to be exact for all polynomials of degree $\leq 2n - 1$, then it is also exact for all polynomials of degree $\leq n - 1$ and therefore, by Theorem 1 of Chapter 6, it must be interpolatory.

Let $Q(x)$ be any polynomial of degree $\leq n - 1$. The product $f(x) = \omega(x)Q(x)$ is a polynomial of degree $\leq 2n - 1$ and equation (7.1.1) must be exact for it. But $f(x_k) = 0$ $(k = 1, 2, \ldots, n)$ and hence

$$\int_a^b p(x)\,\omega(x)\,Q(x)\,dx = 0.$$

This shows the necessity of the orthogonality condition.

We now prove the sufficiency of the conditions. Let $f(x)$ be an arbitrary polynomial of degree $\leq 2n - 1$. We can divide $f(x)$ by $\omega(x)$ and represent $f(x)$ in the form

$$f(x) = Q(x)\,\omega(x) + \rho(x)$$

where $Q(x)$ and $\rho(x)$ are polynomials of degree $\leq n - 1$. Since $\omega(x_k) = 0$ we have

$$f(x_k) = \rho(x_k), \qquad (k = 1, 2, \ldots, n)$$

$$\int_a^b p(x)f(x)dx = \int_a^b p(x)\,\omega(x)\,Q(x)dx + \int_a^b p(x)\,\rho(x)dx.$$

The first of the integrals on the right hand side is zero by the assumed orthogonality. Because the degree of $\rho(x)$ is not greater than $n - 1$, and because formula (7.1.1) is assumed to be interpolatory, then the equation

$$\int_a^b p(x)\rho(x)dx = \sum_{k=1}^n A_k\rho(x_k)$$

must be exact. Since $f(x_k) = \rho(x_k)$ we must also have

$$\int_a^b p(x)f(x)dx = \sum_{k=1}^n A_k f(x_k)$$

and formula (7.1.1) will indeed be exact for an arbitrary polynomial of degree $\leq 2n - 1$. This completes the proof.

The possibility of constructing formulas with degree of precision $2n - 1$ is related to the existence of polynomials $\omega(x)$ of degree n which possess the above stated orthogonality property. If the weight function $p(x)$ changes sign on $[a, b]$ then such a polynomial $\omega(x)$ may not exist. If such a polynomial does exist its roots might not satisfy the above requirements.

In the remainder of this section we will assume that the weight function $p(x)$ is nonnegative on $[a, b]$:

$$p(x) \geq 0, \quad \text{for} \quad x \in [a, b].$$

In this case, as was shown in Section 2.1, the polynomial $\omega(x)$ of nth degree, which is orthogonal on $[a, b]$ with respect to $p(x)$ to all polynomials of lower degree, does exist for all n. The roots of $\omega(x)$ are real, distinct and lie inside the segment $[a, b]$.

These remarks are summarized by the following statement:

If $p(x) \geq 0$ for $x \in [a, b]$, then a quadrature formula (7.1.1), which is exact for all polynomials of degree $\leq 2n - 1$, exists for all n.

Up until now it has not been established that $2n - 1$ is the highest degree for which formula (7.1.1) is exact. If $p(x)$ changes sign on $[a, b]$ then this may not be true. But, if $p(x)$ does not change sign then it is easy to prove the following:

Theorem 2. *If $p(x) \geq 0$ then no matter how we choose the x_k and A_k equation (7.1.1) can not be exact for all polynomials of degree $2n$.*

Proof. For the polynomial $P(x) = \omega^2(x)$, which has degree $2n$, the integral $\int_a^b p(x)\omega^2(x)dx > 0$ because the function $p(x)$ is nonnegative and not identically zero. The quadrature sum $\sum A_k P(x_k)$ is zero because $P(x_k) = 0$. Hence equation (7.1.1) can not be exact for $P(x) = \omega^2(x)$.

Now we will discuss the construction of quadrature formulas which have the highest degree of precision. Let us consider the system of polynomials $P_n(x)$ $(n = 1, 2, \ldots)$ which are orthogonal on $[a, b]$ with respect to the weight function $p(x)$. In order to be definite let us assume that this system is normalized. The nth degree polynomial of this system $P_n(x)$ can differ from $\omega(x)$ by only a constant multiple. The roots of $P_n(x)$ will thus be the nodes x_k $(k = 1, 2, \ldots, n)$ which are to be used in the quadrature formula.

The coefficients A_k are determined by equation (6.1.4) or equivalently by

$$A_k = \int_a^b p(x) \frac{P_n(x)}{(x - x_k) P_n'(x_k)} dx. \tag{7.1.2}$$

In order to calculate A_k by (7.1.2) we make use of the Christoffel-Darboux identity (2.1.11) by substituting in that equation $t = x_k$. After dividing by $x - x_k$ we obtain

$$\sum_{s=0}^{n-1} P_s(x) P_s(x_k) = -\frac{a_n}{a_{n+1}} \frac{P_n(x) P_{n+1}(x_k)}{x - x_k}$$

where a_n is the coefficient of x^n in $P_n(x)$.

Let us multiply this last equation by $p(x)$ and integrate over $[a, b]$. The integral $P_s(x_k) \int_a^b p(x) P_s(x) dx$ is zero for $s \geq 1$, by the orthogonality of $P_s(x)$, and is 1 for $s = 0$, by the normality of $P_0(x)$. After carrying out the integration we have

$$1 = -\frac{a_n}{a_{n+1}} P_{n+1}(x_k) \int_a^b p(x) \frac{P_n(x)}{x - x_k} dx.$$

Hence we obtain

$$A_k = -\frac{a_{n+1}}{a_n} \frac{1}{P_n'(x_k) P_{n+1}(x_k)}. \tag{7.1.3}$$

This expression for A_k can be changed slightly by making use of the recursion relation (2.1.10) for orthonormal polynomials. Let us substitute the root x_k of $P_n(x)$ in place of x in (2.1.10). This gives

$$\frac{a_n}{a_{n+1}} P_{n+1}(x_k) + \frac{a_{n-1}}{a_n} P_{n-1}(x_k) = 0.$$

From this relationship we can write (7.1.3) in the form

$$A_k = \frac{a_n}{a_{n-1}} \frac{1}{P_n{}'(x_k) P_{n-1}(x_k)}. \tag{7.1.4}$$

An important fact is that a quadrature formula of the highest degree of precision has all positive coefficients:

Theorem 3. *If the quadrature formula* (7.1.1) *is exact for all possible polynomials of degree* $\leq 2n - 2$ *then all of the coefficients* A_k *are positive.*

Proof. Consider the function $f(x) = \left[\dfrac{\omega(x)}{x - x_i}\right]^2$. This is a polynomial of degree $2n - 2$ and hence equation (7.1.1) must be exact for it. But

$$f(x_k) = \begin{cases} 0 & \text{for} \quad k \neq i \\ \omega'^2(x_i) & \text{for} \quad k = i \end{cases}$$

which means that

$$\int_a^b p(x) \left[\frac{\omega(x)}{x - x_i}\right]^2 dx = A_i \omega'^2(x_i)$$

or

$$A_i = \int_a^b p(x) \left[\frac{\omega(x)}{(x - x_i) \omega'(x_i)}\right]^2 dx > 0,$$

which then proves the theorem.

We will now study the remainder of the quadrature. The segment of integration $[a, b]$ can be any finite or infinite segment. Let us assume that the product $p(x)f(x)$ is summable on $[a, b]$.

Theorem 4. *If* $f(x)$ *has a continuous derivative of order* $2n$ *on* $[a, b]$ *then there exists a point* η *in* $[a, b]$ *for which the remainder of the quadrature formula of the highest degree of precision is*

$$R(f) = \frac{f^{(2n)}(\eta)}{(2n)!} \int_a^b p(x) \omega^2(x)dx. \tag{7.1.5}$$

Proof. Let us construct the interpolating polynomial $H(x)$ of degree $\leq 2n - 1$ which satisfies the conditions

$$H(x_k) = f(x_k), \qquad H'(x_k) = f'(x_k).$$

By Theorem 6 of Chapter 3, the remainder $r(x) = f(x) - H(x)$ of the interpolation can be expressed as

$$r(x) = \frac{f^{(2n)}(\xi)}{(2n)!}\omega^2(x)$$

where ξ belongs to the segment which contains x and the nodes x_k. Thus

$$\int_a^b p(x)f(x)dx = \int_a^b p(x)H(x)dx + \frac{1}{(2n)!}\int_a^b p(x)f^{(2n)}(\xi)\omega^2(x)\,dx.$$

Because the quadrature formula is exact for all polynomials of degree $\leq 2n - 1$ it is exact for $H(x)$:

$$\int_a^b p(x)H(x)dx = \sum_{k=1}^{n} A_k H(x_k) = \sum_{k=1}^{n} A_k f(x_k)$$

and hence we obtain as the remainder of the quadrature

$$R(f) = \frac{1}{(2n)!}\int_a^b f^{(2n)}(\xi)\,p(x)\,\omega^2(x)dx.$$

By the usual reasoning[1] it can be shown that there exists a point $\eta \in [a, b]$ for which (7.1.5) is valid. This completes the proof.

We mention one other integral representation for the remainder. Everything we discussed in Section 5.3 holds true for the remainder of an arbitrary quadrature formula. Let us assume that $f(x)$ has a continuous derivative of order $2n$ on $[a, b]$. Then, with $r = 2n$, equation (5.3.6) gives an integral representation for $R(f)$:

$$R(f) = \int_a^b f^{(2n)}(t)\,K(t)\,dt. \tag{7.1.6}$$

---

[1]If $n = \inf_{[a,b]} f^{(2n)}$ and $M = \sup_{[a,b]} f^{(2n)}$ then

$$m\int_a^b p(x)\omega^2(x)\,dx \leq \int_a^b p(x)f^{(2n)}(\xi)\omega^2(x)\,dx \leq M\int_a^b p(x)\omega^2(x)\,dx.$$

Therefore

$$\int_a^b f^{(2n)}(\xi)p(x)\omega^2(x)\,dx = T\int_a^b p(x)\omega^2(x)\,dx$$

where $m \leq T \leq M$. Thus it is easy to establish the existence of the point η.

If the segment $[a, b]$ is finite then such a representation is certainly possible.

From (7.1.5) we see that if $f^{(2n)}(x)$ is different from zero throughout $[a, b]$ then $R(f)$ is not zero and has the same sign as $f^{(2n)}(x)$. Because this is true for an arbitrary function $f^{(2n)}(x)$, which possesses the properties we have assumed, then the kernel $K(t)$ of (7.1.6) must be nonnegative throughout $[a, b]$.

We will now establish a theorem on the convergence of the quadrature formula. This result could also be obtained as a corollary to a more general result of Chapter 12. We prove this theorem now, however, because we are able to use a much simpler argument than that in Chapter 12.

Let $p(x)$ be a weight function which is nonnegative on $[a, b]$ and let $\omega_n(x)$ $(n = 0, 1, \ldots)$ be the corresponding orthogonal system of polynomials. Also, let $x_k^{(n)}$ $(k = 1, 2, \ldots, n)$ be the roots of the polynomial $\omega_n(x)$ and let $A_k^{(n)}$ $(k = 1, 2, \ldots, n)$ be the coefficients of the quadrature formula of the highest degree of precision.

Theorem 5. *If the segment $[a, b]$ is finite and if $f(x)$ is continuous on $[a, b]$ then*

$$\lim_{n \to \infty} \sum_{k=1}^{n} A_k^{(n)} f(x_k^{(n)}) = \int_a^b p(x) f(x)\, dx. \qquad (7.1.7)$$

Proof. Since $f(x)$ is continuous on $[a, b]$ for any $\epsilon > 0$ we can find a polynomial $P(x)$ with the property that for any $x \in [a, b]$ we have

$$|f(x) - P(x)| < \epsilon. \qquad (7.1.8)$$

Then

$$\left| \int_a^b p(x) f(x)\, dx - \sum_{k=1}^{n} A_k^{(n)} f(x_k^{(n)}) \right| \leq$$

$$\leq \left| \int_a^b p(x) f(x)\, dx - \int_a^b p(x) P(x)\, dx \right| +$$

$$+ \left| \int_a^b p(x) P(x)\, dx - \sum_{k=1}^{n} A_k^{(n)} P(x_k^{(n)}) \right| +$$

$$+ \left| \sum_{k=1}^{n} A_k^{(n)} P(x_k^{(n)}) - \sum_{k=1}^{n} A_k^{(n)} f(x_k^{(n)}) \right|.$$

But, by (7.1.8),

$$\left| \int_a^b p(x)f(x)\,dx \ - \ \int_a^b p(x)P(x)\,dx \right| < \epsilon \int_a^b p(x)\,dx$$

and

$$\left| \sum_{k=1}^n A_k{}^{(n)} P(x_k{}^{(n)}) \ - \ \sum_{k=1}^n A_k{}^{(n)} f(x_k{}^{(n)}) \right| \leq$$

$$\leq \epsilon \sum_{k=1}^n A_k{}^{(n)} \ = \ \epsilon \int_a^b p(x)\,dx.$$

Now if m is the degree of the polynomial $P(x)$, then for $2n - 1 \geq m$ we will have

$$\int_a^b p(x)P(x)\,dx \ = \ \sum_{k=1}^n A_k{}^{(n)} P(x_k{}^{(n)}),$$

and for such an n

$$\left| \int_a^b p(x)f(x)\,dx \ - \ \sum_{k=1}^n A_k{}^{(n)} f(x_k{}^{(n)}) \right| < 2\,\epsilon \int_a^b p(x)\,dx,$$

which proves (7.1.7).

7.2. CONSTANT WEIGHT FUNCTION

The formulas of Gauss are historically the first formulas of the highest algebraic degree of precision. These formulas are used to approximate the integral

$$\int_a^b f(x)\,dx \tag{7.2.1}$$

where $[a, b]$ is a finite segment; here $p(x) \equiv 1$.

By a linear transformation we can transform an arbitrary segment $[a, b]$ into any standard segment we choose. In order to make use of the symmetry of the nodes x_k and coefficients A_k we will take the standard segment to be $[-1, +1]$. Thus, we will assume that (7.2.1) has been transformed into the form

$$\int_{-1}^{+1} f(x)\,dx. \tag{7.2.2}$$

The system of polynomials which are orthogonal on $[-1, +1]$ with respect to the constant weight function are the Legendre polynomials

$$P_n(x) = \frac{1}{2^n \, n!} \frac{d^n (x^2 - 1)^n}{dx^n}.$$

The quadrature formula of the highest degree of precision $2n - 1$

$$\int_{-1}^{+1} f(x) \, dx \approx \sum_{k=1}^{n} A_k^{(n)} f(x_k^{(n)}) \qquad (7.2.3)$$

has for its n nodes the roots of the Legendre polynomial of degree n:

$$P_n(x_k^{(n)}) = 0.$$

The coefficients $A_k^{(n)}$ of this formula can be found from either equation (7.1.3) or (7.1.4); we must remember, however, that in those equations we used orthonormal polynomials. The orthonormal Legendre polynomials are the polynomials $p_n(x) = \sqrt{\dfrac{2n+1}{2}} \, P_n(x)$. The leading coefficients of these are $a_n = \sqrt{\dfrac{2n+1}{2}} \dfrac{(2n)!}{2^n (n!)^2}$ (see equations (2.2.10) and (2.2.11)). A simple calculation then gives

$$A_k^{(n)} = -\frac{2}{(n+1) P_n'(x_k^{(n)}) P_{n+1}(x_k^{(n)})} = \frac{2}{n P_n'(x_k^{(n)}) P_{n-1}(x_k^{(n)})}. \qquad (7.2.4)$$

This can be simplified by use of the following relationship which is known from the theory of Legendre polynomials[2]:

$$(1 - x^2) P_n'(x) = (n+1)[x P_n(x) - P_{n+1}(x)] = n[P_{n-1}(x) - x P_n(x)].$$

If we substitute $x_k^{(n)}$ for x in this equation we obtain

$$[1 - (x_k^{(n)})^2] P_n'(x_k^{(n)}) = -(n+1) P_{n+1}(x_k^{(n)}) = n P_{n-1}(x_k^{(n)}).$$

This permits us to eliminate either P_n', P_{n+1}, or P_{n-1} from (7.2.4). We can obtain, for example,

$$A_k^{(n)} = \frac{2}{[1 - (x_k^{(n)})^2] [P_n'(x_k^{(n)})]^2}. \qquad (7.2.5)$$

In Appendix A we give values of the nodes and coefficients for (7.2.3) for[3] $n = 2(1)16(4)40(8)48$.

---

[2] See, for example: E. W. Hobson, *The Theory of Spherical and Ellipsoidal Harmonics*, Cambridge, 1931.

[3] Writing $n = 2(1)16\ldots$ means that n takes values from 2 to 16 in steps of 1 and so forth. The original Russian edition of this book gave only the 15 decimal place values of the $x_k^{(n)}$ and $A_k^{(n)}$ for $n = 1(1)16$ tabulated by: A. N. Lowan, N. Davids and A. Levenson, "Table of the zeros of the Legendre polynomials of order 1 to 16 and the weight coefficients for the Gauss' mechanical quadrature formula," *Bull. Amer. Math. Soc.*, Vol. 48, 1942, pp. 739–43.

If the integrand $f(x)$ has a continuous derivative of order $2n$ on $[-1, +1]$ then we can use equation (7.1.5) to find the remainder of the Gauss quadrature formula. In (7.1.5) we must now use $p(x) \equiv 1$ and take as $\omega(x)$ the polynomial of degree n, with leading coefficient unity, which is orthogonal with respect to $p(x) \equiv 1$ on $[-1, +1]$, to all polynomials of degree $\leq n - 1$. The polynomial $\omega(x)$ differs from the Legendre polynomial $P_n(x)$ by a constant multiple:

$$\omega(x) = \frac{2^n (n!)^2}{(2n)!} P_n(x).$$

Now, since

$$\int_{-1}^{+1} P_n^2(x)\, dx = \frac{2}{2n+1},$$

we have the following representation for the remainder of the Gauss formula (7.2.3)

$$R(f) = \frac{2^{2n+1}}{(2n+1)(2n)!} \left[\frac{(n!)^2}{(2n)!} \right]^2 f^{(2n)}(\eta), \qquad (7.2.6)$$

where η is a point in the segment $[-1, +1]$.

Example 1. Suppose we wish to calculate the integral

$$J = \int_0^1 \frac{dt}{1+t} = \ln 2 \approx 0.69314718.$$

Let us use the 5-point Gauss formula. In order to use the nodes and coefficients which are tabulated in Appendix A we must transform the segment of integration $[0, 1]$ to the segment $[-1, +1]$. This is accomplished by the transformation

$$t = \frac{1}{2}(1 + x).$$

We then obtain

$$J = \int_{-1}^{+1} \frac{dx}{3+x}.$$

The approximate value of J, using the 5-point Gauss formula, is then

$$J \approx A_1^{(5)}(3 + x_1^{(5)})^{-1} + A_2^{(5)}(3 + x_2^{(5)})^{-1} + \cdots + A_5^{(5)}(3 + x_5^{(5)})^{-1}.$$

After substituting the values from the table we obtain, to eight significant figures,

$$J \approx 0.69314717.$$

We could approximately evaluate J in its original form as an integral over the segment $[0, 1]$ by transforming the Gauss formula for the segment $[-1, +1]$ into the corresponding formula for the segment $[0, 1]$. This would be done as follows:

$$u_k^{(5)} = \frac{1}{2} [1 + x_k^{(5)}], \qquad B_k^{(5)} = \frac{1}{2} A_k^{(5)} \qquad (k = 1, 2, \ldots, 5)$$

$$u_1^{(5)} = 0.04691 \cdots \qquad B_1^{(5)} = 0.11846 \cdots$$
$$u_2^{(5)} = 0.23076 \cdots \qquad B_2^{(5)} = 0.23931 \cdots$$
$$u_3^{(5)} = 0.50000 \cdots \qquad B_3^{(5)} = 0.28444 \cdots$$
$$u_4^{(5)} = 0.76923 \cdots \qquad B_4^{(5)} = 0.23931 \cdots$$
$$u_5^{(5)} = 0.95308 \cdots \qquad B_5^{(5)} = 0.11846 \cdots$$

Now we can use these nodes and coefficients to calculate J in the original form

$$J \approx B_1^{(5)}(1 + u_1^{(5)})^{-1} + B_2^{(5)}(1 + u_2^{(5)})^{-1} + \cdots + B_5^{(5)}(1 + u_5^{(5)})^{-1}.$$

Example 2 The integral equation

$$y(x) = f(x) + \int_a^b K(x, s) y(s) ds$$

is often solved approximately by replacing it with a linear system[4]. Such a system can be constructed, for example, if we replace the integral by a quadrature sum:

$$y(x) = f(x) + \sum_{j=1}^{n} A_j K(x, x_j) y(x_j) + R(x).$$

If we substitute, in turn, $x = x_1, x_2, \ldots, x_n$ into this equation we obtain the linear system of equations

$$y(x_i) = f(x_i) + \sum_{j=1}^{n} A_j K(x_i, x_j) y(x_j) + R(x_i), \qquad (i = 1, 2, \ldots, n).$$

If we ignore the remainder terms $R(x_i)$ then this is a system of n equations which have as unknowns the n approximate values $\tilde{y}(x_i)$ of the unknown function $y(x)$:

$$\tilde{y}(x_i) = f(x_i) + \sum_{j=1}^{n} A_j K(x_i, x_j) \tilde{y}(x_j), \qquad (i = 1, 2, \ldots, n). \qquad (7.2.7)$$

---

[4]For a more complete discussion of the use of quadrature formulas in the approximate solution of integral equations, than is given in this example, see: L. V. Kantorovich and V. I. Krylov, *Approximate Methods of Higher Analysis*, Interscience and Noordhoff, 1958. (Translated from the Russian, *Priblizhennye metody vysshego analiza*, Moscow, 1952).

The magnitude of the remainders $R(x_i)$ depend on the precision of the quadrature formula and we can expect that the more precise the formula the more accurate will be the solution of the integral equation.

The solution of the linear system (7.2.7) becomes increasingly difficult as the number of equations increases. Therefore, if we wish to find the approximate solution of an integral equation by replacing it by a linear system it is desirable to use a quadrature formula of the highest degree of precision.

Let us consider the integral equation

$$y(x) - \frac{1}{2} \int_0^1 e^{xt} y(t) \, dt = \frac{1}{2x} (e^x - 1)$$

and let us use the Gauss 2-point formula to find its approximate solution. The nodes and coefficients of this formula for the segment [0, 1] are:

$$A_1^{(2)} = A_2^{(2)} = \frac{1}{2}, \quad x_1^{(2)} \approx 0.2113, \quad x_2^{(2)} \approx 0.7887.$$

The system (7.2.7) has the form

$$(1 - \frac{1}{2} K_{1,1}) \tilde{y}_1 - \frac{1}{2} K_{1,2} \tilde{y}_2 = f_1$$

$$- \frac{1}{2} K_{2,1} \tilde{y}_1 + (1 - \frac{1}{2} K_{2,2}) \tilde{y}_2 = f_2$$

where

$$\tilde{y}_i = \tilde{y}(x_i^{(2)}), \quad K_{i,j} = K(x_i^{(2)}, x_j^{(2)}), \quad K(t, x) = \frac{1}{2} e^{xt}, \quad f_i = f(x_i^{(2)}).$$

After computing the coefficients this system becomes

$$0.7386 \, \tilde{y}_1 - 0.2954 \, \tilde{y}_2 = 0.4434$$
$$- 0.2954 \, \tilde{y}_1 + 0.5343 \, \tilde{y}_2 = 0.2384.$$

Solving these equations we find

$$\tilde{y}_1 = \tilde{y}(0.2113) = 0.9997, \quad \tilde{y}_2 = \tilde{y}(0.7887) = 0.9990.$$

The exact solution of the equation, as can be easily verified by substitution, is $y(x) = 1$.

7.3. INTEGRALS OF THE FORM $\int_a^b (b - x)^\alpha (x - a)^\beta f(x) \, dx$

AND THEIR APPLICATION TO THE CALCULATION OF MULTIPLE INTEGRALS

Let $[a, b]$ be an arbitrary finite segment and let us be given the corresponding weight function $p(x) = (b - x)^\alpha (x - a)^\beta$, $\alpha, \beta > -1$. In order to study the integral $\int_a^b (b - x)^\alpha (x - a)^\beta f(x)\, dx$ and for the construction of quadrature formulas for its approximation, one usually transforms the segment $[a, b]$ into the segment $[-1, +1]$ by the linear transformation

$$x = \frac{1}{2}[a + b + t(b - a)], \qquad -1 \le t \le +1.$$

We will assume that such a transformation has been carried out and will restrict our attention to the integral

$$\int_{-1}^{+1} (1 - x)^\alpha (1 + x)^\beta f(x)\, dx. \tag{7.3.1}$$

The orthogonal system of polynomials which correspond to the segment $[-1, +1]$ and the weight function $(1 - x)^\alpha (1 + x)^\beta$ is the system of Jacobi polynomials $P_n^{(\alpha, \beta)}(x)$ $(n = 0, 1, 2, \ldots)$. A quadrature formula with n nodes

$$\int_{-1}^{+1} (1 - x)^\alpha (1 + x)^\beta f(x)\, dx \approx \sum_{k=1}^n A_k f(x_k), \tag{7.3.2}$$

which has the highest degree of precision $2n - 1$, must have for its nodes x_k the roots of the Jacobi polynomial of degree n

$$P_n^{(\alpha, \beta)}(x_k) = 0.$$

The coefficients[5] A_k can be found from either equation (7.1.3) or (7.1.4).

The normalized Jacobi polynomials are [by (2.2.2), (2.2.5) and (2.2.7)]:

$$p_n^{(\alpha, \beta)}(x) = \delta_n^{-\frac{1}{2}} P_n^{(\alpha, \beta)}(x)$$

where

$$\delta_n = \frac{2^{\alpha+\beta+1} \Gamma(\alpha + n + 1) \Gamma(\beta + n + 1)}{(\alpha + \beta + 2n + 1)\, n!\, \Gamma(\alpha + \beta + n + 1)}.$$

The leading coefficients of the normalized Jacobi polynomials are

$$a_n = \delta_n^{-\frac{1}{2}} \frac{\Gamma(\alpha + \beta + 2n + 1)}{2^n n!\, \Gamma(\alpha + \beta + n + 1)}.$$

---

[5]Trans. note: We omit the superscript (n) from the symbols $x_k^{(n)}$ and $A_k^{(n)}$ whenever it is clear to which values of n they correspond.

We then find

$$A_k = \frac{(\alpha + \beta + 2n)\, 2^{\alpha+\beta}\, \Gamma(\alpha + n)\, \Gamma(\beta + n)}{n!\, \Gamma(\alpha + \beta + n + 1)\, P_n^{(\alpha,\beta)'}(x_k)\, P_{n-1}^{(\alpha,\beta)}(x_k)}. \tag{7.3.3}$$

This expression for the coefficients can be simplified somewhat if we make use of the relationship[6]

$$(\alpha + \beta + 2n)(1 - x^2)\, \frac{d}{dx} P_n^{(\alpha,\beta)}(x) =$$

$$= -n[(\alpha + \beta + 2n)\, x + \beta - \alpha]\, P_n^{(\alpha,\beta)}(x) + 2(\alpha + n)(\beta + n)\, P_{n-1}^{(\alpha,\beta)}(x).$$

Substituting $x = x_k$ we obtain

$$(\alpha + \beta + 2n)(1 - x_k^2)\, P_n^{(\alpha,\beta)'}(x_k) = 2(\alpha + n)(\beta + n)\, P_{n-1}^{(\alpha,\beta)}(x_k)$$

which permits us to write A_k in the form

$$A_k = \frac{2^{\alpha+\beta+1}\, \Gamma(\alpha + n + 1)\, \Gamma(\beta + n + 1)}{n!\, \Gamma(\alpha + \beta + n + 1)(1 - x_k^2)[P_n^{(\alpha,\beta)'}(x_k)]^2}. \tag{7.3.4}$$

The leading coefficient of the polynomial $P_n^{(\alpha,\beta)}(x)$ has the value (2.2.2). Therefore the polynomial

$$\omega(x) = \frac{2^n\, n!\, \Gamma(\alpha + \beta + n + 1)}{\Gamma(\alpha + \beta + 2n + 1)}\, P_n^{(\alpha,\beta)}(x)$$

has unity for its leading coefficient. If $f(x)$ has a continuous derivative of order $2n$ on the segment $[-1, +1]$ then the remainder term of formula (7.3.2) is

$$R(f) = \frac{f^{(2n)}(\eta)}{(2n)!} \left[\frac{2^n\, n!\, \Gamma(\alpha + \beta + n + 1)}{\Gamma(\alpha + \beta + 2n + 1)} \right]^2 \times$$

$$\times \int_{-1}^{+1} (1 - x)^\alpha (1 + x)^\beta [P_n^{(\alpha,\beta)}(x)]^2\, dx = \frac{f^{(2n)}(\eta)}{(2n)!} \times$$

$$\times \frac{2^{\alpha+\beta+2n+1}\, n!\, \Gamma(\alpha + n + 1)\, \Gamma(\beta + n + 1)\, \Gamma(\alpha + \beta + n + 1)}{(\alpha + \beta + 2n + 1)[\Gamma(\alpha + \beta + 2n + 1)]^2} \tag{7.3.5}$$

where $-1 < \eta < +1$.

Let us now consider some special cases of quadrature formulas for use with Jacobi weight functions.

---

[6]See, for example: G. Szegö, *Orthogonal Polynomials*, Amer. Math. Soc. Colloquim Publ., Vol. 23, 1959.

1. Quadrature formulas on $[-1, +1]$.

A. For $\alpha = \beta = -1/2$ the weight function is $(1 - x^2)^{-\frac{1}{2}}$ and the corresponding Jacobi polynomials are a multiple of the Chebyshev polynomials of the first kind (see (2.3.4)):

$$P_n^{(-\frac{1}{2}, -\frac{1}{2})}(x) = C_n T_n(x) = C_n \cos(n \text{ arc cos } x).$$

The roots of T_n are the nodes to be used in the quadrature formula; these are

$$x_k = \cos \frac{2k - 1}{2n} \pi \qquad (k = 1, 2, \ldots, n).$$

The coefficients A_k are easily computed. Since

$$T_n'(x_k) = \frac{n \sin(n \text{ arc cos } x_k)}{\sqrt{1 - x_k^2}} = \frac{(-1)^{k-1} n}{\sqrt{1 - x_k^2}}$$

then

$$(1 - x_k^2)[P_n^{(-\frac{1}{2}, -\frac{1}{2})}(x_k)]^2 = C_n^2(1 - x_k^2)[T_n'(x_k)]^2 = C_n^2 n^2$$

and

$$A_k = \frac{2^n \left[\Gamma\left(n + \frac{1}{2}\right)\right]^2}{n! \Gamma(n) C_n^2 n^2}.$$

The righthand side of this expression is independent of k and hence, for a fixed n, $A_1 = \cdots = A_n$. Let A denote the common value of the A_k. The easiest way to find the value of A is to use the fact that the quadrature formula is exact for $f(x) \equiv 1$ and hence

$$\sum_{k=1}^{n} A_k = nA = \int_{-1}^{+1} \frac{dx}{\sqrt{1 - x^2}} = \pi.$$

Hence

$$A = \frac{\pi}{n}.$$

The quadrature formulas of the highest degree of precision for the weight function $(1 - x^2)^{-\frac{1}{2}}$ have the form[7]

---

[7]This formula was found by F. G. Mehler in 1864. See the reference at the end of this chapter.

$$\int_{-1}^{+1} \frac{f(x)}{\sqrt{1-x^2}}\, dx = \frac{\pi}{n} \sum_{k=1}^{n} f\left(\cos \frac{2k-1}{2n}\pi\right) + R(f). \qquad (7.3.6)$$

Using (7.3.5) we obtain the following expression for the remainder

$$R(f) = \frac{\pi}{2^{2n-1}} \frac{f^{(2n)}(\eta)}{(2n)!}, \qquad -1 < \eta < +1.$$

B. Let $\alpha = \beta = \dfrac{1}{2}$ and $p(x) = \sqrt{1-x^2}$. The Jacobi polynomials

$P_n^{(\frac{1}{2},\frac{1}{2})}(x)$ are a multiple of the Chebyshev polynomials of the second kind [see (2.3.7)]:

$$P_n^{(\frac{1}{2},\frac{1}{2})}(x) = \frac{(2n+1)!}{2^{2n}\, n!\, (n+1)!}\, U_n(x)$$

$$U_n(x) = \frac{\sin\left[(n+1)\, \text{arc}\, \cos x\right]}{\sqrt{1-x^2}}.$$

The roots of $P_n^{(\frac{1}{2},\frac{1}{2})}(x)$ are $x_k = \cos \dfrac{k}{n+1}\pi$ $(k = 1, 2, \ldots, n)$.
The coefficients A_k can be computed from (7.3.4):

$$A_k = \frac{\pi}{n+1} \sin^2 \frac{k\pi}{n+1}.$$

The quadrature formulas have the form

$$\int_{-1}^{+1} \sqrt{1-x^2}\, f(x)\, dx =$$

$$= \frac{\pi}{n+1} \sum_{k=1}^{n} \sin^2 \frac{k\pi}{n+1}\, f\left(\cos \frac{k\pi}{n+1}\right) + R(f). \qquad (7.3.7)$$

The remainder $R(f)$ can be found from (7.3.5)

$$R(f) = \frac{\pi}{2^{2n}} \frac{f^{(2n)}(\eta)}{(2n)!}, \qquad -1 < \eta < +1.$$

C. Let $\alpha = \dfrac{1}{2}$, $\beta = -\dfrac{1}{2}$ so that $p(x) = \sqrt{\dfrac{1-x}{1+x}}$. As in the two previous cases the Jacobi polynomials $P_n^{(\frac{1}{2}, -\frac{1}{2})}(x)$ can be simply expressed in terms of trigonometric functions. If $Q(x)$ is an arbitrary polynomial of degree less than n then the following orthogonality condition must be satisfied:

$$\int_{-1}^{+1} \sqrt{\frac{1-x}{1+x}} \; P_n^{(\frac{1}{2}, -\frac{1}{2})}(x) \, Q(x) \, dx =$$

$$= \int_{-1}^{+1} (1-x) P_n^{(\frac{1}{2}, -\frac{1}{2})}(x) \, Q(x) \; \frac{dx}{\sqrt{1-x^2}} = 0.$$

Let us consider the polynomial $S(x) = (1-x) P_n^{(\frac{1}{2}, -\frac{1}{2})}(x)$. This is a polynomial of degree $n + 1$ and it is orthogonal on the segment $[-1, +1]$ with respect to the weight function $(1 - x^2)^{-\frac{1}{2}}$ to each polynomial $Q(x)$ of degree less than n. If $S(x)$ is expanded in terms of Chebyshev polynomials of the first kind $T_k(x)$ $(k = 0, 1, \ldots, n + 1)$ then all the coefficients of the polynomials $T_k(x)$, for $k \leq n - 1$, in this expansion must be zero by the orthogonality properties of $S(x)$. Hence this expansion must have the form $S(x) = C_n T_n(x) + C_{n+1} T_{n+1}(x)$. Since $S(x)$ is divisible by $1 - x$ we must have

$$S(1) = C_n T_n(1) + C_{n+1} T_{n+1}(1) = C_n + C_{n+1} = 0.$$

Therefore $C_{n+1} = -C_n$ and

$$P_n^{(\frac{1}{2}, -\frac{1}{2})}(x) = C_n \; \frac{T_n(x) - T_{n+1}(x)}{1-x} \; .$$

If we equate the leading coefficients from (2.2.2) and (2.3.2) we find

$$C_n = \frac{(2n)!}{2^{2n}(n!)^2} \; .$$

Setting $x = \cos \theta$ and using the fact that $T_k(x) = \cos (k \arccos x) = \cos k\theta$ we obtain

$$P_n^{(\frac{1}{2}, -\frac{1}{2})}(x) = \frac{(2n)!}{2^{2n}(n!)^2} \; \frac{\sin (2n + 1)\theta/2}{\sin \theta/2} \; .$$

The roots of this polynomial are

$$x_k = \cos \frac{2k}{2n + 1} \pi \qquad (k = 1, 2, \ldots, n).$$

The coefficients of the quadrature formula can be computed from (7.3.4):

$$A_k = \frac{4\pi}{2n+1} \sin^2 \frac{k\pi}{2n+1} .$$

Thus the quadrature formulas for use with the weight function $\sqrt{\dfrac{1-x}{1+x}}$ have the form

$$\int_{-1}^{+1} \sqrt{\frac{1-x}{1+x}} \, f(x) \, dx =$$

$$= \frac{4\pi}{2n+1} \sum_{k=1}^{n} \sin^2 \frac{k\pi}{2n+1} f\left(\cos \frac{2k\pi}{2n+1}\right) + R(f). \qquad (7.3.8)$$

The remainder term in this formula is

$$R(f) = \frac{\pi}{2^{2n}} \frac{f^{(2n)}(\eta)}{(2n)!}, \qquad -1 < \eta < +1.$$

2. Quadrature formulas on [0, 1].

A. The first case we consider is $\alpha = 0$, $\beta = \dfrac{1}{2}$; this corresponds to the integral $\int_0^1 \sqrt{x} \, f(x) dx$. The polynomials $Q_n(x)$ which are orthogonal on the segment [0, 1] with respect to the weight function $\sqrt{x}$ are closely related to the Legendre polynomials $P_m(x)$. Let us put $m = 2n+1$ and consider the Legendre polynomials of odd order $P_{2n+1}(y)$ ($n = 0, 1, 2, \ldots$). These are odd functions of y and the ratio $P_{2n+1}(y)/y$ depends only on y^2. Let us replace y^2 by x. We will show that $Q_n(x)$ can be taken as the polynomial

$$Q_n(x) = \frac{P_{2n+1}(\sqrt{x})}{\sqrt{x}} .$$

Using the substitution $x = y^2$ we obtain

$$\int_0^1 \sqrt{x} \, Q_n(x) Q_m(x) \, dx = \int_0^1 P_{2n+1}(\sqrt{x}) P_{2m+1}(\sqrt{x}) \, \frac{dx}{\sqrt{x}} =$$

$$= 2 \int_0^1 P_{2n+1}(y) P_{2m+1}(y) \, dy = \int_{-1}^{+1} P_{2n+1}(y) P_{2m+1}(y) \, dy = 0.$$

This proves the orthogonality of $Q_n(x)$.

In the quadrature formula of the highest degree of precision

$$\int_0^1 \sqrt{x}\, f(x)\, dx = \sum_{k=1}^n A_k f(x_k) + R(f) \qquad (7.3.9)$$

the nodes x_k are the squares of the positive roots y_k of the Legendre polynomial $P_{2n+1}(y)$:

$$x_k = y_k^2.$$

We can also show that the coefficients A_k

$$A_k = \int_0^1 \sqrt{x}\, \frac{Q_n(x)}{(x - x_k)\, Q'(x_k)}\, dx$$

are simply related to the coefficients of the Gauss formula (7.2.3) with $2n + 1$ nodes. Using the relationships

$$x_k = y_k^2, \qquad Q_n'(x_k) = \frac{P_{2n+1}'(y_k)}{2y_k^2}$$

we obtain

$$A_k = 2y_k^2 \int_0^1 \frac{P_{2n+1}(y)}{P_{2n+1}'(y_k)} \frac{2y}{y^2 - y_k^2}\, dy.$$

This integral can be written as the sum of two integrals since

$$\frac{2y}{y^2 - y_k^2} = \frac{1}{y - y_k} + \frac{1}{y + y_k}.$$

If, in the second of these two integrals, we replace y by $-y$ we obtain

$$A_k = 2y_k^2 \int_{-1}^{+1} \frac{P_{2n+1}(y)}{(y - y_k)\, P_{2n+1}'(y_k)}\, dy. \qquad (7.3.10)$$

Let us write the coefficients of the Gauss formula (7.2.3) with $2n + 1$ nodes as $A_k^{(2n+1)}$ ($k = -n, -n + 1, \ldots, -1, 0, 1, \ldots, n$). Then the integral in (7.3.10) is equal to $A_k^{(2n+1)}$. Therefore

$$A_k = 2y_k^2 A_k^{(2n+1)} \qquad (k = 1, 2, \ldots, n).$$

The remainder $R(f)$ of formula (7.3.9) can be found from the general expression (7.1.5) if we use the fact that the leading coefficient of $Q_n(x)$ is the same as the leading coefficient of $P_{2n+1}(y)$:

$$\omega(x) = \frac{2^{2n+1}[(2n+1)!]^2}{(4n+2)!}\, Q_n(x) = \frac{2^{2n+1}[(2n+1)!]^2}{(4n+2)!}\, \frac{P_{2n+1}(\sqrt{x})}{\sqrt{x}}.$$

Thus we obtain

$$R(f) = \frac{f^{(2n)}(\eta)}{(2n)!}\, \frac{2}{4n+3}\left\{ \frac{2^{2n+1}[(2n+1)!]^2}{(4n+2)!} \right\}^2, \qquad 0 < \eta < 1.$$

Here we give values of the x_k and A_k in formula (7.3.9) for $n = 1(1)8$:

Quadrature Formulas for the Integral $\displaystyle\int_0^1 \sqrt{x}\, f(x)\, dx.$

$x_k^{(n)}$		$A_k^{(n)}$	
	$n = 1$		
0.60000	00000	0.66666	66667
	$n = 2$		
0.28994	91979	0.27755	59982
0.82116	19131	0.38911	06684
	$n = 3$		
0.16471	02869	0.12578	26743
0.54986	84991	0.30760	23676
0.90080	58292	0.23328	16246
	$n = 4$		
0.10514	02826	0.06568	05199
0.37622	45144	0.19609	62654
0.69894	80124	0.25252	73457
0.93733	42493	0.15236	25356
	$n = 5$		
0.07265	35129	0.03818	73467
0.26946	07913	0.12567	31527
0.53312	19512	0.19863	08015
0.78688	00558	0.19763	33763
0.95693	13076	0.10654	19894
	$n = 6$		
0.05311	10354	0.02403	62680
0.20114	57477	0.08360	26285
0.41261	26738	0.14701	05789
0.64252	74355	0.17846	00808
0.84198	68221	0.15513	01778
0.96861	62852	0.07842	69326
	$n = 7$		
0.04047	90635	0.01606	46414
0.15535	52844	0.05784	21902
0.32600	92219	0.10841	05888
0.52478	10495	0.14648	80937
0.71945	44081	0.15419	23470
0.87848	14120	0.12363	05295
0.97612	92156	0.06003	82760
	$n = 8$		
0.03185	66030	0.01124	93760
0.12336	37516	0.04145	12327

$x_k^{(n)}$		$A_k^{(n)}$	
	$n = 8\,(contd.)$		
0.26285	15868	0.08098	23455
0.43253	13536	0.11690	14328
0.61076	41382	0.13666	92830
0.77482	09677	0.13177	55814
0.90378	39476	0.10024	68648
0.98123	97722	0.04739	05504

B. In a manner similar to the last case we can construct formulas of the highest degree of precision for the segment [0, 1] and the weight function $x^{-\frac{1}{2}}$

$$\int_0^1 x^{-\frac{1}{2}} f(x)\,dx = \sum_{k=1}^n A_k f(x_k) + R(f). \qquad (7.3.11)$$

The polynomials $S_n(x)$ which are orthogonal on [0, 1] with respect to $x^{-\frac{1}{2}}$ are related to the Legendre polynomials $P_k(x)$ by

$$S_n(x) = P_{2n}(\sqrt{x}).$$

Thus the abscissas x_k in (7.3.11) are the squares of the positive roots y_k of the Legendre polynomial $P_{2n}(y)$:

$$y_k = y_k^2 \qquad (k = 1, 2, \ldots, n).$$

Let us write the coefficients of the Gauss formula (7.2.3) with $2n$ nodes as $A_k^{(2n)}$ $(k = -n, \ldots, -1, +1, \ldots, n)$. The coefficients A_k in (7.3.11) are then

$$A_k = 2 A_k^{(2n)} \qquad (k = 1, 2, \ldots, n).$$

The remainder has the form

$$R(f) = \frac{f^{(2n)}(\eta)}{(2n)!} \frac{2}{4n+1} \left\{ \frac{2^{2n}[(2n)!]^2}{(4n)!} \right\}^2, \qquad 0 < \eta < 1.$$

Values of the x_k and A_k in formula (7.3.11) are tabulated here for $n = 1(1)8$:

Quadrature Formulas for the Integral $\int_0^1 \frac{1}{\sqrt{x}} f(x)\,dx.$

$x_k^{(n)}$		$A_k^{(n)}$	
	$n = 1$		
0.33333	33333	2.00000	00000
	$n = 2$		
0.11558	71100	1.30429	03097
0.74155	57471	0.69570	96903

$x_k^{(n)}$	$A_k^{(n)}$
$n=3$	
0.05693 91160	0.93582 78691
0.43719 78527	0.72152 31461
0.86949 93948	0.34264 89847
$n=4$	
0.03364 82681	0.72536 75667
0.27618 43139	0.62741 32917
0.63467 74762	0.44476 20689
0.92215 66084	0.20245 70726
$n=5$	
0.02216 35688	0.59104 84494
0.18783 15676	0.53853 34386
0.46159 73614	0.43817 27250
0.74833 46283	0.29890 26983
0.94849 39262	0.13334 26886
$n=6$	
0.01568 34066	0.49829 40916
0.13530 00116	0.46698 50731
0.34494 23794	0.40633 48534
0.59275 01277	0.32015 66571
0.81742 80132	0.21387 86520
0.96346 12786	0.09435 06728
$n=7$	
0.01167 58719	0.43052 77069
0.10183 27040	0.41039 69274
0.26548 11572	0.37107 67950
0.47237 15370	0.31440 63343
0.68426 20156	0.24303 71414
0.86199 13331	0.16031 61743
0.97275 57512	0.07023 89207
$n=8$	
0.00902 73770	0.37890 12209
0.07930 05598	0.36520 68301
0.20977 93686	0.33831 30388
0.38177 10533	0.29919 19776
0.57063 58201	0.24925 79425
0.74931 73785	0.19031 70234
0.89222 19741	0.12450 70479
0.97891 42101	0.05430 49188

3. Application to multiple integrals.

One method often used in practice is to separate the variables, if possible, of the multiple integral and to apply quadrature formulas for functions of a single variable in turn to each of the variables separately. As an example consider the integral

$$I = \iint_\sigma f(x,\,y)\,d\sigma$$

where the region σ is a rectangle $a \leq x \leq b$, $c \leq y \leq d$. The integral I can be written as two single integrals

$$I = \int_a^b \left[\int_c^d f(x, y) \, dy \right] dx.$$

Here we can replace the integral with respect to y by a quadrature sum with m nodes y_i and coefficients B_i $(i = 1, \ldots, m)$ and the integral with respect to x by a quadrature sum with n nodes x_j and coefficients A_j $(j = 1, \ldots, n)$. This leads to the following integration formula for I:

$$I \approx \sum_{j=1}^n \sum_{i=1}^m A_j B_i f(x_j, y_i)$$

This formula requires us to evaluate the integrand $f(x, y)$ at mn points which is a relatively large number compared to the individual numbers m and n.

This method can also be applied to regions other than rectangles. In every case it leads to a relatively large number of points in the integration formula. The problem becomes even more acute when the above method is applied to triple and higher dimensional integrals. This method, however, does give useful formulas especially for two and three dimensions and they are especially valuable for relatively smooth functions so that formulas with extremely high accuracy do not have to be used.

We now consider certain special cases of this method.[8]

4. Double integrals in polar coordinates.

Let us consider

$$I = \iint_\sigma F(r, \phi) \, r \, dr \, d\phi,$$

and assume that the region of integration σ is defined by the inequalities

$$\alpha \leq \phi \leq \beta, \qquad 0 \leq r \leq R = R(\phi).$$

---

[8]Trans. note: The author's discussion of methods for combining quadrature formulas for single integrals is, up to this point and in the remainder of this section, mostly descriptive in nature; he does not show for what class of functions the resulting formulas will be exact. Recent papers cited in the references at the end of this chapter by the following authors give some exact results of this nature: Hammer and Wymore; Hammer, Marlowe and Stroud; Peirce; Hetherington; and Secrest and Stroud.

Other formulas for multiple integrals, not of the type discussed in this section, which use fewer points for the same algebraic degree of precision are also known in a few cases. For references to such formulas see: A. H. Stroud, "A bibliography on approximate integration," *Math. Comp.*, Vol. 15, 1961, pp. 52–80.

If we introduce the parameter ρ by setting $r = \rho R$, $0 \le \rho \le 1$, then the integral I can be written in the form

$$I = \int_\alpha^\beta \left[\int_0^1 F(\rho R, \phi)\, \rho\, d\rho \right] R^2\, d\phi.$$

Hence we see that calculation of a double integral in polar coordinates reduces to a consideration of the integral

$$\int_0^1 f(x)\, x\, dx. \tag{7.3.12}$$

If we wish to construct a quadrature formula

$$\int_0^1 f(x)\, x\, dx = \sum_{k=1}^n A_k f(x_k) + R(f)$$

of the highest degree of precision we must take its nodes as the roots of the polynomial $\Pi_n(x)$ which is orthogonal on the segment $[0, 1]$ with respect to $p(x) = x$ to all polynomials of degree $\le n - 1$. The coefficients A_k can be calculated by the usual equations (7.1.3) or (7.1.4).

To find the x_k and A_k we can use the previously obtained results for the weight function $(1 - z)^\alpha (1 + z)^\beta$. By making the change of variable $x = \dfrac{1}{2}(1 - z)$, $-1 \le z \le 1$, (7.3.12) becomes

$$\int_0^1 f(x)\, x\, dx = \frac{1}{4} \int_{-1}^1 F(z)(1 - z)\, dz$$
$$F(z) = f\left(\frac{1-z}{2}\right). \tag{7.3.13}$$

Under this transformation $\Pi_n(x)$ is transformed into a polynomial of degree n in z which is orthogonal on the segment $[-1, 1]$ with respect to the weight $1 - z$ to all polynomials of degree $\le n - 1$ and will differ from the Jacobi polynomial $P_n^{(1,0)}(z)$ by only a constant factor

$$\Pi_n(x) = c P_n^{(1,0)}(z).$$

Hence we see that the nodes x_k of formula (7.3.7) are related to the roots z_k of $P_n^{(1,0)}(z)$ by the relationship

$$x_k = \frac{1 - z_k}{2} \qquad (k = 1, \ldots, n).$$

From (7.3.8), (7.3.2) and (7.3.4), for $\alpha = 1$, $\beta = 0$, we have the following general expression for the coefficients A_k:

$$A_k = \frac{1}{(1 - z_k^2) [P_n^{(1,0)'}(z_k)]^2}.$$

Values of the x_k and A_k are given below for $n = 1(1)6$[9]:

Quadrature Formulas for the Integral $\displaystyle\int_0^1 x f(x) \, dx$.

$x_k^{(n)}$			$A_k^{(n)}$		
		$n = 1$			
0.66666	66666	67	0.50000	00000	00
		$n = 1$			
0.35505	10257	22	0.18195	86182	56
0.84494	89742	78	0.31804	13817	44
		$n = 3$			
0.21234	05382	39	0.06982	69799	01
0.59053	31355	59	0.22924	11063	60
0.91141	20404	87	0.20093	19137	39
		$n = 4$			
0.13975	98643	44	0.03118	09709	50
0.41640	95676	31	0.12984	75476	08
0.72315	69863	62	0.20346	45680	10
0.94289	58038	85	0.13550	69134	31
		$n = 5$			
0.09853	50857	99	0.01574	79145	22
0.30453	57266	46	0.07390	88700	73
0.56202	51897	53	0.14638	69870	85
0.80198	65821	26	0.16717	46380	94
0.96019	01429	49	0.09678	15902	27
		$n = 6$			
0.07305	43286	80	0.00873	83018	14
0.23076	61379	70	0.04395	51655	51
0.44132	84812	28	0.09866	11508	91
0.66301	53097	19	0.14079	25537	88
0.85192	14003	32	0.13554	24972	32
0.97068	35728	40	0.07231	03307	26

5. Triple integrals in spherical coordinates.

To calculate

$$I = \iiint_\sigma f(r, \theta, \phi) r^2 \sin\theta \, dr \, d\theta \, d\phi$$

we can reduce it to single integrals in each of the variables r, θ, ϕ. For the integration with respect to r we will have an integral of the form

$$\int_0^1 f(x) x^2 \, dx. \tag{7.3.14}$$

---

[9]Trans. note: These values are from: H. Fishman, "Numerical integration constants," *Math. Tables Aids Comput.*, Vol. 11, 1957, pp. 1-9.

As in the last case for polar coordinates we can show that in the quadrature formula for (7.3.14) of the highest degree of precision

$$\int_0^1 f(x)\, x^2\, dx = \sum_{k=1}^n A_k f(x_k) + R\,(f) \qquad (7.3.15)$$

the nodes x_k must be related to the roots $z_1, z_2, \ldots, z_n$ of the Jacobi polynomial $P_n^{(2,0)}(z)$ by the relation

$$x_k = \frac{1 - z_k}{2}$$

and the coefficients A_k must have the values

$$A_k = \frac{1}{(1 - z_k^2)\, [P_n^{(2,0)\prime}(z_k)]^2}.$$

6. Double integrals in Cartesian coordinates.

Consider

$$I = \iint_\sigma f(x,\, y)\, dxdy. \qquad (7.3.16)$$

We will assume that $f(x,\, y)$ is continuous and relatively smooth in σ.

Under certain assumptions about the region σ the integral I can be reduced to two single integrals

$$I = \int_a^b F\,(x)\, dx \qquad (7.3.17)$$

$$F\,(x) = \int_{y_1(x)}^{y_2(x)} f(x,\, y)\, dy \qquad (7.3.18)$$

where $y_1(x)$, $y_2(x)$, a and b are known quantities. We will assume that the integral (7.3.18) can be calculated for all values of x for which we are interested and will concern ourselves with the problem of evaluating (7.3.17). The function $F\,(x)$ depends both on the integrand $f(x,\, y)$ and on the region σ.

It can be expected that among the quadrature formulas of highest degree of precision the Gauss formulas will not always give the best result since they are intended for use with a specific weight function and do not take into account the influence of σ on the function $F\,(x)$.

We now make some remarks about an appropriate weight function for (7.3.17). Construct a line through the region σ which passes through the point x and which is parallel to the y axis. The part of this line which lies in σ has length $y_2(x) - y_1(x)$. The longer this line the greater will be the influence of a narrow strip of σ along this line on the formation of the double integral. Therefore to calculate the integral I we use the weight function

$$p(x) = y_2(x) - y_1(x) \qquad (7.3.19)$$

and write I in the form

$$I = \int_a^b [y_2(x) - y_1(x)]\, \Phi(x)\, dx \qquad (7.3.20)$$

$$\Phi(x) = \frac{F(x)}{y_2(x) - y_1(x)}.$$

In many cases the weight function (7.3.19) will account sufficiently well for the influence of the region on I and for sufficiently smooth functions $f(x, y)$ will give good results. But this has the disadvantage that each region would have its own special class of quadrature formulas. However, the selection of the weight function for the integral I can be simplified by the following considerations. Consider the integral

$$I_1 = \int_a^b p(x) f(x)\, dx$$

and suppose that to evaluate it we wish to apply the quadrature formula of the highest degree of precision with n nodes

$$I_1 = \int_a^b p(x) f(x)\, dx \approx \sum_{k=1}^n A_k f(x_k). \qquad (7.3.21)$$

The nodes of this formula are the zeros of the nth degree polynomial of the system of orthogonal polynomials for the weight function $p(x)$. The accuracy of the quadrature formula (7.3.21) will, in general, depend on how closely the function $f(x)$ can be approximated by a polynomial of degree $2n - 1$.

Suppose now that $p(x)$ can be represented as a product

$$p(x) = \rho(x) q(x)$$

where $q(x)$ is positive throughout the interval $[a, b]$. Let us combine the function $q(x)$ with the integrand $f(x)$: $q(x) f(x) = F(x)$ and consider the integral I_1 with weight function $\rho(x)$

$$I_1 = \int_a^b \rho(x) F(x)\, dx.$$

Using the roots x_k of the polynomial of degree n which belongs to the system of orthogonal polynomials with weight function $\rho(x)$ we can construct the quadrature formula of the highest degree of precision

$$I_1 = \int_a^b \rho(x) F(x)\,dx = \sum_{k=1}^n B_k F(x_k) \tag{7.3.22}$$

If $q(x)$ is a slowly varying function which has derivatives of high order or if it is an analytic function with singular points far from the segment $[a, b]$ then we can expect that the function $f(x)$ and $F(x) = q(x) f(x)$ can both be closely approximated by polynomials of degree $2n - 1$. We can hope, therefore, that formulas (7.3.21) and (7.3.22), which both serve for calculating the integral I_1, will have about the same error and that only a small error is introduced in passing from (7.3.21) with weight function $p(x)$ to (7.3.22) with weight function $\rho(x)$.

In order to calculate the integral (7.3.17) this permits us to pass from the "natural" weight function $p(x) = y_2(x) - y_1(x)$ to a simpler weight function. In many cases this can be done without a significant loss of accuracy. The simpler weight function can be chosen so that it can be used for many regions.

Suppose the interval of integration $[a, b]$ is finite and assume it is possible to select exponents α and β so that the ratio

$$q(x) = \frac{y_2(x) - y_1(x)}{(x-a)^\beta (b-x)^\alpha} \qquad a \le x \le b$$

is bounded from above and from below by positive numbers

$$0 < m \le q(x) \le M < \infty.$$

Then we can use the weight function $(x-a)^\beta (b-x)^\alpha$ to calculate (7.3.17):

$$I = \int_a^b (x-a)^\beta (b-x)^\alpha \Psi(x)\,dx$$

where

$$\Psi(x) = (x-a)^{-\beta} (b-x)^{-\alpha} F(x).$$

For example, if the region of integration has the form shown in Figure 3 where the boundary λ of the region has at the point A with coordinate $x = a$ a tangent of the first order[10] we can take $\alpha = 0$, $\beta = \dfrac{1}{2}$ and use as the

---

[10] Mme. H. Berthod-Zaborowski and H. Mineur, "Sur le calcul numérique des intégrales doubles," C. R. Acad. Sci. Paris, Vol. 229, 1949, pp. 919–21. Taking y as the independent variable then at the point A the boundary curve λ can be written in the form $x = a + c_2 (y - y_0)^2 + c_3 (y - y_0)^3 + \ldots$; where $c_2 \ne 0$.

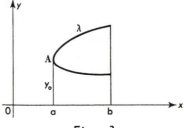

Figure 3.

weight function

$$p(x) = \sqrt{x - a}.$$

The integral

$$I = \int_a^b \sqrt{x - a}\ \Psi(x)\,dx, \qquad \Psi(x) = (x - a)^{-\frac{1}{2}} F(x)$$

can be calculated by formula (7.3.9).

If the region σ has the form shown in Fig. 4 where the boundary λ has tangents of the first order at $x = a$ and $x = b$ then we can use the weight function

$$p(x) = \sqrt{(x - a)(b - x)}.$$

To calculate the integral

$$I = \int_a^b \sqrt{(x - a)(b - x)}\ \Psi(x)\,dx$$

$$\Psi(x) = [(x - a)(b - x)]^{-\frac{1}{2}} F(x)$$

we can use (7.3.11).

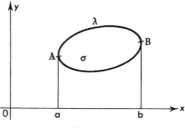

Figure 4.

7.4. THE INTEGRAL $\int_{-\infty}^{+\infty} e^{-x^2} f(x)\, dx.$

The system of polynomials which are orthogonal on the entire real axis $-\infty < x < +\infty$ with respect to the weight function e^{-x^2} is the system of Chebyshev-Hermite polynomials

$$H_n(x) = (-1)^n e^{x^2} \frac{d^n}{dx^n} e^{-x^2}.$$

A quadrature formula of the highest degree of precision

$$\int_{-\infty}^{+\infty} e^{-x^2} f(x)\, dx = \sum_{k=1}^{n} A_k f(x_k) + R(f) \qquad (7.4.1)$$

must have as its nodes the roots of the polynomial $H_n(x)$:

$$H_n(x_k) = 0, \qquad (k = 1, 2, \ldots, n).$$

The coefficients A_k can be found from (7.1.3) by using the leading coefficients (2.4.4) of the normalized Chebyshev-Hermite polynomials (2.4.3):

$$A_k = -\frac{2^{n+1} n!\, \pi^{\frac{1}{2}}}{H_n'(x_k) H_{n+1}(x_k)}.$$

If we substitute $x = x_k$ in (2.4.2) we obtain $H_{n+1}(x_k) = -H_n'(x_k)$ and thus this expression for the A_k can be written as

$$A_k = \frac{2^{n+1} n!\, \pi^{\frac{1}{2}}}{[H_n'(x_k)]^2} \qquad (7.4.2)$$

To find the remainder in (7.4.1) we use the polynomial $\omega(x) = 2^{-n} H_n(x)$; then

$$\int_{-\infty}^{+\infty} p(x)\, \omega^2(x)\, dx = 2^{-2n} \int_{-\infty}^{+\infty} e^{-x^2} H_n^2(x)\, dx = 2^{-n} n!\, \pi^{\frac{1}{2}},$$

and by (7.1.5)

$$R(f) = \frac{n!\, \pi^{\frac{1}{2}}}{2^n} \frac{f^{(2n)}(\eta)}{(2n)!}.$$

In Appendix B we give values of the x_k and A_k in formula (7.4.1) for[11] $n = 1 (1) 20$.

As an example[12], let us evaluate numerically the integral

$$\int_{-\infty}^{+\infty} e^{-x^2} J_0(x)\, dx = \sqrt{\pi}\ e^{-1/8} I_0(1/8) \approx 1.5703011006678$$

where $J_0(x)$ is the Bessel function of order zero and $I_0(x)$ is the modified Bessel function of order zero[13]. Applying the quadrature formula (7.4.1) with 10 nodes we obtain

$$\sum_{k=1}^{10} A_k J_0(x_k) = 1.5703011006676$$

which differs by only two in the last place from the true value which was found from the series expansion for $I_0(x)$.

7.5. INTEGRALS OF THE FORM $\int_0^\infty x^a e^{-x} f(x)\, dx$.

The system of polynomials which are orthogonal on the semi-infinite axis $0 \leq x < \infty$ with respect to the weight function $x^a e^{-x}$ is the system of Chebyshev-Laguerre polynomials

$$L_n^{(a)}(x) = (-1)^n x^{-a} e^x \frac{d^n}{dx^n} x^{a+n} e^{-x}.$$

A quadrature formula of the highest degree of precision

$$\int_0^\infty x^a e^{-x} f(x)\, dx = \sum_{k=1}^n A_k f(x_k) + R(f) \qquad (7.5.1)$$

must have as its nodes the roots of the Laguerre polynomial $L_n^{(a)}(x)$.

The normalized Laguerre polynomials are given by (2.5.4) and their

---

[11]Trans. note: In the original Russian edition of this book the author gave the values of the x_k and A_k for $n = 1 (1) 10$ given by: R. E. Greenwood and J. J. Miller, "Zeros of the Hermite polynomials and weights for Gauss' mechanical quadrature formula," *Bull. Amer. Math. Soc.*, Vol. 54, 1948, p. 765–769.

[12]This example is from the paper by H. E. Salzer, R. Zucker, and R. Capuano cited in Appendix B.

[13]See: G. N. Watson, *A Treatise on the Theory of Bessel Functions*, 2nd ed., Cambridge Univ. Press, 1944, p. 394.

leading coefficients by (2.5.5). Therefore the coefficients A_k can be found from (7.1.3) to be

$$A_k = -\frac{n!\,\Gamma\,(\alpha+n+1)}{L_n^{(\alpha)\prime}\,(x_k)\,L_{n+1}^{(\alpha)}\,(x_k)}\,.$$

Using the relationship

$$L_{n+1}^{(\alpha)}\,(x) = (x-\alpha-n-1)\,L_n^{(\alpha)}\,(x) - x L_n^{(\alpha)\prime}\,(x),$$

from the theory of Laguerre polynomials, we obtain

$$L_{n+1}^{(\alpha)}\,(x_k) = -\,x_k\,L_n^{(\alpha)\prime}\,(x_k)$$

and therefore

$$A_k = \frac{n!\,\Gamma\,(\alpha+n+1)}{x_k\,[L_n^{(\alpha)\prime}\,(x_k)]^2}.$$

Values of the x_k and A_k for $\alpha = 0$ for $n = 1\,(1)\,16\,(4)\,32$ are given in Appendix C[14].

1. Consider the integral

$$I = \int_0^\infty e^{-x}\,\frac{x}{1-e^{-2x}}\,dx = \frac{\pi^2}{8} \approx 1.2337\,.$$

Let us calculate the integral by using formula (7.5.1) for $\alpha = 0$ with 5 nodes. Using the x_k and A_k tabulated in Appendix C for $n = 5$ we obtain

$$I \approx A_1 f(x_1) + \cdots + A_5 f(x_5) = 1.2338\,.$$

2. We now calculate the integral

$$I = \int_0^\infty \frac{x\,dx}{e^x + e^{-x} - 1} = \int_0^\infty x e^{-x}\,[1 + e^{-2x} - e^{-x}]^{-1}\,dx \approx 1.17$$

by using the formula with two nodes for the weight function

$$p\,(x) = x e^{-x}$$

which corresponds to (7.5.1) with $\alpha = 1$. The second degree polynomial orthogonal with respect to $x e^{-x}$ is found from (2.5.2) to be

$$L_2^{(1)}\,(x) = x^2 - 6x + 6.$$

---

[14]Trans. note: In the original edition of this book the author gave the values tabulated by: H. E. Salzer and R. Zucker, "Table of the zeros and weight factors of the first fifteen Laguerre polynomials," *Bull. Amer. Math. Soc.*, Vol. 55, 1949, pp. 1004–12.

The roots of this polynomial are $x_1 = 3 - \sqrt{3}$ and $x_2 = 3 + \sqrt{3}$ and the co-efficients can be calculated from (7.5.2):

$$A_1 = \frac{3 + \sqrt{3}}{6}, \qquad A_2 = \frac{3 - \sqrt{3}}{6}.$$

We then obtain

$$I \approx A_1 f(x_1) + A_2 f(x_2) = 1.20.$$

REFERENCES

B. Bronwin, "On the determination of the coefficients in any series of sines and cosines of multiples of a variable angle from particular values of that series," *Philosophical Magazine*, (3) Vol. 34, 1849, pp. 260-68.

E. B. Christoffel, "Über die Gaussische Quadratur und eine Verallgemeinerungderselben," *J. Reine Angew. Math.*, Vol. 55, 1858, pp. 61-82.

C. F. Gauss, "Methodus nova integralium valores per approximationen inveniendi," *Werke*, Vol. 3, 1866, pp. 163-96.

Ia. L. Geronimus, *Theory of Orthogonal Polynomials*, Gostekhizdat, Moscow, 1950 (Russian).

P. C. Hammer, O. J. Marlowe, and A. H. Stroud, "Numerical integration over simplexes and cones," *Math. Tables Aids Comput.*, Vol. 10, 1956, pp. 130-37.

P. C. Hammer and A. W. Wymore, "Numerical evaluation of multiple integrals I," *Math. Tables Aids Comput.*, Vol. 11, 1957, pp. 59-67.

R. G. Hetherington, "Numerical integration over hypershells," Thesis, Univ. of Wisc., 1961,

A. A. Markov, *Calculus of Finite Differences*, Moscow, 1911 (Russian).

F. G. Mehler, "Bemerkungen zur Theorie der mechanischen Quadraturen," *J. Reine Angew. Math.*, Vol. 63, 1864, pp. 152-57.

W. H. Peirce, "Numerical integration over the planar annulus," *J. Soc. Indust. Appl. Math.*, Vol. 5, 1957, pp. 66-73.

W. H. Peirce, "Numerical integration over the spherical shell," *Math. Tables Aids Comput.*, Vol. 11, 1957, pp. 244-49.

K. A. Posse, "On functions analogous to Legendre functions," Comm. Société Math. Kharkow, 1885, 15 pp.

K. A. Posse, *Sur quelques applications des fractions continues algébriques*, Petersburg, 1886.

D. Secrest and A. H. Stroud, "Approximate integration over certain spherically symmetric regions," *Math. Comput.*, to appear in the future.

N. Ia. Sonin, "On the approximate evaluation of definite integrals and on the related integral functions," *Warshawskia Universitetskia Izvestia.* Vol. 1, 1887, pp. 1-76 (Russian).

T. J. Stieltjes, "Quelques recherches sur la theorie des quadratures dites mécanique," *Ann. Sci. École Norm. Sup.*, (3) Vol. 1, 1884, pp. 409-26.

CHAPTER 8

Quadrature Formulas with Least Estimate of the Remainder

8.1. MINIMIZATION OF THE REMAINDER OF QUADRATURE FORMULAS

In Chapter 7 we studied quadrature formulas of the highest algebraic degree of precision. It is reasonable to suppose that such formulas will give a small error provided that the integrand $f(x)$ can be closely approximated by a polynomial of moderate degree, in particular if $f(x)$ is an analytic function in a sufficiently wide region about the segment of integration $[a, b]$. Many years of experimentation has shown that these formulas give excellent precision in comparison with other types of quadrature formulas.

However these formulas are not universal, and in some practical cases they are known to give worse results than some of the elementary formulas: the midpoint formula, the trapezoidal formula, Simpson's formula, and others. This usually happens when the function $f(x)$ has a low order of differentiability or is an analytic function with singular points close to the segment of integration.

In the theory of quadrature there arose the need for the construction of formulas for the integration of functions which belong to a predetermined class, in particular to a class of functions of low order of differentiability.

Let us briefly recall the comments we made on this problem in Section 5.1. Let us be given a class of functions F. For each function $f \epsilon F$ the remainder $R(f)$ of the quadrature is defined as

$$R(f) = \int_a^b p(x) f(x) \, dx - \sum_{k=1}^n A_k f(x_k). \qquad (8.1.1)$$

A number which can be used to characterize the precision of the quadrature formula for all functions of F is

$$R = \sup_f \left| R(f) \right| = \sup_f \left| \int_a^b p(x) f(x)\, dx - \sum_{k=1}^n A_k f(x_k) \right|. \qquad (8.1.2)$$

The value of R depends on the x_k and the A_k, and we wish to select the nodes and coefficients so that R has the smallest possible value. The x_k and A_k are usually subjected to certain restraints which are related to the class F and to the way in which the functions of F are given. Two examples of such restraints are:

1. If the functions f are given in tabular form for a certain set of values of x, then it would be desirable to restrict the choice of the x_k to values for which the function is tabulated.

2. In order to construct quadrature formulas with the least estimate of the remainder for the class of functions with continuous r^{th} derivative for which $\left| f^{(r)} \right| \le M_r$ we must require that the quadrature formula be exact for all polynomials of degree $\le r - 1$. This is the same as requiring

$$\sum_{k=1}^n A_k x_k^m = \int_a^b p(x)\, x^m\, dx, \qquad (m = 0, 1, \ldots, r-1). \qquad (8.1.3)$$

In this chapter we assume that the segment of integration is finite. This assumption will be necessary for the particular cases which we will consider. With this assumption we can always consider that the segment $[a, b]$ has been transformed into the segment $[0, 1]$.

8.2. MINIMIZATION OF THE REMAINDER IN THE CLASS $L_q^{(r)}$

We will say that $f(x)$ belongs to the class $L_q^{(r)}$ ($q \ge 1$) if $f(x)$ has an absolutely continuous derivative of order $r - 1$ on $[0, 1]$ and $f^{(r)}(x)$ is q^{th} power summable on $[0, 1]$.

Each function $f \,\epsilon\, L_q^{(r)}$ can be represented in the form

$$f(x) = \sum_{i=0}^{r-1} \frac{f^{(i)}(0)}{i!} x^i + \int_0^1 f^{(r)}(t)\, E(x-t)\, \frac{(x-t)^{r-1}}{(r-1)!}\, dt \qquad (8.2.1)$$

where the $f^{(i)}(0)$ are numbers and $f^{(r)}(t)$ is a measurable and q^{th} power summable function on $[0, 1]$. The converse is also true: for any numbers $f^{(i)}(0)$ and any $f^{(r)}(t) \,\epsilon\, L_q$ the function defined by (8.2.1) belongs to $L_q^{(r)}$.

Consider the integral $\int_0^1 \rho(x) f(x)\, dx$, where $f(x) \,\epsilon\, L_q^{(r)}$. At first it will be sufficient to assume that the weight function $\rho(x)$ is measurable and summable on $[0, 1]$.

Suppose we use the quadrature formula

$$\int_0^1 \rho(x) f(x)\, dx \approx \sum_{k=1}^n A_k f(x_k) \tag{8.2.2}$$

to calculate this integral approximately. We wish to construct a formula which will be the "best" for all functions $f(x) \, \epsilon \, L_q^{(r)}$ ($q \geq 1$) assuming that (8.2.2) is exact for all polynomials of degree $< r$. If we use the representation (8.2.1) for the functions of $L_q^{(r)}$, then the remainder $R(f)$ of the quadrature has the form:

$$R(f) = \int_0^1 \rho(x) f(x)\, dx - \sum_{k=1}^n A_k f(x_k) = \int_0^1 f^{(r)}(t) K(t)\, dt \tag{8.2.3}$$

$$K(t) = \int_t^1 \rho(x) \frac{(x-t)^{r-1}}{(r-1)!}\, dx - \sum_{k=1}^n A_k E(x_k - t) \frac{(x_k - t)^{r-1}}{(r-1)!}. \tag{8.2.4}$$

Consider now the class F of functions $f(x)$ which satisfy the condition

$$\left(\int_0^1 |f^{(r)}(t)|^q dt \right)^{\frac{1}{q}} \leq M_r.$$

By Hölder's inequality we have

$$|R(f)| \leq \left(\int_0^1 |f^{(r)}|^q dt \right)^{\frac{1}{q}} \left(\int_0^1 |K(t)|^p dt \right)^{\frac{1}{p}} \leq M_r \left(\int_0^1 |K(t)|^p dt \right)^{\frac{1}{p}}$$

for $\dfrac{1}{p} + \dfrac{1}{q} = 1$. The function

$$f^{(r)}(t) = M_r \left(\int_0^1 |K(t)|^p dt \right)^{-\frac{1}{q}} |K(t)|^{\frac{p}{q}} \text{ sign } K(t)$$

belongs to the class F and, as is easily seen, for this function the above inequality becomes an equality. Therefore the right side will be an upper bound for $|R(f)|$ on the class F:

$$R = \sup_F |R(f)| = M_r \left(\int_0^1 |K(t)|^p dt \right)^{\frac{1}{p}}. \tag{8.2.5}$$

Thus we see that the dependence of R on x_k and A_k occurs only in the term $\displaystyle\int_0^1 |K(t)|^p dt$. Our aim will be to select the x_k and A_k so that the integral $\displaystyle\int_0^1 |K(t)|^p dt$ will be a minimum. If such x_k and A_k exist then they will furnish a least value for R for each M_r and the corresponding quadrature formula can be considered "the best" for the entire class $L_q^{(r)}$.[1]

The problem of minimizing $\int_0^1 |K(t)|^p dt$ can be interpreted as the problem of best approximating the function $\int_t^1 \rho(x) \frac{(x-t)^{r-1}}{(r-1)!} dx$ in the metric L_p (see Section 4.1) by means of functions of the form

$$\sum_{k=1}^n A_k E(x_k - t) \frac{(x_k - t)^{r-1}}{(r-1)!}.$$

For arbitrary $\rho(x)$, r and n this problem can not be solved in closed form. We will restrict ourselves to certain special cases when the solution can be found by simple methods.

First of all we need to become familiar with certain facts from the theory of approximation of functions. Let us be given on the segment $[0, 1]$ a certain function $f \in L_p$. In addition, let us suppose that the functions $\phi_k \in L_p$ $(k = 1, 2, \ldots, n)$ are linearly independent on $[0, 1]$. This means that the equation

$$\int_0^1 \left| \sum_{k=1}^n a_k \phi_k \right|^p dx = 0$$

is possible only when all the a_k are zero. This is equivalent to the statement that the equation $\sum_{k=1}^n a_k \phi_k(x) = 0$ can be fulfilled on a set of points of measure greater than zero if and only if $a_k = 0$ $(k = 1, 2, \ldots, n)$.

The error ϵ in the approximation of f by a linear combination $s = \sum_{k=1}^n a_k \phi_k$ is defined by

$$\epsilon^p = \int_0^1 |f - s|^p dx = I.$$

We now discuss the conditions under which ϵ^p will be a minimum. From

---

[1]$R(f)$ is a linear functional defined for functions $f^{(r)} \in L_q$. The integral $\left(\int_0^1 |K|^p dt \right)^{\frac{1}{p}}$ is the norm of $R(f)$ in the space L_q. In the terminology of functional analysis our problem is to construct a quadrature formula (8.2.2) with the least norm for the remainder.

a theorem of calculus we can assert that the values of a_k which give a minimum for I must satisfy the equations

$$\frac{\partial I}{\partial a_i} = p \int_0^1 |f - s|^{p-1} \operatorname{sign} (f - s) \phi_i \, dx = 0 \qquad i = 1, 2, \ldots, n. \qquad (8.2.6)$$

We now show that the linear combination s which satisfies (8.2.6) indeed gives the best approximation to f. Let us take any other linear combination $s^* = \displaystyle\sum_{k=1}^n a_k^* \phi_k$. We must show that $I \leq I^* = \displaystyle\int_0^1 |f - s^*|^p \, dx$. We have

$$I = \int_0^1 |f - s|^p \, dx = \int_0^1 |f - s|^{p-1} (f - s) \operatorname{sign} (f - s) \, dx =$$

$$= \int_0^1 |f - s|^{p-1} (f - (s - s^*) - s^*) \operatorname{sign} (f - s) dx.$$

By (8.2.6)

$$I = \int_0^1 |f - s|^{p-1} (f - s^*) \operatorname{sign} (f - s) \, dx. \qquad (8.2.7)$$

This integral can not be made smaller if $\operatorname{sign} (f - s)$ is replaced by $\operatorname{sign} (f - s^*)$. Therefore

$$I \leq \int_0^1 |f - s|^{p-1} |f - s^*| \, dx. \qquad (8.2.8)$$

Applying Hölder's inequality[2]

---

[2]See, for example, I. P. Natanson, *Theory of Functions of a Real Variable*, Ungar, New York, 1955, Chap. 7, Sec. 6. If $F \epsilon L_p$ and $G \epsilon L_q$, $\dfrac{1}{p} + \dfrac{1}{q} = 1$, then the product FG is summable and

$$\int_0^1 FG \, dx \leq \left(\int_0^1 |F|^p \, dx \right)^{\frac{1}{p}} \left(\int_0^1 |G|^q \, dx \right)^{\frac{1}{q}}. \qquad (a)$$

For the following presentation it is essential to note that equality can occur only when the following two conditions are satisfied:

1. $\dfrac{|F|^p}{\displaystyle\int_0^1 |F|^p \, dx} = \dfrac{|G|^q}{\displaystyle\int_0^1 |G|^q \, dx}$,

2. The signs of F and G coincide almost everywhere on $[0, 1]$. To apply (a) to (8.2.9) we take $F = |f - s^*|$ and $G = |f - s|^{p-1}$.

$$I \leq \left(\int_0^1 |f - s*|^p \, dx \right)^{\frac{1}{p}} \left(\int_0^1 |f - s|^p \, dx \right)^{\frac{p-1}{p}} = I^* \frac{1}{p} I^{\frac{p-1}{p}} . \quad (8.2.9)$$

This gives $I^{\frac{1}{p}} \leq I^* \frac{1}{p}$ and thus $I \leq I^*$.

Finally we show that the linear combination s, which minimizes I, is unique. It is necessary to verify that if $I = I^*$ then $a_k = a_k^*$ $(k = 1, 2, \ldots, n)$. This is clear if $I = 0$ because if $f = s$ for almost all x then

$$I^* = \int_0^1 |f - s^*|^p \, dx = \int_0^1 |s - s^*|^p \, dx = 0.$$

Therefore, for almost all x, $s = s^*$ and since the ϕ_k are linearly independent $a_k = a_k^*$. Thus we can suppose that $|f - s|$ is positive on a set of points of measure greater than zero.

From the argument leading to (8.2.9) we see that $I = I^*$ only if two conditions are satisfied:

1. For almost all x we must have

$$|f - s|^{p-1} (f - s^*) \operatorname{sign} (f - s) = |f - s|^{p-1} |f - s^*| . \quad (8.2.10)$$

This is necessary if (8.2.8) is to be an equality.

2. In (8.2.9) equality can only occur when almost everywhere

$$\frac{|f - s|^p}{\displaystyle\int_0^1 |f - s|^p \, dx} = \frac{|f - s^*|^p}{\displaystyle\int_0^1 |f - s^*|^p \, dx} .$$

Since $I = \int_0^1 |f - s|^p \, dx = I^* = \int_0^1 |f - s^*|^p \, dx$ then almost everywhere we must have

$$|f - s| = |f - s^*| . \quad (8.2.11)$$

But $|f - s| > 0$ on a set of positive measure and from (8.2.10) and (8.2.11) it follows that on a set of positive measure

$$f - s = f - s^* \quad \text{or} \quad s = s^* .$$

Since the ϕ_k are linearly independent this is only possible when $a_k = a_k^*$ $(k = 1, 2, \ldots, n)$.

We now assume $\rho(x) \equiv 1$ and let us consider the quadrature formula

$$\int_0^1 f(x) \, dx = \sum_{k=1}^n A_k f(x_k) + R(f) . \quad (8.2.12)$$

Now let $f(x)$ be absolutely continuous and $f'(x)$ be q^{th} power summable on $[0, 1]$. This corresponds to the case $r = 1$. We require that (8.2.12) be exact for $f(x) \equiv 1$ which imposes the condition $\sum_{k=1}^{n} A_k = 1$ on the co-

efficients. In the class $L_q^{(1)}$ the remainder $R(f)$ has the precise estimate

$$R = \sup_f R(f) = M_1 \left(\int_0^1 |K(t)|^p \, dt \right)^{\frac{1}{p}}$$

$$M_1^q \geq \int_0^1 |f'(x)|^q \, dx$$

$$K(t) = 1 - t - \sum_{k=1}^{n} A_k E(x_k - t).$$

The kernel $K(t)$ of the remainder is a piece-wise linear function with leading coefficient equal to -1, for which the nodes x_k are points of discontinuity. At the node x_k the function $K(t)$ has a jump of A_k. If the x_k lie inside the segment $[0, 1]$ then at $t = 0$ and $t = 1$ the kernel is zero. A typical graph of $K(t)$ is illustrated in Fig. 5.

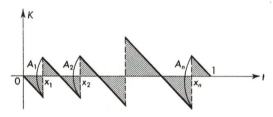

Figure 5.

The problem of minimizing the integral $\int_0^1 |K(t)|^p \, dt$ has the following

geometric meaning: it is necessary to determine for what arrangement of points of discontinuity x_k $(k = 1, 2, \ldots, n)$ and for what values of the jumps A_k $(k = 1, 2, \ldots, n)$, subject to the restraint $\Sigma A_k = 1$, will the cross-hatched area in Fig. 5 have the least mean p^{th} power. The answer is easy to foresee: the minimum will be achieved when the area consists of $2n$ equal triangles.

The nodes x_k must be located at the points $x_k = \dfrac{2k-1}{2n}$ $(k = 1, 2, \ldots,$

$n)$. The coefficients A_k must all be equal and since their sum is unity $A_k = \dfrac{1}{n}$ $(k = 1, 2, \ldots, n)$. This result can be easily verified by a calculation which we will not carry out.

The corresponding quadrature formula is

$$\int_0^1 f(x)\,dx = \frac{1}{n} \sum_{k=1}^{n} f\left(\frac{2k-1}{2n}\right) + R(f) \tag{8.2.13}$$

which is well known as the repeated midpoint formula. Its remainder in the class $L_q^{(1)}$ is

$$|R(f)| \le \frac{M_1}{2n \sqrt[p]{p+1}}, \qquad M_1 = \left(\int_0^1 |f'|^q\,dt\right)^{\frac{1}{q}}, \quad \frac{1}{p} + \frac{1}{q} = 1.$$

Let us now take $r = 2$ and consider the class $L_q^{(2)}$ of functions with absolutely continuous first derivative for which $f^{(2)}(x)$ is q^{th} power summable.

We require that the quadrature formula (8.2.12) be exact whenever $f(x)$ is a polynomial of degree zero or one. This is equivalent to the following two restraints on the x_k and A_k:

$$\sum_{k=1}^{n} A_k = \int_0^1 1\,dx = 1$$

$$\sum_{k=1}^{n} A_k x_k = \int_0^1 x\,dx = \frac{1}{2} \tag{8.2.14}$$

Under these conditions the remainder $R(f)$ has the following precise estimate in the class $L_q^{(2)}$

$$|R(f)| \le M_2 \left(\int_0^1 |K(t)|^p\,dt\right)^{\frac{1}{p}}, \qquad M_2 \ge \left(\int_0^1 |f^{(2)}(x)|^q\,dx\right)^{\frac{1}{q}}$$

$$K(t) = \frac{(1-t)^2}{2} - \sum_{k=1}^{n} A_k E(x_k - t)(x_k - t). \tag{8.2.15}$$

For later use we tabulate the value of $K(t)$ on each of the segments $[0, x_1], [x_1, x_2], \dots, [x_n, 1]$:

$$K(t) = \begin{cases} \dfrac{t^2}{2} & \text{for } 0 \le t \le x_1 \\[2ex] \dfrac{(1-t)^2}{2} - \displaystyle\sum_{k=i+1}^{n} A_k (x_k - t) & \text{for } x_i \le t \le x_{i+1} \\[2ex] \dfrac{(1-t)^2}{2} & \text{for } x_n \le t \le 1; \end{cases}$$

$K(t)$ is a continuous function of t on $[0, 1]$. The first derivative $K'(t)$ has discontinuities of the first kind at the points x_k and the size of the jumps of $K'(t)$ are

$$K'(x_k + 0) - K'(x_k - 0) = -A_k. \tag{8.2.16}$$

On each of the indicated segments $K(t)$ is a quadratic polynomial with leading coefficient $\frac{1}{2}t^2$. A typical graph of $K(t)$ is given in Fig. 6.

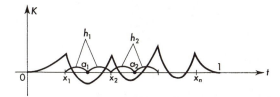

F igure 6.

Let us now turn to the problem of minimizing the integral

$$U = \int_0^1 |K(t)|^p dt$$

with the restraints (8.2.14). We will assume that the minimum value exists and will use the method of Lagrangian multipliers to find it. The result obtained by this method will later be justified. Let us consider the function

$$G = U + \lambda_1 \left(\sum_{k=1}^n A_k - 1 \right) + \lambda_2 \left(\sum_{k=1}^n A_k x_k - \frac{1}{2} \right)$$

and let us set equal to zero the partial derivatives of this function with respect to the x_i and A_i:

$$\frac{\partial G}{\partial x_i} = -A_i p \int_0^1 |K(t)|^{p-1} S(t) E(x_i - t) \, dt + \lambda_2 A_i = 0 \tag{8.2.17}$$

$$S(t) = \text{sign } K(t);$$

$$\frac{\partial G}{\partial A_i} = -p \int_0^1 |K(t)|^{p-1} S(t) E(x_i - t)(x_i - t) \, dt +$$

$$+ \lambda_1 + \lambda_2 x_i = 0. \tag{8.2.18}$$

Here the A_i are assumed to be different from zero because otherwise the quadrature sum would contain less than n nodes. The term A_i can then be cancelled from (8.2.17) to give:

$$\int_0^{x_i} |K(t)|^{p-1} S(t)\, dt = \frac{\lambda_2}{p}.$$

Since i takes the values $1, 2, \ldots, n$ we have

$$\int_0^{x_1} |K(t)|^{p-1} S(t)\, dt = \frac{\lambda_2}{p}$$

$$\int_{x_i}^{x_{i+1}} |K(t)|^{p-1} S(t)\, dt = 0, \qquad i = 1, 2, \ldots, n-1. \qquad (8.2.19)$$

By using (8.2.17) equation (8.2.18) can be reduced to the form

$$p \int_0^{x_i} |K(t)|^{p-1} S(t)\, t\, dt + \lambda_1 = 0.$$

Hence we see that

$$\int_0^{x_1} |K(t)|^{p-1} S(t)\, t\, dt = -\frac{\lambda_1}{p}$$

$$\int_{x_i}^{x_{i+1}} |K(t)|^{p-1} S(t)\, t\, dt = 0, \qquad i = 1, 2, \ldots, n-1. \qquad (8.2.20)$$

From (8.2.19) and (8.2.20) it follows that on each of the segments $[x_i, x_{i+1}]$ $(i = 1, 2, \ldots, n-1)$ the function $K(t)$ is a polynomial of second degree with leading coefficient $\frac{1}{2} t^2$ which deviates least from zero in the metric L_p. In order to find these polynomials let us first find the polynomial of the form $T_2(x) = x^2 + mx + r$ which deviates least from zero on the segment $[-1, +1]$:

$$\int_{-1}^1 |T_2(x)|^p\, dx = \text{minimum}.$$

We can see at once that for this polynomial $m = 0$. Indeed, if we replace x by $-x$ in the above integral we find that $T_2(-x)$ also deviates least from zero. But then

$$T_2(x) = T_2(-x)$$

and consequently $m = 0$.

Let us now find the constant term r.

The following conditions must be satisfied by T_2

$$\int_{-1}^{1} |T_2|^{p-1} \operatorname{sign} T_2 \, dx = 0$$

(8.2.21)

$$\int_{-1}^{1} |T_2|^{p-1} x \operatorname{sign} T_2 \, dx = 0.$$

The second of these equations is identically satisfied. From the first it follows that sign T_2 must change sign inside $[-1, +1]$. We suppose, therefore, that $r = -l^2$, $0 < l < 1$ and $T_2(x) = x^2 - l^2$, and write the first condition in the form

$$- \int_0^l (l^2 - x^2)^{p-1} \, dx + \int_l^1 (x^2 - l^2)^{p-1} \, dx = 0.$$

If we set $x = l\sqrt{t}$ then we can reduce this equation to the form

$$\int_1^{l^{-2}} t^{-\frac{1}{2}} (t - 1)^{p-1} \, dt = \frac{\sqrt{\pi}\, \Gamma(p)}{\Gamma\left(p + \dfrac{1}{2}\right)}.$$

(8.2.22)

From this equation we can find l. As l increases from 0 up to 1 the left side of (8.2.22) decreases from ∞ to 0 and hence this equation has one and only one solution.

In order to transform this result to the segment $[x_i, x_{i+1}]$ we write $a_i = \frac{1}{2}(x_i + x_{i+1})$ and $h_i = \frac{1}{2}(x_{i+1} - x_i)$. Then the second degree polynomial with leading coefficient $\frac{1}{2}t^2$ which deviates least from zero in the metric L_p on $[x_i, x_{i+1}]$ is

$$K(t) = \frac{h_i^2}{2} T_2\left(\frac{t - a_i}{h_i}\right) = \frac{h_i}{2}\left[\left(\frac{t - a_i}{h_i}\right)^2 - l^2\right], \qquad x_i \le t \le x_{i+1}.$$

At the point $t = x_i = a_i - h_i$ this polynomial has the value

$$K(x_i) = \frac{h_i^2}{2}(1 - l^2).$$

Similarly if we write $K(t)$ for the segment $[x_{i-1}, x_i]$ we see that its value at the point $t = x_i$ is

$$K(x_i) = \frac{h_{i-1}^2}{2}(1 - l^2).$$

Because $K(t)$ is continuous these values must be equal; thus

$$h_{i-1} = h_i \qquad (i = 2, 3, \ldots, n).$$

The common value of the h_i we will denote by h. Then for each of the x_i we have

$$K(x_i) = \frac{h^2}{2}(1 - l^2).$$

Now consider the segment $0 \leq t \leq x_1$. Here $K(t) = \frac{1}{2}t^2$ and at $t = x_1$ we must have

$$K(x_1) = \frac{x_1^2}{2} = \frac{h^2}{2}(1 - l^2) \qquad \text{or} \qquad x_1 = h\sqrt{1 - l^2}.$$

Finally, a consideration of $K(t)$ on the segment $[x_n, 1]$ gives for its length

$$1 - x_n = h\sqrt{1 - l^2}.$$

Since the sum of the lengths of the segments $[0, x_1], [x_1, x_2], \ldots,$ $[x_n, 1]$ is equal to 1 we must have

$$2h\sqrt{1 - l^2} + (n - 1)2h = 1$$

and

$$h = \frac{1}{2}\left[n - 1 + \sqrt{1 - l^2}\right]^{-1}$$

$$x_k = x_1 + 2h(k - 1) = h\left[2(k - 1) + \sqrt{1 - l^2}\right]. \qquad (8.2.23)$$

To calculate the coefficients A_k we use equation (8.2.16). On the segment $x_i \leq t \leq x_{i+1}$

$$K(t) = \frac{1}{2}h^2\left[\left(\frac{t - a_i}{h}\right)^2 - l^2\right],$$

$$K'(t) = t - a_i$$

$$K'(x_i + 0) = K'(a_i - h + 0) = -h$$

$$K'(x_{i+1} - 0) = K'(a_i + h - 0) = +h.$$

Therefore

$$A_i = 2h, \qquad (i = 2, 3, \ldots, n - 1). \qquad (8.2.24)$$

A similar calculation for the nodes x_1 and x_n gives

$$A_1 = A_n = \left(1 + \sqrt{1 - l^2}\right)h. \qquad (8.2.25)$$

Let us find the value of $\displaystyle\int_0^1 |K(t)|^p \, dt$. We have

$$\int_0^1 |K(t)|^p \, dt = \int_0^{x_1} \left(\frac{t^2}{2}\right)^p dt + \sum_{i=1}^{n-1} \int_{x_i}^{x_{i+1}} \left|\frac{h^2}{2} T_2\left(\frac{t-a_i}{h}\right)\right|^p dt +$$

$$+ \int_{x_n}^1 \left[\frac{1}{2}(1-t)^2\right]^p dt.$$

It is easily verified that

$$\int_0^{x_1} \frac{t^{2p}}{2^p} \, dt = \int_{x_n}^1 \frac{(1-t)^{2p}}{2^p} \, dt = \frac{h^{2p+1}(1-l^2)^{p+\frac{1}{2}}}{(2p+1)\,2^p}$$

$$\int_{x_i}^{x_{i+1}} \left|\frac{h^2}{2} T_2\left(\frac{t-a_i}{h}\right)\right|^p dt = \frac{h^{2p+1}}{2^p} \int_{-1}^1 |T_2(x)|^p \, dx$$

$$I = \int_{-1}^1 |T_2(x)|^p \, dx = |T_2|^p x \Big|_{-1}^{+1} -$$

$$- p \int_{-1}^1 |T_2(x)|^{p-1} x \, T_2' \, \text{sign}\,(T_2) \, dx =$$

$$= 2(1-l^2)^p - 2p \int_{-1}^1 |T_2|^{p-1} x^2 \, \text{sign}\,(T_2) \, dx.$$

If, in this last integral, we replace x^2 by $x^2 - l^2 + l^2 = T_2 + l^2$ we obtain

$$I = 2(1-l^2)^p - 2pI - 2pl^2 \int_{-1}^1 |T_2|^{p-1} \, \text{sign}\,(T_2) \, dx.$$

But, by (8.2.21), the integral on the right side of this equation is zero and consequently

$$I = \frac{2(1-l^2)^p}{2p+1},$$

$$\int_0^1 |K(t)|^p \, dt = \frac{h^{2p+1}}{(2p+1)\,2^{p-1}} \left[(1-l^2)^{p+\frac{1}{2}} + (n-1)(1-l^2)^p\right] =$$

$$= \frac{h^{2p+1}(1-l^2)^p}{(2p+1)\,2^{p-1}} \left[\sqrt{1-l^2} + n - 1\right] =$$

$$= \frac{h^{2p}(1-l^2)^p}{(2p+1)\,2^p}. \tag{8.2.26}$$

The remainder $R(f)$ in formula (8.2.12) with nodes (8.2.23) and coefficients (8.2.24) and (8.2.25) will have the following estimate for a function $f \in L_p^{(2)}$:

$$|R(f)| \le M_2 \, \frac{h^2(1 - l^2)}{2\sqrt[p]{2p + 1}}, \qquad M_2 = \left(\int_0^1 |f''(x)|^q \, dx \right)^{\frac{1}{q}} \qquad (8.2.27)$$

We will now show that the nodes x_k and coefficients A_k indeed give the least value of the integral (8.2.26).

Let x_k^* and A_k^* be any other nodes and coefficients and $K^*(t)$ the corresponding kernel. We must show that

$$\int_0^1 |K^*(t)|^p \, dt \ge \frac{h^{2p}(1 - l^2)^p}{(2p + 1) 2^p} \, .$$

We have

$$\int_0^1 |K^*|^p \, dt = \int_0^{x_1^*} \left(\frac{t^2}{2} \right)^p dt + \sum_{i=1}^{n-1} \int_{x_i^*}^{x_{i+1}^*} |K^*|^p \, dt +$$

$$+ \int_{x_n^*}^1 \left[\frac{(1 - t)^2}{2} \right]^p dt =$$

$$= \frac{x_1^{*\,2p+1} + (1 - x_n^*)^{2p+1}}{(2p + 1) 2^p} + \sum_{i=1}^{n-1} \int_{x_i^*}^{x_{i+1}^*} |K^*|^p \, dt.$$

On each of the segments $[x_i^*, x_{i+1}^*]$ the kernel $K^*(t)$ is a certain quadratic polynomial with leading coefficient $\frac{1}{2} t^2$. Let us replace $K^*(t)$ by the second degree polynomial with the same leading coefficient which deviates least from zero on $[x_i^*, x_{i+1}^*]$ in the metric L_p. If we denote $a_i^* = \frac{1}{2}(x_i^* + x_{i+1}^*)$ and $h_i^* = \frac{1}{2}(x_{i+1}^* - x_i^*)$ then such a polynomial will be $\dfrac{h_i^{*\,2}}{2} T_2 \left(\dfrac{t - a_i^*}{h_i^*} \right).$ The last equation then becomes an inequality:

$$\int_0^1 |K^*|^p \, dt \ge \frac{x_1^{*\,2p+1} + (1 - x_n^*)^{2p+1}}{(2p + 1) 2^p} +$$

$$+ \sum_{i=1}^{n-1} \left(\frac{h_i^{*\,2}}{2} \right)^p \int_{x_i^*}^{x_{i+1}^*} \left| T_2 \left(\frac{t - a_i^*}{h_i^*} \right) \right|^p dt.$$

Equality is possible only when $K^*(t)$ is the polynomial which deviates least from zero on each segment $[x_i^*, x_{i+1}^*]$. But in that case we will have $x_k^* = x_k$ and $A_k^* = A_k$.

Using our previous notation the integrals in the summation were shown to have the common value $\dfrac{2(1 - l^2)^p}{2p + 1}$. Therefore

$$\int_0^1 |K^*|^p \, dt \geq \frac{x_1^{*\,2p+1} + (1 - x_n^*)^{2p+1}}{(2p + 1)\, 2^p} +$$

$$+ \frac{(1 - l^2)^p}{(2p + 1)\, 2^{3p}} \sum_{i=1}^{n-1} (x_{i+1}^* - x_i^*)^{2p+1}. \tag{8.2.28}$$

If in the sum $u_n = \displaystyle\sum_{i=1}^{n-1} (x_{i+1}^* - x_i^*)^{2p+1}$, we fix x_1^* and x_n^* then, as a function of $x_2^*, \ldots, x_{n-1}^*$, u_n is a minimum when all the segments $[x_i^*, x_{i+1}^*]$ have the same length. This can be shown by means of induction. Suppose $n = 3$ and consider

$$u_3 = (x_3^* - x_2^*)^{2p+1} + (x_2^* - x_1^*)^{2p+1}.$$

Then

$$\frac{\partial u_3}{\partial x_2^*} = (2p + 1)[(x_2^* - x_1^*)^{2p} - (x_3^* - x_2^*)^{2p}] = 0$$

shows that

$$x_2^* - x_1^* = x_3^* - x_2^* \tag{8.2.29}$$

and because

$$\frac{\partial^2 u_3}{\partial x_2^{*2}} = (2p + 1)\, 2p[(x_2^* - x_1^*)^{2p-1} + (x_3^* - x_2^*)^{2p-1}] > 0$$

then (8.2.29) indeed gives a minimum.

Assuming that the assertion is true for u_{n-1} we can verify it for u_n:

$$u_n = \sum_{i=1}^{n-2} (x_{i+1}^* - x_i^*)^{2p+1} + (x_n^* - x_{n-1}^*)^{2p+1} \geq$$

$$\geq (n - 2) \left(\frac{x_{n-1}^* - x_1^*}{n - 2} \right)^{2p+1} + (x_n^* - x_{n-1}^*)^{2p+1} = v.$$

Let us find the minimum of v as a function of x_{n-1}^*. From

$$\frac{\partial v}{\partial x_{n-1}^*} = (2p+1)\left[\left(\frac{x_{n-1}^* - x_1^*}{n-2}\right)^{2p} - (x_n^* - x_{n-1}^*)^{2p}\right] = 0$$

it follows that the segment $[x_{n-1}^*, x_n^*]$ must have the same length as all the other segments $[x_i^*, x_{i+1}^*]$:

$$x_n^* - x_{n-1}^* = \frac{x_{n-1}^* - x_1^*}{n-2}. \tag{8.2.30}$$

Since

$$\frac{\partial^2 v}{\partial x_{n-1}^{*2}} = (2p+1)2p\left[\left(\frac{x_{n-1}^* - x_1^*}{n-2}\right)^{2p-1} + (x_n^* - x_{n-1}^*)^{2p-1}\right] > 0$$

equation (8.2.30) indeed gives a minimum and therefore

$$u_n \geq (n-1)\left(\frac{x_n^* - x_1^*}{n-1}\right)^{2p+1}.$$

Substituting in (8.2.28) the minimum value for u_n we obtain

$$\int_0^1 |K^*|^p\, dt \geq \frac{x_1^{*\,2p+1} + (1-x_n^*)^{2p+1}}{(2p+1)2^p} +$$

$$+ \frac{(1-l^2)^p(x_n^* - x_1^*)^{2p+1}}{(2p+1)2^{3p}(n-1)^{2p}} = w.$$

By an argument similar to the preceding we can show that the minimum value of w is achieved for

$$x_1^* = 1 - x_n^* = \frac{\sqrt{1-l^2}}{2(n-1+\sqrt{1-l^2})} = h\sqrt{1-l^2}$$

and that

$$\min w = \frac{(1-l^2)^p h^{2p}}{2^p(2p+1)}.$$

We finally obtain

$$\int_0^1 |K^*|^p\, dt \geq \frac{(1-l^2)^p h^{2p}}{2^p(2p+1)} = \int_0^1 |K|^p\, dt.$$

From the above argument we see that equality can only be achieved when the x_k^* and A_k^* coincide with the values given by (8.2.23), (8.2.24), and (8.2.25).

As a supplementary remark we show that

$$\lim_{n \to \infty} \sum_{k=1}^{n} A_k f(x_k) = \int_0^1 f(x)dx$$

whenever $f(x)$ is Riemann integrable on $[0, 1]$. To do this it is sufficient to show that the quadrature sum is a Riemann sum which is equivalent to

$$A_1 + \cdots + A_{k-1} \le x_k \le A_1 + \cdots + A_k \qquad (k = 1, 2, \ldots, n).$$

This inequality is very easily verified. If we substitute for the x_k and A_k the values which we have found for them we obtain the following valid inequality:

$$\left(1 + \sqrt{1 - l^2}\right)h + \left(2(k - 2) + \sqrt{1 - l^2}\right)h \le \left(2(k - 1) + \sqrt{1 - l^2}\right)h \le$$
$$\le \left(1 + \sqrt{1 - l^2}\right)h + \left(2(k - 1) + \sqrt{1 - l^2}\right)h.$$

8.3. MINIMIZATION OF THE REMAINDER IN THE CLASS C_r

In Section 5.3 we defined C_r as the class of functions $f(x)$ which have a continuous derivative of order r on $[0, 1]$. The characteristic representation for a function $f(x) \in C_r$ is given by

$$f(x) = \sum_{i=0}^{r-1} \frac{f^{(r)}(0)}{i!} x^i + \int_0^1 f^{(r)}(t) E(x - t) \frac{(x - t)^{r-1}}{(r - 1)!} dt, \qquad (8.3.1)$$

where the $f^{(i)}(0)$ are arbitrary real numbers and $f^{(r)}(t)$ is an arbitrary continuous function on $[0, 1]$.

A quadrature formula

$$\int_0^1 f(x) \, dx \approx \sum_{k=1}^{n} A_k f(x_k) \qquad (8.3.2)$$

which has the least estimate of the remainder in C_r must be exact whenever $f(x)$ is a polynomial of degree $<r$. Then the remainder in (8.3.2) can be represented in the form

$$R(f) = \int_0^1 f^{(r)}(t) K(t) dt \qquad (8.3.3)$$

where

$$K(t) = \frac{(1 - t)^r}{r!} - \sum_{k=1}^{n} A_k E(x_k - t) \frac{(x_k - t)^{r-1}}{(r - 1)!}.$$

Consider the class F, of functions $f(x) \in C_r$, which satisfy the condition $|f^{(r)}(x)| \le M_r$. For functions of F we have

$$|R(f)| \le M_r \int_0^1 |K(t)| dt.$$

We can easily see that the right side of this inequality is an upper bound for $|R(f)|$ on F. This follows if we take a function $f(t)$ for which

$$f^{(r)}(t) = M_r \operatorname{sign} K(t).$$

For such a function

$$R(f) = M_r \int_0^1 |K(t)| dt.$$

Such a function does not belong to F because $f^{(r)}(t)$ is not continuous, but this function, together with its first and second derivatives, can be approximated to any degree of precision in the metric L by means of a function of F. Therefore in the above inequality for $|R(f)|$ the right side can not be decreased:

$$R = \sup_F |R(f)| = M_r \int_0^1 |K(t)| dt. \qquad (8.3.4)$$

We must minimize $\int_0^1 |K(t)| dt$ subject to the restraining conditions

$$\sum_{k=1}^{n} A_k x_k^i = \frac{1}{i + 1} \qquad (i = 0, 1, \ldots, r - 1). \qquad (8.3.5)$$

As in the preceding section we will solve this problem for only the two simplest cases.

Let $r = 1$ and consider the class of functions with a continuous derivative on $[0, 1]$. In this case we must require that the quadrature formula will be exact whenever $f(x)$ is a constant function. This is equivalent to requiring

$$\sum_{k=1}^{n} A_k = 1.$$

The kernel $K(t)$ is

$$K(t) = 1 - t - \sum_{k=1}^{n} A_k E(x_k - t).$$

A typical graph of such a kernel is given in Fig. 5. The integral $\int_0^1 |K(t)|\,dt$ is numerically equal to the area which is shaded in the figure. This area will be the smallest when all of the $2n$ triangles have the same size. Therefore the formula which gives the least estimate of the remainder in the class F is, for each M_1, the repeated midpoint formula (8.2.13).

The smallest value of the shaded area in Fig. 5 is $\dfrac{1}{4n}$ and hence the remainder $R(f)$ of formula (8.2.13) in the class C_1 has the estimate

$$|R(f)| \le M_1 \frac{1}{4n}, \qquad |f'(x)| \le M_1.$$

Let us now consider the class of functions C_2 which have two continuous derivatives on $[0, 1]$. The nodes and coefficients must satisfy the two conditions (8.2.14) and hence the quadrature formula must be exact for any linear function.

The kernel of the remainder $K(t)$ is given by (8.2.15). We will obtain a representation for this kernel on the segment $[x_k, x_{k+1}]$. Let us assume that the minimum of $u = \int_0^1 |K(t)|\,dt$ exists. We construct the auxiliary function

$$G = u + \lambda_1 \left(\sum_{k=1}^{n} A_k - 1 \right) + \lambda_2 \left(\sum_{k=1}^{n} A_k x_k - \frac{1}{2} \right)$$

and set the partial derivatives of G with respect to the x_i and A_i equal to zero:

$$\frac{\partial G}{\partial x_i} = -A_i \int_0^1 S(t) E(x_i - t)\,dt + \lambda_2 A_i = 0 \qquad (8.3.6)$$

$$S(t) = \text{sign } K(t)$$

$$\frac{\partial G}{\partial A_i} = -\int_0^1 S(t) E(x_i - t)(x_i - t)\,dt + \lambda_1 + \lambda_2 x_i = 0 \qquad (8.3.7)$$

$$(i = 1, 2, \ldots, n).$$

From these equations we see that on each of the segments $[x_i,\ x_{i+1}]$

$$\int_{x_i}^{x_{i+1}} S(t)\,dt = 0, \qquad \int_{x_i}^{x_{i+1}} t\,S(t)\,dt = 0 \qquad (i = 1, 2, \ldots, n - 1).$$

Thus on each of these segments the kernel $K(t)$ is a second degree polynomial with leading coefficient $\frac{1}{2}t^2$ which deviates least from zero on $[x_i,\ x_{i+1}]$ in the metric L.

In Section 2.3 we showed that among all polynomials of degree n with leading coefficient equal to unity the polynomial which deviates least from zero on $[-1, 1]$ in the metric L is

$$P_n(x) = \frac{1}{2^n} U_n(x) = \frac{\sin\,[(n + 1)\ \text{arc}\ \cos\ x]}{2^n \sqrt{1 - x^2}}.$$

For $n = 2$ this is the polynomial

$$P_2(x) = x^2 - \frac{1}{4}.$$

Transforming the segment $[-1, 1]$ into the segment $[x_i,\ x_{i+1}]$ by the linear transformation

$$t = a_i + h_i x, \qquad a_i = \frac{1}{2}(x_i + x_{i+1}), \qquad h_i = \frac{1}{2}(x_{i+1} - x_i)$$

and making the leading coefficient equal to $\frac{1}{2}$ we obtain

$$K(t) = \frac{1}{2} h_i^2 P_2\left(\frac{t - a_i}{h_i}\right), \qquad x_i \leq t \leq x_{i+1}.$$

If we start from this representation for $K(t)$ and use an argument similar to that of the preceding section we can prove that this kernel indeed gives a minimum value for u.

For the quadrature formula which provides the least estimate for the remainder in C_2 we have proven:

1. The nodes and coefficients are

$$x_k = \frac{\sqrt{3} + 4(k - 1)}{2}\,h \qquad h = [\sqrt{3} + 2(n - 1)]^{-1}$$

$$A_1 = A_n = \frac{2 + \sqrt{3}}{2}\,h \qquad A_k = 2h\ (k = 2, \ldots, n - 1).$$

2. These x_k and A_k minimize the integral $\displaystyle\int_0^1 |K(t)|\,dt$ and they are unique.

3. *The remainder $R(f)$ has the estimate*

$$|R(f)| \le M_2 \frac{h^2}{8}, \qquad |f''(x)| \le M_2 \quad \text{for} \quad x \in [0, 1].$$

4. *The quadrature sum* $\sum\limits_{k=1}^{n} A_k f(x_k)$ *is a Riemann sum and hence for any Riemann integrable function*

$$\lim_{n \to \infty} \sum_{k=1}^{n} A_k f(x_k) = \int_0^1 f(x) \, dx.$$

8.4. THE PROBLEM OF MINIMIZING THE ESTIMATE OF THE REMAINDER FOR QUADRATURE WITH FIXED NODES

We consider the problem of constructing quadrature formulas with given nodes and with minimal estimate of the remainder. We consider the case which occurs most often in applications: equally spaced nodes and a constant weight function. Let us assume that the segment of integration $[0, 1]$ is divided into n equal parts of length $h = 1/n$.

The quadrature formula

$$\int_0^1 f(x) \, dx \approx \sum_{k=0}^{n} A_k f\left(\frac{k}{n}\right) \tag{8.4.1}$$

has $n + 1$ coefficients A_k which are to be determined. If we require that (8.4.1) be exact for all polynomials of degree $\le n$ then, as we saw in Chapter 6, the coefficients A_k are completely defined and the formulas are the Newton-Cotes formulas. Let us assume then that (8.4.1) is exact for polynomials of degree $r - 1 < n$. This imposes the following restraints on the A_k:

$$\sum_{k=0}^{n} A_k = 1 \tag{8.4.2}$$

$$\sum_{k=1}^{n} A_k k^i = \frac{n^i}{i + 1}, \qquad (i = 1, 2, \ldots, r - 1).$$

If $f^{(r-1)}(x)$ is absolutely continuous on $[0, 1]$ then the remainder of the quadrature can be represented in the form:

$$R(f) = \int_0^1 f^{(r)}(t) K(t) dt$$

$$(8.4.3)$$

$$K(t) = \frac{(1-t)^r}{r!} - \sum_{k=1}^n A_k E\left(\frac{k}{n} - t\right) \frac{\left(\frac{k}{n} - t\right)^{r-1}}{(r-1)!} \, .$$

Among the $n+1$ coefficients A_k there are $n+1-r$ independent relations which are available for decreasing the estimate of the remainder of the formula (8.4.1).

In two cases we will find quadrature formulas which minimize the estimate of $R(f)$.

Let us take first of all the functions of the class $L_q^{(1)}$ for which

$$\left(\int_0^1 |f'(t)|^q dt\right)^{\frac{1}{q}} \le M_1$$

If we assume that the formula is exact when $f(x)$ is a constant function then the coefficients A_k must satisfy the first of the conditions (8.4.2) and we have the following estimate for the remainder

$$|R(f)| \le \left(\int_0^1 |f'|^q dt\right)^{\frac{1}{q}} \left(\int_0^1 |K|^p dt\right)^{\frac{1}{p}} \le$$

$$\le M_1 \left(\int_0^1 |K|^p dt\right)^{\frac{1}{p}} = \sup_f |R(f)|.$$

The integral $\displaystyle\int_0^1 |K(t)|^p dt$ depends only on the A_k and these coefficients must be chosen to minimize this integral. The kernel $K(t)$ is given by

$$K(t) = 1 - t - \sum_{k=1}^n A_k E\left(\frac{k}{n} - t\right).$$

On each of the segments $\left[\dfrac{i-1}{n}, \dfrac{i}{n}\right]$ $K(t)$ is a linear function of t:

$$K(t) = 1 - t - \sum_{k=1}^n A_k.$$

At the points $t = i/n$ $(i = 1, 2, \ldots, n - 1)$ $K(t)$ is discontinuous with a jump of A_i and at the ends of the segment $[0, 1]$ the kernel has the values A_0 and $-A_n$ respectively.

A typical graph of $K(t)$ is illustrated in Fig. 7.

Figure 7.

We must determine the A_k, subject to the restriction $\displaystyle\sum_{k=0}^{n} A_k = 1$, so that the pth power of the shaded area in the figure will have the least average value. A simple calculation shows that this will occur when the shaded area consists of $2n$ equal right triangles.

Thus it immediately follows that

$$A_0 = A_n = \frac{1}{2n}, \qquad A_1 = A_2 = \cdots = A_{n-1} = \frac{1}{n}.$$

This is the well-known repeated trapezoidal rule:

$$\int_0^1 f(x)\, dx = \frac{1}{n}\left[\frac{1}{2} f(0) + f\left(\frac{1}{n}\right) + \cdots + f\left(\frac{n-1}{n}\right) + \frac{1}{2} f(1)\right] + R(f)$$

and its remainder $R(f)$ has the estimate

$$|R(f)| \le \frac{M_1}{2n(p+1)^{\frac{1}{p}}}, \qquad M_1 = \left(\int_0^1 |f'(t)|^q\, dt\right)^{\frac{1}{q}}.$$

Now we consider quadrature formulas with least estimate of the remainder in classes of functions of higher degrees of differentiability. We restrict ourselves exclusively to the class $L_2^{(r)}(r \ge 2)$. In this case the problem of determining the coefficients A_k has a simple solution.

We assume that the quadrature formula is exact for polynomials of degree $< r$ which is equivalent to equations (8.4.2) being satisfied. The remainder has the representation (8.4.3). In the class of functions $f(x)$ which satisfy

$$\left(\int_0^1 [f^{(r)}(t)]^2 \, dt \right)^{\frac{1}{2}} \le M_r$$

the remainder has the estimate

$$R(f) \le M_r \left(\int_0^1 [K(t)]^2 \, dt \right)^{\frac{1}{2}} = \sup_f |R(f)|$$

The integral $I = \int_0^1 [K(t)]^2 \, dt$ is only a function of the A_k and, as before, the problem is to minimize I. This problem is one of minimizing a second degree polynomial in the A_k with the linear restraints (8.4.2). The integral I does not depend on A_0 since A_0 only enters in the first of the equations (8.4.2). This equation is not needed to find the minimum of I because it does not impose any restraint on the A_k ($k = 1, 2, \ldots, n$) and we will use this equation to calculate A_0 when we have calculated the other A_k ($k \ge 1$). The other restraints

$$\sum_{k=1}^n A_k k^i = \frac{n^i}{i+1} \qquad (i = 1, 2, \ldots, r-1)$$

are independent and can be written as functions of any $r-1$ of the A_k, for example as functions of $A_1, \ldots, A_{r-1}$.

In the integral I the terms of second degree in the A_k are obtained from the integral

$$\sigma(A_1, \ldots, A_n) = \frac{1}{(r-1)!} \int_0^1 \left[\sum_{k=1}^n A_k E\left(\frac{k}{n} - t\right)\left(\frac{k}{n} - t\right)^{r-1} \right]^2 dt.$$

The quadratic form $\sigma(A_1, \ldots, A_n)$ is positive definite since, clearly, $\sigma(A_1, \ldots, A_n) \ge 0$ and $\sigma(A_1, \ldots, A_n) = 0$ can only occur when for each $t \in [0, 1]$

$$\sum_{k=1}^n A_k E\left(\frac{k}{n} - t\right)\left(\frac{k}{n} - t\right)^{r-1} = 0$$

and this is possible only when $A_k = 0$ ($k = 1, 2, \ldots, n$).

From this it follows, by the usual algebraic argument, that the problem of minimizing

$$I = \int_0^1 [K(t)]^2 \, dt$$

subject to the restraints (8.4.2) has a unique solution. If we write the usual conditions for an extremum of I then we obtain a system of linear equations which determine the A_k. Values of the A_k and I have been calculated by Sard and Meyers for $r = 2$, $m = 1\,(1)\,20$; $r = 3$, $m = 2\,(1)\,12$; and $r = 4$, $m = 2\,(1)\,9$. We give these values in the following tables.

$$r = 2$$

m	1	2	3	4	5	6	7	8	9
δ	2	16	30	112	190	624	994	3 104	4 770
$A_0\delta = A_m\delta$	1	3	4	11	15	41	56	153	209
$A_1\delta = A_{m-1}\delta$		10	11	32	43	118	161	440	601
$A_2\delta = A_{m-2}\delta$				26	37	100	137	374	511
$A_3\delta = A_{m-3}\delta$						106	143	392	535
$A_4\delta = A_{m-4}\delta$								386	529
$m^5 I$	$\dfrac{1}{120}$	$\dfrac{1}{160}$	$\dfrac{1}{120}$	$\dfrac{1}{105}$	$\dfrac{5}{456}$	$\dfrac{77}{6\,240}$	$\dfrac{39}{2\,840}$	$\dfrac{22}{1\,455}$	$\dfrac{7}{424}$

m	10	11	12	13	14	15
δ	14 480	21 758	64 848	95 966	282 352	413 250
$A_0\delta = A_m\delta$	571	780	2 131	2 911	7 953	10 864
$A_1\delta = A_{m-1}\delta$	1 642	2 243	6 128	8 371	22 870	31 241
$A_2\delta = A_{m-2}\delta$	1 396	1 907	5 210	7 117	19 444	26 561
$A_3\delta = A_{m-3}\delta$	1 462	1 997	5 456	7 453	20 362	27 815
$A_4\delta = A_{m-4}\delta$	1 444	1 973	5 390	7 363	20 116	27 479
$A_5\delta = A_{m-5}\delta$	1 450	1 979	5 408	7 387	20 182	27 569
$A_6\delta = A_{m-6}\delta$			5 402	7 381	20 164	27 545
$A_7\delta = A_{m-7}\delta$					20 170	27 551
$m^5 I$	$\dfrac{311}{17\,376}$	$\dfrac{763}{39\,560}$	$\dfrac{419}{20\,265}$	$\dfrac{9\,773}{442\,920}$	$\dfrac{28\,381}{1\,210\,080}$	$\dfrac{8\,213}{330\,600}$

m	16	17	18	19	20
δ	1 204 288	1 747 906	5 056 272	7 290 718	20 966 960
$A_0\ \delta = A_m\delta$	29 681	40 545	110 771	151 316	413 403
$A_1\ \delta = A_{m-1}\delta$	85 352	116 593	318 538	435 131	1 188 800
$A_2\ \delta = A_{m-2}\delta$	72 566	99 127	270 820	369 947	1 010 714
$A_3\ \delta = A_{m-3}\delta$	75 992	103 807	283 606	387 413	1 045 432
$A_4\ \delta = A_{m-4}\delta$	75 074	102 553	280 180	382 733	1 045 646
$A_5\ \delta = A_{m-5}\delta$	75 320	102 889	281 098	383 987	1 049 072
$A_6\ \delta = A_{m-6}\delta$	75 254	102 799	280 852	383 651	1 048 154
$A_7\ \delta = A_{m-7}\delta$	75 272	102 823	280 918	383 741	1 048 400
$A_8\ \delta = A_{m-8}\delta$	75 266	102 817	280 900	383 717	1 048 334
$A_9\ \delta = A_{m-9}\delta$			280 906	383 723	1 048 352
$A_{10}\delta = A_{m-10}\delta$					1 048 346
$m^5 I$	$\dfrac{2\,468}{94\,085}$	$\dfrac{170\,393}{6\,169\,080}$	$\dfrac{162\,977}{5\,618\,080}$	$\dfrac{699\,869}{23\,023\,320}$	$\dfrac{8\,331}{262\,087}$

$$r = 3$$

m	2	3	4	5	6	7	8
δ	6	24	240	1 560	930	607 152	643 104
$A_0\delta = A_m\delta$	1	3	21	112	55	30 927	28 603
$A_1\delta = A_{m-1}\delta$	4	9	76	379	192	106 573	99 124
$A_2\delta = A_{m-2}\delta$			46	289	132	76 573	69 874
$A_3\delta = A_{m-3}\delta$					172	89 503	85 684
$A_4\delta = A_{m-4}\delta$							76 534
$m^7 I$	1	11	11	73	11	134 081	3 961
	1 890	8 960	12 600	69 888	10 850	124 899 840	3 617 460

m	9	10	11	12
δ	4 700 880	34 572 870	2 789 581 080	143 254 032
$A_0\delta = A_m\delta$	186 016	1 230 777	90 294 905	4 250 217
$A_1\delta = A_{m-1}\delta$	643 081	4 259 404	312 347 051	14 705 148
$A_2\delta = A_{m-2}\delta$	457 051	3 016 564	221 544 971	10 423 398
$A_3\delta = A_{m-3}\delta$	549 131	3 656 464	267 523 241	12 607 228
$A_4\delta = A_{m-4}\delta$	515 161	3 358 804	247 986 521	11 640 978
$A_5\delta = A_{m-5}\delta$		3 528 844	255 093 851	12 084 348
$A_6\delta = A_{m-6}\delta$				11 831 398
$m^7 I$	662 807	507 029	3 062 211 497	1 028 343
	584 998 400	435 618 162	2 556 270 662 400	835 648 520

$$r = 4$$

m	2	3	4	5	6
δ	6	24	28 992	432 840	19 740 084
$A_0\delta = A_m\delta$	1	3	2 349	29 392	1 082 811
$A_1\delta = A_{m-1}\delta$	4	9	9 932	110 209	4 409 946
$A_2\delta = A_{m-2}\delta$			4 430	76 819	2 225 043
$A_3\delta = A_{m-3}\delta$					4 304 484
$m^9 I$	1	13	6 557	61 633	210 047
	9 072	17 920	36 529 920	193 912 320	921 203 920

m	7	8	9
δ	167 985 552	12 298 253 184	291 277 352 304
$A_0\delta = A_m\delta$	8 013 897	509 110 987	10 764 281 184
$A_1\delta = A_{m-1}\delta$	31 412 443	2 040 010 996	42 647 140 119
$A_2\delta = A_{m-2}\delta$	18 665 443	1 105 566 730	24 253 340 709
$A_3\delta = A_{m-3}\delta$	25 900 993	1 867 200 148	37 040 022 813
$A_4\delta = A_{m-4}\delta$		1 254 475 462	30 933 891 327
$m^9 I$	56 097 271	2 876 254 589	18 892 720 083
	207 342 167 040	11 621 849 258 880	72 495 696 573 440

REFERENCES

S. M. Nikol'skii, "To the question concerning estimates of approximation with quadrature formulas," *Uspehi Mat. Nauk*, Vol. 5, 1950, pp. 165–77 (Russian).

S. M. Nikol'skii, *Quadrature Formulas*, Fizmatgiz, Moscow, 1958 (Russian).

A. Sard, "Best approximate integration formulas; best approximation formulas," *Amer. J. Math.*, Vol. 71, 1949, pp. 80–91.

A. Sard and L. F. Meyers, "Best approximate integration formulas," *J. Math. Phys.*, Vol. 29, 1950, pp. 118–23.

T. A. Shaidaeva, "The most exact quadrature formulas for certain classes of functions," Dissertation, Leningrad State Univ., 1954 (Russian).

T. A. Shaidaeva, "Quadrature formulas with least estimate of the remainder for certain classes of functions," *Trudy Mat. Inst. Steklov*, Vol. 53, 1959, pp. 313–41 (Russian).

CHAPTER 9

Quadrature Formulas Containing Preassigned Nodes

9.1. GENERAL THEOREMS

In applied problems it is sometimes necessary to construct quadrature formulas in which some of the nodes are given beforehand and the other nodes are free and may be chosen by any criterion we may desire.

Consider, for example, the boundary value problem on the segment $[a, b]$ for the second order differential equation

$$L(y) + \lambda \rho(x) y = \frac{d}{dx}\left[p(x)\frac{dy}{dx}\right] + (\lambda \rho(x) - q(x)) y = -f(x) \quad (9.1.1)$$

with the boundary conditions

$$y(a) = 0, \qquad y(b) = 0. \quad (9.1.2)$$

If we know Green's function for the operator $L(y)$ under the conditions (9.1.2) then the solution of the boundary value problem can be reduced to the solution of the integral equation[1]

$$y(x) = F(x) + \lambda \int_a^b G(x, \xi) \, \rho(\xi) \, y(\xi) \, d\xi$$

$$F(x) = \int_a^b G(x, \xi) \, f(\xi) \, d\xi. \quad (9.1.3)$$

---

[1]See, for example, V. I. Smirnov, *Course of Higher Mathematics*, Vol. 4 Gostekhizdat, Moscow, 1954, pp. 519–21 (Russian).

Suppose we wish to approximate the solution of this equation by apply-ing a quadrature formula to the integrals in (9.1.3). It is natural to use the fact that the value of $y(x)$ is known on the ends of the segment $[a, b]$ and to use a quadrature formula of the form

$$\int_a^b f(x)\,dx \approx Af(a) + Bf(b) + \sum_{k=1}^n A_k f(x_k)$$

which contains the two fixed nodes a and b. The other nodes x_k ($k = 1, 2, \ldots, n$) are determined by some other method.

The above is a "two-point" boundary value problem. In other problems we may wish to use a quadrature formula which contains more than two fixed nodes.

Consider the quadrature formula

$$\int_a^b p(x)f(x)\,dx \approx \sum_{k=1}^n A_k f(x_k) + \sum_{j=1}^m B_j f(a_j) \qquad (9.1.4)$$

in which the m nodes $a_1, \ldots, a_m$ are fixed. It contains the $2n + m$ pa-rameters x_k, A_k ($k = 1, \ldots, n$) and B_j ($j = 1, \ldots, m$). We will show how to choose these parameters so that (9.1.4) is exact for polynomials of as high degree as possible.

Let us introduce the two polynomials

$$\Omega(x) = (x - a_1)\cdots(x - a_m)$$
$$\omega(x) = (x - x_1)\cdots(x - x_n).$$

By counting the choices of the coefficients A_k and B_j we see that formula (9.1.4) can be made exact for polynomials of degree $\leq n + m - 1$. This can be accomplished by requiring that the formula be interpolatory. In order to make the formula exact for polynomials of higher degree we have at our disposal only the choice of the nodes x_k.

Theorem 1. *In order that formula* (9.1.4) *be exact for all polynomials of degree* $\leq 2n + m - 1$ *it is necessary and sufficient that* (1) *it be inter-polatory, and* (2) *the polynomial* $\omega(x)$ *be orthogonal on the segment* $[a, b]$ *with respect to the weight function* $p(x)\Omega(x)$ *to every polynomial* $Q(x)$ *of degree* $< n$.

Proof. The necessity of the first condition is obvious since if formula (9.1.4) is exact for all polynomials of degree $\leq n + m - 1$ then it must be interpolatory. The necessity of the second condition can be verified if we put $f(x) = \Omega(x)\omega(x)Q(x)$. Then $f(x)$ is a polynomial of degree $\leq 2n + m - 1$ and for it (9.1.4) must be exact. Since $f(x)$ is zero at the points a_j and x_k the quadrature sum for this function is also zero and therefore

$$\int_a^b p(x)\,\Omega(x)\,\omega(x)\,Q(x)\,dx = 0. \tag{9.1.5}$$

Now let $f(x)$ be an arbitrary polynomial of degree $\leq 2n + m - 1$. It can be written in the form $f(x) = \Omega(x)\,\omega(x)\,Q(x) + r(x)$ where $Q(x)$ and $r(x)$ are polynomials of degrees $\leq n - 1$ and $\leq n + m - 1$ respectively. Here it is clear that $f(a_j) = r(a_j)$ $(j = 1, \ldots, m)$ and $f(x_k) = r(x_k)$ $(k = 1, \ldots, n)$.

If the orthogonality condition (9.1.5) is satisfied and if formula (9.1.4) is interpolatory then the following relationship will be satisfied

$$\int_a^b p(x)\,f(x)\,dx = \int_a^b p(x)\,\Omega(x)\,\omega(x)\,Q(x)\,dx + \int_a^b p(x)\,r(x)\,dx =$$

$$= \int_a^b p(x)\,r(x)\,dx = \sum_{k=1}^n A_k\,r(x_k) + \sum_{j=1}^m B_j\,r(a_j) =$$

$$= \sum_{k=1}^n A_k\,f(x_k) + \sum_{j=1}^m B_j\,f(a_j)$$

This proves the theorem.

The construction of the quadrature formula (9.1.4) which is exact for all algebraic polynomials of degree $\leq 2n + m - 1$ thus reduces to finding the polynomial of degree n which is orthogonal on $[a, b]$, with respect to the weight function $p(x)\,\Omega(x)$, to all polynomials of degree $< n$. The roots of $\omega(x)$ must be real, distinct and lie inside the segment $[a, b]$. They must also be distinct from the fixed nodes a_j $(j = 1, \ldots, m)$.

Let us assume that the polynomial $\omega(x)$ which satisfies the conditions of Theorem 1 exists. Then we can construct formula (9.1.4) so that it is exact for all polynomials of degree $\leq 2n + m - 1$. We will make one more remark about the degree of precision of this formula. To do this we first need to construct a representation for the remainder. Let us construct the interpolating polynomial $H(x)$ of degree $\leq 2n + m - 1$ for $f(x)$ on $[a, b]$ which satisfies the conditions

$$H(a_j) = f(a_j) \qquad (j = 1, \ldots, m)$$

$$H(x_k) = f(x_k), \quad H'(x_k) = f'(x_k) \qquad (k = 1, \ldots, n).$$

If $f(x)$ has a derivative of order $2n + m$ throughout the segment $[a, b]$ then the remainder of the interpolation $r(x) = f(x) - H(x)$ can be represented as

$$r(x) = \Omega(x)\,\omega^2(x)\,\frac{f^{(2n+m)}(\xi)}{(2n+m)!} \qquad a < \xi < b.$$

The remainder of the quadrature $R(f)$ satisfies $R(f) = R(H) + R(r)$. Since (9.1.4) is exact for all polynomials of degree $2n + m - 1$ then $R(H) = 0$. Also, at all the nodes a_j and x_k the remainder $r(x)$ is zero and thus the quadrature sum for $r(x)$ vanishes

$$\sum_{k=1}^{n} A_k r(x_k) + \sum_{j=1}^{m} B_j r(a_j) = 0.$$

Consequently

$$R(f) = R(r) = \int_a^b p(x)r(x)\,dx = \int_a^b p(x)\,\Omega(x)\,\omega^2(x)\,\frac{f^{(2n+m)}(\xi)}{(2n+m)!}\,dx.$$

Thus we see that if

$$I = \int_a^b p(x)\,\Omega(x)\,\omega^2(x)\,dx \neq 0$$

then the degree of precision of (9.1.4) is $2n + m - 1$. This is true since if $f(x)$ is a polynomial of degree $2n + m$ then $f^{(2n+m)}(x)$ is a constant different from zero and for such a function

$$R(f) = \frac{f^{(2n+m)}}{(2n+m)!} \int_a^b p(x)\,\Omega(x)\,\omega^2(x)\,dx \neq 0.$$

If $I = 0$ then the algebraic degree of precision of (9.1.4) will be greater than $2n + m - 1$. We could derive a criterion to determine the exact degree of precision in these exceptional cases but we do not choose to do so.

Since formula (9.1.4) is interpolatory the coefficients A_k and B_j have the following values:

$$A_k = \int_a^b p(x)\,\frac{\omega(x)\,\Omega(x)}{(x - x_k)\,\omega'(x_k)\,\Omega(x_k)}\,dx \tag{9.1.6}$$

$$B_j = \int_a^b p(x)\,\frac{\omega(x)\,\Omega(x)}{(x - a_j)\,\omega(a_j)\,\Omega'(a_j)}\,dx. \tag{9.1.7}$$

We can give for the coefficients A_k a representation which is easier to use for computations than (9.1.6). Let us assume that there exists a unique system of polynomials $\pi_s(x)$ $(s = 1, 2, \ldots)$ which form an orthonormal system with respect to the weight function $\rho(x) = p(x)\,\Omega(x)$ on $[a, b]$ where $\pi_s(x)$ has degree s. The polynomial $\pi_n(x)$ differs from $\omega(x)$ by only a constant factor so that

$$A_k = \frac{1}{\pi_n'(x_k)\,\Omega(x_k)} \int_a^b \rho(x)\,\frac{\pi_n(x)}{x - x_k}\,dx.$$

The integral in this expression was calculated in Section 7.1 in terms of a different notation. We obtained the following two expressions for this integral:

$$\int_a^b \rho(x)\,\frac{\pi_n(x)}{x - x_k}\,dx = \frac{a_{n+1}}{a_n \pi_{n+1}(x_k)} = \frac{a_n}{a_{n-1}\pi_{n-1}(x_k)},$$

where a_n is the leading coefficient of the polynomial $\pi_n(x)$:

$$\pi_n(x) = a_n x^n + \cdots$$

Therefore

$$A_k = -\frac{a_{n+1}}{a_n \pi_n'(x_k)\,\pi_{n+1}(x_k)\,\Omega(x_k)} = \frac{a_n}{a_{n-1}\pi_n'(x_k)\,\pi_{n-1}(x_k)\,\Omega(x_k)}. \qquad (9.1.8)$$

If we compare (9.1.8) with the expressions (7.1.3) and (7.1.4) for the coefficients of the formula of the highest algebraic degree of precision then it is clear that the A_k in (9.1.4) differ only by the factor $\dfrac{1}{\Omega(x_k)}$ from the corresponding coefficients in the quadrature formula with weight function $\rho(x) = p(x)\,\Omega(x)$

$$\int_a^b \rho(x)f(x)\,dx \approx \sum_{k=1}^n A_k^* f(x_k)$$

which is exact for polynomials of degree $\leq 2n - 1$.

To construct formula (9.1.4) for each n we must construct the system of polynomials which are orthogonal on $[a, b]$ with respect to the weight function $p(x)\,\Omega(x)$. In certain cases we can make use of a result on the representation of such a system of polynomials which are orthogonal with respect to a nonnegative function times a polynomial. We will formulate this result with the degree of generality which is required in the remainder of this chapter.

For simplicity of notation we assume that the orthogonal polynomials have leading coefficient of unity. We denote such a system by $P_s^*(x)$ or $\pi_s^*(x)$ to distinguish it from the corresponding system of orthonormal polynomials.

Together with the weight function $p(x)$ we also consider the weight function $\rho(x) = p(x)\,\Omega(x)$ where $\Omega(x) = (x - a_1)\cdots(x - a_m)$ is any polynomial with distinct roots $a_1, a_2, \ldots, a_m$.

We will assume that there exists a system of polynomials $P_s^*(x) =$

$x^s + \cdots$ $(s = 0, 1, \ldots)$ which are orthogonal on $[a, b]$ with respect to the weight function $p(x)$ and that

$$\int_a^b p(x) [P_s^*(x)]^2 dx \neq 0.$$

This is equivalent to assuming that there exists a unique system of polynomials $\pi_s^*(x) = x^s + \cdots$ $(s = 0, 1, \ldots)$ which are orthogonal on $[a, b]$ with respect to the weight function $\rho(x)$. We will show that $\pi_s^*(x)$ can be expressed in terms of the $P_s^*(x)$ as follows:

$$\pi_n^*(x) = \frac{1}{\Delta\Omega(x)} \begin{vmatrix} P_{n+m}^*(x) & P_{n+m}^*(a_1) & \cdots & P_{n+m}^*(a_m) \\ P_{n+m-1}^*(x) & P_{n+m-1}^*(a_1) & \cdots & P_{n+m-1}^*(a_m) \\ \cdots\cdots\cdots\cdots\cdots\cdots\cdots\cdots\cdots\cdots\cdots \\ P_n^*(x) & P_n^*(a_1) & \cdots & P_n^*(a_m) \end{vmatrix} = \frac{D_{n+m}(x)}{\Delta\Omega(x)}$$

$$\Delta = \begin{vmatrix} P_{n+m-1}^*(a_1) & \cdots & P_{n+m-1}^*(a_m) \\ \cdots\cdots\cdots\cdots\cdots\cdots\cdots \\ P_n^*(a_1) & \cdots & P_n^*(a_m) \end{vmatrix}.$$

(9.1.9)

The product $\Omega(x)\pi_n^*(x)$ is a polynomial of degree $n + m$ with leading coefficient of unity. This polynomial can be expanded in terms of the polynomials $P_s^*(x)$:

$$\Omega(x)\pi_n^*(x) = P_{n+m}^*(x) + c_1 P_{n+m-1}^*(x) + c_2 P_{n+m-2}^*(x) + \cdots$$

The orthogonality of $\pi_n^*(x)$ with respect to the weight function $p(x)\Omega(x)$ to all polynomials of degree less than n means that in this expression the terms involving $P_s^*(x)$ for $s \leq n - 1$ must be absent and therefore the expansion has the form

$$\Omega(x)\pi_n^*(x) = P_{n+m}^*(x) + c_1 P_{n+m-1}^*(x) + \cdots + c_m P_n^*(x). \quad (9.1.10)$$

When x is replaced by one of the numbers $a_1, a_2, \ldots, a_m$ the left side of this equation is zero and therefore the coefficients $c_1, \ldots, c_m$ must satisfy the system of equations

$$P_{n+m}^*(a_1) + c_1 P_{n+m-1}^*(a_1) + \cdots + c_m P_n^*(a_1) = 0.$$
$$\cdots\cdots\cdots\cdots\cdots\cdots\cdots\cdots\cdots\cdots\cdots\cdots\cdots\cdots\cdots\cdots\cdots\cdots \quad (9.1.11)$$
$$P_{n+m}^*(a_m) + c_1 P_{n+m-1}^*(a_m) + \cdots + c_m P_n^*(a_m) = 0.$$

Thus the right side of (9.1.10) is divisible by $\Omega(x)$ and hence we can write $\pi_n^*(x)$ in the form

$$\pi_n^*(x) = \Omega^{-1}(x)[P_{n+m}^*(x) + c_1 P_{n+m-1}^*(x) + \cdots + c_m P_n^*(x)].$$

Since we assumed that there exists a unique sequence of polynomials $\pi_s^*(x)$ the system (9.1.11) must have, for each n, a unique solution for the unknown coefficients $c_1, \ldots, c_m$. The determinant of the system coincides with Δ and therefore $\Delta \neq 0$. The expression (9.1.9) for $\pi_n^*(x)$ can be obtained in the following way. If we adjoin equation (9.1.10) to the system (9.1.11) then we obtain the system:

$$-\Omega(x)\pi_n^*(x) + P_{n+m}^*(x) + c_1 P_{n+m-1}^*(x) + \cdots + c_m P_n^*(x) = 0$$

$$P_{n+m}^*(a_1) + c_1 P_{n+m-1}^*(a_1) + \cdots + c_m P_n^*(a_1) = 0$$

$$\cdots\cdots\cdots\cdots\cdots\cdots\cdots\cdots\cdots\cdots\cdots\cdots\cdots\cdots\cdots$$

$$P_{n+m}^*(a_m) + c_1 P_{n+m-1}^*(a_m) + \cdots + c_m P_n^*(a_m) = 0.$$

This can be considered as a homogeneous system of $n+1$ equations in the $m+1$ quantities $1, c_1, \ldots, c_m$. By a well-known theorem of algebra we can assert that the determinant of this system must be zero:

$$\begin{vmatrix} -\Omega(x)\pi_n^*(x) + P_{n+m}^*(x) & P_{n+m-1}^*(x) & \cdots & P_n^*(x) \\ P_{n+m}^*(a_1) & P_{n+m-1}^*(a_1) & \cdots & P_n^*(a_1) \\ \cdots\cdots\cdots\cdots\cdots\cdots\cdots\cdots & & & \\ P_{n+m}^*(a_m) & P_{n+m-1}^*(a_m) & \cdots & P_n^*(a_m) \end{vmatrix} = 0.$$

This proves (9.1.9).

9.2. FORMULAS OF SPECIAL FORM

In the quadrature formulas considered in Chapter 7 all of the nodes and coefficients were chosen so that the formulas were exact for polynomials of the highest possible degree.

In attempting to generalize this idea A. A. Markov considered formulas in which all of the coefficients A_k but only part of the nodes x_k are chosen so that the formula has the greatest possible precision. The other nodes are fixed in some way. Markov studied this question for weight functions which do not change sign. Let us assume that the weight function $p(x)$ in (9.1.4) is nonnegative: $p(x) \geq 0$. In order that $\rho(x) = p(x)\Omega(x)$ does not change sign on $[a, b]$ we must also assume that $\Omega(x)$ does not change sign in this interval and thus none of the fixed nodes can lie inside $[a, b]$.

If we do not allow quadrature formulas with nodes outside $[a, b]$ we must then limit ourselves to the cases studied by Markov[2]:

(1) $m = 1$ with a single fixed node $a_1 = a$;

---

[2]Trans. note: These formulas are more often attributed to Radau; case (3) is also attributed to Lobatto. See the references at the end of this chapter.

(2) $m = 1$ with a single fixed node $a_1 = b$ (this case reduces to (1) by the linear transformation $x = a + b - t$; we will not consider this case separately);

(3) $m = 2$ with the two fixed nodes $a_1 = a$, $a_2 = b$.

The assumption that $\rho(x)$ has constant sign on $[a, b]$ means that the polynomial $\omega(x)$ of degree n which is orthogonal on $[a, b]$ with respect to $\rho(x)$ to all polynomials of degree $<n$ exists for each n. The roots x_k of this polynomial are real and distinct and all lie inside $[a, b]$. In each of the above cases the x_k are distinct from the fixed nodes which are situated at the ends of $[a, b]$.

Thus, for the cases considered by Markov, the quadrature formulas (9.1.4) which are exact for all polynomials of degree $\leq 2n + m - 1$ can be constructed for all n. Since $p(x)\Omega(x)\omega^2(x)$ does not change sign inside $[a, b]$ then $\int_a^b p(x)\Omega(x)\omega^2(x)\,dx \neq 0$ and the algebraic degree of precision of such formulas is $2n + m - 1$.

Let us consider the first case: $m = 1$, $a_1 = a$

$$\int_a^b p(x)f(x)\,dx = Af(a) + \sum_{k=1}^n A_k f(x_k) + R(f). \qquad (9.2.1)$$

The highest degree of precision which can be achieved in such a formula is $2n$.

Here $\Omega(x) = x - a$. Let x_k be the roots of the nth degree polynomial $\pi_n(x)$ which is orthogonal on $[a, b]$ with respect to $\rho(x) = (x - a)p(x)$ to all polynomials of degree $<n$. If $P_s(x)$ $(s = 0, 1, \ldots)$ is the orthogonal system of polynomials with respect to $p(x)$ then by (9.1.9) $\pi_n(x)$ can be written in the form

$$\pi_n(x) = \frac{K_n}{x-a}\begin{vmatrix} P_{n+1}(x) & P_{n+1}(a) \\ P_n(x) & P_n(a) \end{vmatrix}$$

$$= \frac{K_n}{x-a}[P_{n+1}(x)P_n(a) - P_n(x)P_{n+1}(a)].$$

where K_n is a nonzero constant. Equation (9.1.8) gives a convenient method to compute the coefficients A_k:

$$A_k = -\frac{a_{n+1}}{a_n(x_k - a)\pi_n'(x_k)\pi_{n+1}(x_k)} = \frac{a_n}{a_{n-1}(x_k - a)\pi_n'(x_k)\pi_{n-1}(x_k)}. \qquad (9.2.2)$$

Using (9.1.7) we find for A

$$A = \pi_n^{-1}(a)\int_a^b p(x)\pi_n(x)\,dx. \qquad (9.2.3)$$

We can show that all the coefficients in formula (9.2.1) are positive. As
an integrand let us take the polynomial of degree $2n - 1$:

$$f(x) = (x - a) \left[\frac{\omega(x)}{x - x_i} \right]^2 .$$

This polynomial has the following values at the nodes of the formula:

$$f(a) = 0, \quad f(x_k) = \begin{cases} 0 & \text{for } k \neq i \\ (x_i - a)[\omega'(x_i)]^2 & \text{for } k = i. \end{cases}$$

Formula (9.2.1) must be exact for this function and thus

$$\int_a^b p(x)(x - a) \left[\frac{\omega(x)}{x - x_i} \right]^2 dx = A_i(x_i - a)[\omega'(x_i)]^2 .$$

Therefore

$$A_i = \frac{1}{(x_i - a)[\omega'(x_i)]^2} \int_a^b p(x)(x - a) \left[\frac{\omega(x)}{x - x_i} \right]^2 dx > 0 .$$

Similarly, if we take $f(x) = \omega^2(x)$ we obtain

$$A = \omega^{-2}(a) \int_a^b p(x)\omega^2(x) dx > 0 .$$

If $f(x)$ has a continuous derivative of order $2n + 1$ then the remainder
$R(f)$ in (9.2.1) can be represented in the form

$$R(f) = \int_a^b p(x)(x - a)\omega^2(x) \frac{f^{(2n+1)}(\xi)}{(2n+1)!} dx ,$$

or, since $p(x)(x - a)\omega^2(x)$ does not change sign on $[a, b]$,

$$R(f) = \frac{f^{(2n+1)}(\eta)}{(2n+1)!} \int_a^b p(x)(x - a)\omega^2(x) dx , \qquad a < \eta < b . \qquad (9.2.4)$$

We will now discuss in more detail the above theory for the weight
function $p(x) \equiv 1$.

We assume that the segment $[a, b]$ has been transformed into the segment $[-1, 1]$ and we consider the formula

$$\int_{-1}^1 f(x) dx = Af(-1) + \sum_{k=1}^n A_k f(x_k) + R(f) \qquad (9.2.5)$$

which has degree of precision equal to $2n$. We have $\Omega(x) = 1 + x$ and the
polynomial $\omega(x) = (x - x_1) \cdots (x - x_n)$ must be orthogonal on $[-1, 1]$ with

respect to $(1 + x)$ to all polynomials of lower degree. Therefore $\omega(x)$ can differ from the Jacobi polynomial $P_n^{(0,1)}(x)$ by only a constant factor:

$$\omega(x) = \frac{2^n n! \, \Gamma(n + 2)}{\Gamma(2n + 2)} P_n^{(0,1)}(x).$$

Thus the nodes x_k must be the roots of $P_n^{(0,1)}(x)$. The coefficients A_k can easily be found if we use the remark following (9.1.8). Quadrature formula (7.3.2) for the Jacobi weight function $(1 - x)^\alpha (1 + x)^\beta$ is exact for all polynomials of degree $\leq 2n - 1$. The coefficients of this formula are given by (7.3.4). To find A_k in (9.2.5) we must multiply the corresponding coefficient (7.3.4), for $a = 0$, $\beta = 1$, by $\dfrac{1}{\Omega(x_k)} = \dfrac{1}{1 + x_k}$.
This gives

$$A_k = \frac{4}{(1 + x_k)(1 - x_k^2)[P_n^{(0,1)\,\prime}(x_k)]^2}. \tag{9.2.6}$$

We can use (9.1.7) to calculate A by substituting $p(x) \equiv 1$, $\Omega(x) = 1 + x$ and $a_j = -1$:

$$A = \int_{-1}^{1} \frac{\omega(x)}{\omega(-1)} \, dx = [P_n^{(0,1)}(-1)]^{-1} \int_{-1}^{1} P_n^{(0,1)}(x) \, dx.$$

This last integral and the factor in front of it are easily found from known properties of Jacobi polynomials; these are $\dfrac{2(-1)^n}{n+1}$ and $(-1)^n (n + 1)$ respectively. Thus

$$A = \frac{2}{(n + 1)^2}. \tag{9.2.7}$$

The remainder $R(f)$ can be computed from (9.2.4):

$$R(f) = \frac{f^{(2n+1)}(\eta)}{(2n + 1)!} \int_{-1}^{1} (1 + x)\omega^2(x) \, dx, \qquad -1 < \eta < 1.$$

The integral in this expression can be found without difficulty

$$\int_{-1}^{1} (1 + x)\omega^2(x) \, dx = \left[\frac{2^n n! \, (n + 1)!}{(2n + 1)!}\right]^2 \int_{-1}^{1} (1 + x)[P_n^{(0,1)}(x)]^2 \, dx =$$

$$= \frac{2}{n + 1} \left[\frac{2^n n! \, (n + 1)!}{(2n + 1)!}\right]^2.$$

Therefore

$$R(f) = \frac{2}{n + 1} \left[\frac{2^n n! \, (n + 1)!}{(2n + 1)!}\right]^2 \frac{f^{(2n+1)}(\eta)}{(2n + 1)!}, \qquad -1 < \eta < 1. \tag{9.2.8}$$

The nodes and coefficients in formula (9.2.5) are given below for $n = 1(1)6$.[3]

x_k	A_k
n = 1	
−1.00000000	0.50000000
0.33333333	1.50000000
n = 2	
−1.00000000	0.22222222
−0.28989794	1.02497166
0.68989794	0.75280612
n = 2	
−1.0000000	0.1250000
−0.5753189	0.6576886
0.1810663	0.7763870
0.8228241	0.4409244
n = 4	
−1.0000000	0.0800000
−0.7204803	0.4462078
0.1671809	0.6236530
0.4463140	0.5627120
0.8857916	0.2874271
n = 5	
−1.0000000	0.0555556
−0.8029298	0.3196408
−0.3909286	0.4853872
0.1240504	0.5209268
0.6039732	0.4169013
0.9203803	0.2015884
n = 6	
−1.0000000	0.0408163
−0.8538913	0.2392274
−0.5384678	0.3809498
−0.1173430	0.4471098
0.3260306	0.4247038
0.7038428	0.3182042
0.9413672	0.1489885

Now we consider case 3 where we are given two fixed nodes at the ends of the segment of integration: $a_1 = a$, $a_2 = b$.

$$\int_a^b p(x) f(x)\, dx = A f(a) + B f(b) + \sum_{k=1}^{n} A_k f(x_k) + R(f). \qquad (9.2.9)$$

The highest degree of precision which can be achieved by such a formula is $2n + 1$. Here $\Omega(x) = (x - a)(x - b)$ and the x_k are the roots of the nth

---

[3] This table was calculated at the Leningrad section of the Mathematical Institute of the Academy of Sciences of the U.S.S.R. by research assistants R. B. Akkerman and K. E. Chernin.

degree polynomial $\pi_n(x)$ which is orthogonal on $[a, b]$ with respect to $\rho(x) = (x - a)(x - b)p(x)$ to all polynomials of degree $<n$.

The polynomials $\pi_n(x)$ are related to the polynomials $P_n(x)$ which are orthogonal with respect to $p(x)$ by the following equation:

$$\pi_n(x) = \frac{K_n}{(x-a)(x-b)} \begin{vmatrix} P_{n+2}(x) & P_{n+2}(a) & P_{n+2}(b) \\ P_{n+1}(x) & P_{n+1}(a) & P_{n+1}(b) \\ P_n(x) & P_n(a) & P_n(b) \end{vmatrix}$$

The coefficients A, B, and A_k can be computed from (9.1.8) and (9.1.7):

$$A_k = -\frac{a_{n+1}}{a_n \pi_n'(x_k)\, \pi_{n+1}(x_k)(x_k - a)(x_k - b)} =$$

$$= \frac{n}{a_{n-1}\pi_n'(x_k)\, \pi_{n-1}(x_k)(x_k - a)(x_k - b)} \tag{9.2.10}$$

$$A = [\omega(a)(a-b)]^{-1} \int_a^b p(x)(x-b)\,\omega(x)\,dx$$

$$B = [\omega(b)(b-a)]^{-1} \int_a^b p(x)(x-a)\,\omega(x)\,dx \tag{9.2.11}$$

$$\omega(x) = (x - x_1)\cdots(x - x_n)$$

If we consider the quadrature formula with respect to the weight function $\rho(x) = p(x)(x - a)(x - b)$:

$$\int_a^b p(x)(x-a)(x-b)f(x)\,dx \approx \sum_{k=1}^n A_k^* f(x_k)$$

which is exact for all polynomials of degree $\leq 2n - 1$, then the coefficients A_k in (9.2.8) differ from the coefficients A_k^* by the factor $\dfrac{1}{(x_k - a)(x_k - b)}$. It is easy to show that the A, B, and A_k are positive. To do this it suffices to apply formula (9.2.9) to the polynomials

$$(b - x)\,\omega^2(x), \quad (x - a)\,\omega^2(x), \quad \text{and} \quad (x - a)(x - b)\left[\frac{\omega(x)}{x - x_i}\right]^2$$

The remainder $R(f)$ of (9.2.9) can be represented in the form

$$R(f) = \frac{f^{(2n+2)}(\eta)}{(2n+2)!} \int_a^b p(x)(x-a)(x-b)\,\omega^2(x)\,dx.$$

Let us apply these results to the particular case of $p(x) \equiv 1$ for the segment $[-1, 1]$:

$$\int_{-1}^{1} f(x)\,dx = Af(-1) + Bf(1) + \sum_{k=1}^{n} A_k f(x_k) + R(f) \qquad (9.2.12)$$

$$\Omega(x) = 1 - x^2.$$

The polynomial $\omega(x)$, which is orthogonal on $[-1, 1]$ with respect to $1 - x^2$ to all polynomials of degree $<n$, differs by only a constant factor from the Jacobi polynomial $P_n^{(1,1)}(x)$:

$$\omega(x) = \frac{2^n n!\,\Gamma(n+3)}{\Gamma(2n+3)} P_n^{(1,1)}(x).$$

As in case 1 we can compute the coefficients and remainder:

$$A_k = \frac{8(n+1)}{(n+2)(1-x_k^2)^2 [P_n^{(1,1)\,\prime}(x_k)]^2}$$

$$A = B = \frac{2}{(n+1)(n+2)}$$

$$R(f) = \frac{8(n+1)}{(2n+3)(n+2)} \left[\frac{2^n n!\,(n+2)!}{(2n+2)!}\right]^2 \frac{f^{(2n+2)}(\eta)}{(2n+2)!}, \quad -1 < \eta < 1.$$

The nodes and coefficients in (9.1.12) are symmetric with respect to $x = 0$ and we tabulate below these values which correspond to $0 \le x_k \le 1$ for $n = 1(1)\,15$.[4]

x_k	A_k
$n = 1$	
1.00000000	0.33333333
0.00000000	1.33333333
$n = 2$	
1.00000000	0.16666667
0.44721360	0.83333333
$n = 3$	
1.00000000	0.10000000
0.65465367	0.54444444
0.00000000	0.71111111
$n = 4$	
1.00000000	0.066666667
0.76505532	0.37847496
0.28523152	0.55485837

---

[4] This table was calculated at the Leningrad section of the Mathematical Institute of the Academy of Sciences of the U.S.S.R. by research assistant R. B. Akkerman.

x_k	A_k
	$n = 5$
1.00000000	0.047619048
0.83022390	0.27682605
0.46884879	0.43174538
0.00000000	0.48761905
	$n = 6$
1.00000000	0.035714286
0.87174015	0.21070423
0.59170018	0.34112268
0.20929922	0.41245881
	$n = 7$
1.00000000	0.027777778
0.89975800	0.16549536
0.67718628	0.27453872
0.36311746	0.34642851
0.00000000	0.37151927
	$n = 8$
1.00000000	0.022222222
0.91953391	0.13330599
0.73877386	0.22488934
0.47792495	0.29204268
0.16527896	0.32753976
	$n = 9$
1.00000000	0.018181818
0.93400143	0.10961227
0.78448347	0.18716989
0.56523533	0.24804811
0.29575814	0.28687913
0.00000000	0.30021759
	$n = 10$
1.00000000	0.015151515
0.94489927	0.091684521
0.81927932	0.15797471
0.63287615	0.21250842
0.39953094	0.25127560
0.13655293	0.27140524
	$n = 11$
1.00000000	0.012820513
0.95330985	0.077801687
0.84634757	0.13498193
0.68618847	0.18364686
0.48290982	0.22076779
0.24928693	0.24401579
0.00000000	0.25193085
	$n = 12$
1.00000000	0.010989011
0.95993505	0.066837283
0.86780105	0.11658665
0.72886860	0.16002185
0.55063940	0.19482615
0.34272401	0.21912625
0.11633187	0.23161279

x_k	A_k
n = 13	
1.00000000	0.0095238095
0.96524592	0.058029922
0.88508205	0.10166004
0.76351967	0.14051171
0.60625322	0.17278965
0.42063805	0.19698723
0.21535396	0.21197360
0.00000000	0.21704810
n = 14	
1.00000000	0.0083333333
0.96956804	0.050850369
0.89920054	0.089393689
0.79200828	0.12425539
0.65238872	0.15402699
0.48605941	0.17749190
0.29983047	0.19369005
0.10132627	0.20195830
n = 15	
1.00000000	0.0073529412
0.97313217	0.044921950
0.91088001	0.079198263
0.81569624	0.11059290
0.69102899	0.13798776
0.54138540	0.16039465
0.37217443	0.17700426
0.18951198	0.18721635
0.00000000	0.19066186

9.3. REMARKS ON INTEGRALS WITH WEIGHT FUNCTIONS THAT CHANGE SIGN

The problem of constructing quadrature formulas with preassigned nodes is related to the problem of transforming weight functions which change sign into weight functions with constant sign.

Let us consider the integral

$$\int_a^b p(x)f(x)\,dx \tag{9.3.1}$$

and assume that $p(x)$ changes sign inside the segment $[a, b]$ at a finite number of points[5] $a_1, a_2, \ldots, a_m$.

We construct for $f(x)$ the interpolating polynomial $P(x)$ of degree $<m$ based on the points a_j:

---

[5] On each of the segments $[a, a_1], [a_1 a_2], \ldots, [a_m, b]$ the function $p(x)$ has constant sign and on adjacent segments it has opposite sign.

$$P(a_j) = f(a_j) \qquad (j = 1, 2, \ldots, m)$$

$$f(x) = P(x) + r(x) \tag{9.3.2}$$

$$P(x) = \sum_{j=1}^{m} \frac{\Omega(x)}{(x - a_j)\,\Omega'(a_j)}\, f(a_j).$$

The remainder $r(x)$ of the interpolation can be represented in the form (see (3.2.9)):

$$r(x) = (x - a_1) \cdots (x - a_m)\, f(a_1, \ldots, a_m, x) = \Omega(x)\, f(a_1, \ldots, a_m, x)$$

where $f(a_1, \ldots, a_m, x)$ is the divided difference corresponding to the nodes $a_1, \ldots, a_m, x$.

The integral (9.3.1) can be divided into two parts in the following way:

$$\int_a^b p(x) f(x)\, dx = \int_a^b p(x) P(x)\, dx + \int_a^b p(x) \Omega(x)\, f(a_1, \ldots, a_m, x)\, dx =$$

$$= \sum_{j=1}^{m} a_j f(a_j) + \int_a^b \rho(x) f(a_1, \ldots, a_m, x)\, dx \tag{9.3.3}$$

$$a_j = \int_a^b p(x) \frac{\Omega(x)}{(x - a_j)\,\Omega'(a_j)}\, dx. \tag{9.3.4}$$

We will now be interested in the last integral in (9.3.3). The function $\rho(x) = p(x)\,\Omega(x)$ in this integral does not change sign on $[a, b]$ because each of the factors changes sign at these points. We take $\rho(x)$ as a new weight function. To calculate the integral

$$\int_a^b \rho(x) f(a_1, \ldots, a_m, x)\, dx$$

we can use any of the methods which we employed for weight functions of constant sign. In particular we can construct for this integral a quadrature formula of the highest algebraic degree of precision. As in the preceding section let us denote by $\pi_n(x)$ the nth degree polynomial of the orthogonal system belonging to the weight function $\rho(x)' = p(x)\,\Omega(x)$. Let us consider the quadrature formula with n nodes which is exact for polynomials of degree $\leq 2n - 1$:

$$\int_a^b \rho(x) f(a_1, \ldots, a_m, x)\, dx \approx \sum_{k=1}^{n} \beta_k f(a_1, \ldots, a_m, x_k) \tag{9.3.5}$$

$$\pi_n(x_k) = 0 \qquad (k = 1, 2, \ldots, n)$$

$$\beta_k = \int_a^b \rho(x) \frac{\pi_n(x)}{(x - x_k)\pi_n'(x_k)} \, dx =$$

$$= \int_a^b p(x) \frac{\Omega(x)\omega(x)}{(x - x_k)\omega'(x_k)} \, dx$$

$$\omega(x) = (x - x_1) \cdots (x - x_n).$$

We then obtain the following formula for the integral (9.3.1):

$$\int_a^b p(x)f(x)\,dx \approx \sum_{j=1}^m a_j f(a_j) + \sum_{k=1}^n \beta_k f(a_1, \ldots, a_m, x_k). \qquad (9.3.6)$$

It is easy to see that the algebraic degree of precision of this formula is $2n + m - 1$.

To prove this let $f(x)$ be any polynomial of degree $\leq 2n + m - 1$. In this case (9.3.3) is an identity whenever the terms on the right side of this equation are defined. The divided difference $f(a_1, \ldots, a_m, x)$ is a polynomial of degree m less than the degree of $f(x)$ and it does not exceed $2n - 1$. Because (9.3.5) has degree of precision $2n - 1$ then it will be exact for $f(a_1, \ldots, a_m, x)$ and therefore (9.3.6) will also be exact.

On the other hand if $f(x)$ is taken to be the polynomial

$$f(x) = \Omega(x)\omega^2(x)$$

of degree $2n + m$ then (9.3.6) can not be exact. Indeed, for this function the interpolating polynomial $P(x)$ is identically zero and from (9.3.2) we see that $f(a_1, \ldots, a_m, x) = \omega^2(x)$. All the terms on the right side of (9.3.6) vanish and the integral on the left side is nonzero since $p(x)\Omega(x)$ does not change sign on $[a, b]$:

$$\int_a^b p(x)f(x)\,dx = \int_a^b p(x)\Omega(x)\omega^2(x)\,dx \neq 0.$$

We will now investigate the relationship between $f(a_1, \ldots, a_m, x_k)$ and $f(x)$. If the roots x_k $(k = 1, \ldots, n)$ of the polynomial $\pi_n(x)$, which are the nodes in (9.3.6), are different from the a_j $(j = 1, \ldots, m)$ then the divided difference $f(a_1, \ldots, a_m, x_k)$ is

$$f(a_1, \ldots, a_m, x_k) = \frac{f(x_k) - P(x_k)}{\Omega(x_k)}. \qquad (9.3.7)$$

In this case $f(a_1, \ldots, a_m, x)$ depends only on the following values of $f(x)$: $f(x_k)$, $f(a_j)$ $(j = 1, \ldots, m)$.

If in (9.3.6) we substitute for $f(a_1, \ldots, a_m, x_k)$ the values (9.3.7) then, by collecting the terms in $f(x_k)$ and $f(a_j)$, we see that (9.3.6) can be written in the form (9.1.4). In this case formula (9.3.6) is a particular case of (9.1.4) for which the a_j are the points at which the weight function $p(x)$ changes sign.

If the node x_k coincides with one of the nodes a_j then $\Omega(x_k) = 0$ and (9.3.7) is meaningless. In this case (9.3.7) must be replaced by

$$f(a_1,\ldots,a_m, x_k) = \frac{f'(x_k) - P'(x_k)}{\Omega'(x_k)}$$

which can be obtained by applying l'Hospital's rule. The divided difference $f(a_1,\ldots,a_m, x_k)$ then depends on $f'(x_k)$ as well as on the $f(a_j)$ $(j = 1,\ldots,m)$. In this case the quadrature formula (9.3.6) will contain, in addition to values of the integrand $f(x)$, the value $f'(x_k)$ where x_k is a point at which $p(x)$ changes sign.

As an example, let us take $m = 1$ and assume that $p(x)$ changes sign inside $[a, b]$ at only the point a_1. The interpolating polynomial $P(x)$ will then be the constant function $P(x) = f(a_1)$ and (9.3.3) becomes

$$\int_a^b p(x) f(x)\,dx \approx f(a_1) \int_a^b p(x)\,dx + \int_a^b p(x)(x - a_1) f(a_1, x)\,dx \qquad (9.3.8)$$

$$f(a_1, x) = \frac{f(x) - f(a_1)}{x - a_1}.$$

If all the x_k are different from a_1 then formula (9.3.6) becomes

$$\int_a^b p(x) f(x)\,dx \approx f(a_1) \int_a^b p(x)\,dx + \sum_{k=1}^n \beta_k \frac{f(x_k) - f(a_1)}{x_k - a_1}.$$

If one of the x_k, for example x_1, coincides with a_1 then (9.3.6) will have the form

$$\int_a^b p(x) f(x)\,dx \approx f(a_1) \int_a^b p(x)\,dx + \beta_1 f'(a_1) + \sum_{k=2}^n \beta_k \frac{f(x_k) - f(a_1)}{x_k - a_1}.$$

Let us obtain a formula of this form to calculate the integral

$$\int_{-1}^1 x e^x dx = 2e^{-1} \approx 0.73576.$$

Here we take $p(x) = x$ and hence $p(x)$ changes sign at $x = 0$. For this integral equation (9.3.8) is

$$\int_{-1}^1 x e^x\,dx = e^0 \int_{-1}^1 x\,dx + \int_{-1}^1 x^2 f(0, x)\,dx = \int_{-1}^1 x^2 f(0, x)\,dx,$$

$$f(0, x) = \frac{e^x - 1}{x}.$$

To calculate this last integral we will use the quadrature formula of the highest algebraic degree of precision with two nodes. The second degree polynomial $\pi_2(x)$ which is orthogonal on $[-1, 1]$ with respect to $\rho(x) = x^2$ is $\pi_2(x) = k(5x^2 - 3)$ which has roots

$$x = \pm \frac{\sqrt{15}}{5} \approx \pm 0.7745967.$$

The formula will then be

$$\int_{-1}^{1} x^2 f(0, x)\, dx \approx \beta_1 f(0, x_1) + \beta_2 f(0, x_2).$$

Since the weight function $\rho(x) = x^2$ is symmetric with respect to $x = 0$ it follows that β_1 and β_2 must be equal and thus

$$\beta_1 + \beta_2 = \int_{-1}^{1} x^2 dx = \frac{2}{3}$$

$$\beta_1 = \beta_2 = \frac{1}{3}$$

$$\int_{-1}^{1} x^2 f(0, x)\, dx \approx \frac{1}{3}\left[\frac{e^{x_1} - 1}{x_1} + \frac{e^{x_2} - 1}{x_2}\right] \approx 0.73536.$$

This result is exact to within 0.06% of the true value.

REFERENCES

J. Bouzitat, "Sur l'intégration numérique approchée par la méthode de Gauss généralisée et sur une extension de cette méthode," *C. R. Acad. Sci. Paris*, Vol. 229, 1949, pp. 1201–1203.

A. Markoff, "Sur la méthode de Gauss pour le calcul approché des intégrales," *Math. Ann.*, Vol. 25, 1885, pp. 427–432.

H. Mineur, *Techniques de Calcul Numerique*, Beranger, Paris, 1952 (Note by J. Bouzitat, pp. 557–605).

R. Radau, "Étude sur les formules d'approximation qui servent a calculer la valeur numérique d'une intégrale définie," *J. Math. Pures Appl.*, (3) Vol. 6, 1880, pp. 283–336.

P. Rabinowitz, "Abscisses and weights for Lobatto quadrature of high order," *Math. Comp.*, Vol. 14, 1960, pp. 47–52.

G. W. Struble, "Orthogonal polynomials: variable-signed weight functions," Thesis, University of Wisconsin, 1961.

CHAPTER 10

Quadrature Formulas
with Equal Coefficients

10.1. DETERMINING THE NODES

Quadrature formulas with equal coefficients

$$\int_a^b p(x) f(x) \, dx \approx c_n \sum_{k=1}^n f(x_k) \qquad (10.1.1)$$

are very convenient for computations and in particular for graphical calculations. These formulas have been the subject of many investigations and in this chapter we will develop their theory.

Formula (10.1.1) contains the $n + 1$ parameters $c_n, x_1, \ldots, x_n$ and we can choose these parameters so that the formula will be exact for all possible polynomials of degree $\leq n$.

The requirement that (10.1.1) be exact for $f(x) \equiv 1$ means that we must have

$$\int_a^b p(x) \, dx = n c_n$$

which determines the coefficient c_n:

$$c_n = \frac{1}{n} \int_a^b p(x) \, dx. \qquad (10.1.2)$$

If we also require that (10.1.1) be exact for the monomials $f(x) = x$, $x^2, \ldots, x^n$ then we obtain the following system of equations for the nodes x_k:

$$x_1 + x_2 + \cdots + x_n = c_n^{-1} \int_a^b p(x) x \, dx$$

$$x_1^2 + x_2^2 + \cdots + x_n^2 = c_n^{-1} \int_a^b p(x) x^2 \, dx$$

$$\cdots\cdots\cdots\cdots\cdots\cdots\cdots\cdots\cdots \quad (10.1.3)$$

$$x_1^n + x_2^n + \cdots + x_n^n = c_n^{-1} \int_a^b p(x) x^n \, dx$$

Let $\omega(x)$ be the polynomial of degree n which has the nodes $x_1, \ldots, x_n$ for its roots

$$\omega(x) = (x - x_1)(x - x_2) \cdots (x - x_n). \quad (10.1.4)$$

Using equations (10.1.3) we can easily construct this polynomial. If we write $\omega(x)$ in the form

$$\omega(x) = x^n + A_1 x^{n-1} + A_2 x^{n-2} + \cdots + A_n \quad (10.1.5)$$

then the coefficients $A_1, \ldots, A_n$ are the well-known elementary symmetric functions of the roots. On the other hand the left sides of equations (10.1.3) are the sums of powers of the roots:

$$s_k = x_1^k + x_2^k + \cdots + x_n^k \quad (k = 1, 2, \ldots, n).$$

The right sides of (10.1.3) are the values of these functions for the polynomial (10.1.4).

In the theory of equations the relationship between the elementary symmetric functions A_i $(i = 1, \ldots, n)$ and the functions s_k $(k = 1, \ldots, n)$ is well known. This is given by the following equations which are often called *Newton's equations*[1]

---

[1] The logarithmeric derivative of (10.1.4) leads to the following equation

$$\frac{\omega'(x)}{\omega(x)} = \sum_{i=1}^n \frac{1}{x - x_i}.$$

If $|x| > |x_i|$ then the fraction $\frac{1}{x - x_i}$ can be expanded in a power series in negative powers of x:

$$(x - x_i)^{-1} = \sum_{\nu=0}^\infty \frac{x_i^\nu}{x^{\nu+1}}.$$

Therefore if $|x| > |x_i|$ $(i = 1, \ldots, n)$ then the following expansion is valid:

$$s_1 + A_1 = 0$$

$$s_2 + A_1 s_1 + 2A_2 = 0$$

$$\dots\dots\dots\dots\dots\dots\dots\dots\dots\dots\dots\dots\dots\dots \quad (10.1.6)$$

$$s_n + A_1 s_{n-1} + A_2 s_{n-2} + \cdots + nA_n = 0.$$

From these equations we can sequentially calculate the coefficients A_i $(i = 1, \dots, n)$ from the values of s_k given by (10.1.3). From the A_i we can construct the polynomial $\omega(x)$ and calculating the roots of this polynomial gives the quadrature formula (10.1.1). If (10.1.1) is to be useful the x_k should all be real, distinct and should belong to the segment of integration. The possibility of constructing formula (10.1.1) which is exact for all polynomials of degree $\leq n$ is, therefore, determined by whether or not the roots of $\omega(x)$ satisfy the above requirements where the coefficients A_k in $\omega(x)$ are found from Newton's formulas.

We can construct another expression for the polynomial (10.1.4) by making use of a few results from the theory of analytic functions. Let us apply formula (10.1.1) to the fraction $f(x) = \dfrac{1}{z - x}$ which is the kernel of the Cauchy integral and consider the remainder:

$$R\left(\frac{1}{z-x}\right) = \int_a^b \frac{p(x)}{z-x}\,dx - c_n \sum_{k=1}^{n} \frac{1}{z-x_k} = \int_a^b \frac{p(x)}{z-x}\,dx - c_n \frac{\omega'(z)}{\omega(z)}.$$

We will find the expansion of the remainder in powers of z^{-1} for $|z|$ large. Let ρ be a number so large that the segment of integration $[a, b]$ and all of the nodes x_k lie in the circle $|z| \leq \rho$. Then for $|z| > \rho$

$$\frac{\omega'(x)}{\omega(x)} = \sum_{i=1}^{n} \sum_{\nu=0}^{\infty} \frac{x_i^{\nu}}{x^{\nu+1}} = \sum_{\nu=0}^{\infty} \frac{s_{\nu}}{x^{\nu+1}}$$

$$s_{\nu} = \sum_{i=1}^{n} x_i^{\nu}.$$

Multiplying both sides of this equation by $\omega(x)$ and replacing $\omega(x)$ by its representation (10.1.5) gives

$$nx^{n-1} + (n-1)A_1 x^{n-2} + (n-2)A_2 x^{n-3} + \cdots + A_{n-1} =$$

$$= (x^n + A_1 x^{n-1} + \cdots)\sum_{\nu=0}^{\infty} \frac{s_{\nu}}{x^{\nu+1}}.$$

Equating the coefficients of x^{n-2}, $x^{n-3}, \dots$, we then obtain Newton's equations.

$$\frac{1}{z - x} = \sum_{\nu=0}^{\infty} \frac{x^{\nu}}{z^{\nu+1}}$$

and

$$\int_a^b \frac{p(x)}{z - x} \, dx = \sum_{\nu=0}^{\infty} \frac{1}{z^{\nu+1}} \int_a^b p(x) x^{\nu} \, dx = \sum_{\nu=0}^{\infty} \frac{\mu_{\nu}}{z^{\nu+1}}.$$

Here μ_{ν} denotes the moment of order ν of the weight function $p(x)$. Similarly

$$\frac{1}{z - x_k} = \sum_{\nu=0}^{\infty} \frac{x_k^{\nu}}{z^{\nu+1}}$$

and

$$\frac{\omega'(z)}{\omega(z)} = \sum_{k=1}^{n} \frac{1}{z - x_k} = \sum_{\nu=0}^{\infty} \frac{s_{\nu}}{z^{\nu+1}},$$

$$R\left(\frac{1}{z - x}\right) = \sum_{\nu=0}^{\infty} \frac{\mu_{\nu} - c_n s_{\nu}}{z^{\nu+1}}.$$

$$(10.1.7)$$

Assuming that (10.1.1) is exact for the powers $x, x^2, \ldots, x^n$ then by (10.1.2) and (10.1.3) we have

$$\mu_{\nu} - c_n s_{\nu} = 0, \qquad \nu = 0, 1, \ldots, n$$

and the smallest exponent of $1/z$ in the last expansion will be $n + 2$:

$$\int_a^b \frac{p(x)}{z - x} \, dx - c_n \frac{\omega'(z)}{\omega(z)} = \sum_{\nu=n+1}^{\infty} \frac{\mu_{\nu} - c_n s_{\nu}}{z^{\nu+1}}.$$

Integrating with respect to z and applying a simple transformation we obtain:

$$\omega(z) \exp\left(\sum_{\nu=n+1}^{\infty} \frac{s_{\nu} - c_n^{-1} \mu_{\nu}}{\nu z^{\nu}}\right) = A \, \exp\left(c_n^{-1} \int_a^b p(x) \ln(z - x) \, dx\right) \quad (10.1.8)$$

where A is a certain constant.

Since the expansion of $\exp\left(\displaystyle\sum_{\nu=n+1}^{\infty} \frac{s_{\nu} - c_n^{-1} \mu_{\nu}}{\nu z^{\nu}}\right)$ in powers of $1/z$ differs from unity by only powers of $1/z$ greater than n it is clear that the integer part of the expansion of the right side of (10.1.8) in powers of $1/z$ must coincide with $\omega(z)$ for large $|z|$:

$$\omega(z) = \text{integer part of } A \, \exp\left(c_n^{-1} \int_a^b p(x) \, ln \, (z - x) \, dx\right). \qquad (10.1.9)$$

The constant A could be found by using the fact that the leading term of $\omega(z)$ is z^n. We will not need to calculate this factor since it does not affect the roots of the right side of (10.1.9).

As mentioned above the formula (10.1.1) which is exact for all polynomials of degree $\leq n$ is of interest only when the roots of $\omega(x)$ are real, distinct and lie inside the segment $[a, b]^2$. The polynomial $\omega(x)$ is completely defined by the weight function $p(x)$ and we would like to know for what weight functions this polynomial has the properties we desire. The solution to this problem is not known in general. Below we discuss two weight functions for which the answer is known.

10.2. UNIQUENESS OF THE QUADRATURE FORMULAS OF THE HIGHEST ALGEBRAIC DEGREE OF PRECISION WITH EQUAL COEFFICIENTS

In Chapter 7 we discussed the quadrature formulas of the highest algebraic degree of precision for the weight function $p(x) = (1 - x^2)^{-\frac{1}{2}}$ on $[-1, 1]$. We obtained formula (7.3.2)

$$\int_{-1}^1 (1 - x^2)^{-\frac{1}{2}} f(x) \, dx \approx \frac{\pi}{n} \sum_{k=1}^n f\left(\cos \frac{2k - 1}{2n} \pi\right)$$

which is exact for all polynomials of degree $\leq 2n - 1$. In this formula the number of nodes n is arbitrary. It is remarkable that the coefficients in any one of these formulas are all equal.

We may ask whether these formulas are unique: does there exist on the segment $[-1, 1]$ another weight function $p(x)$ which is different from $(1 - x^2)^{-\frac{1}{2}}$ for which quadrature formulas of the highest algebraic degree of precision exist and which also have equal coefficients?

A negative answer to this question was first given by K. A. Posse and also later by N. Ia. Sonin.

Here we prove a more general theorem due to Ia. L. Geronimus from which the theorem of Posse easily follows.

Let us be given a weight function $p(x)$ which is almost everywhere positive on the segment $[-1, 1]$. Let us take the system of orthogonal polynomials $\omega_n(x) = x^n + \beta_n x^{n-1} + \gamma_n x^{n-2} + \cdots$ $(n = 0, 1, 2, \dots)$ which correspond to this weight function. Let $x_k^{(n)}$ $(k = 1, \dots, n)$ denote the

---

[2]We assume that the integrand is only defined on $[a, b]$ and therefore do not consider formulas with nodes outside $[a, b]$.

roots of $\omega_n(x)$ and consider the quadrature formula with equal coefficients for which the nodes coincide with $x_k^{(n)}$:

$$\int_{-1}^{1} p(x) f(x)\, dx \approx c_n \sum_{k=1}^{n} f(x_k^{(n)}). \tag{10.2.1}$$

Theorem 1. *If for arbitrary values of* $n = 1, 2, \ldots$, *there exists constants* c_n *such that formula* (10.2.1) *is exact for*[3] $f(x) = 1$, $f(x) = x$, $f(x) = x^2$, *then* $p(x)$ *coincides with the Chebyshev weight function* $(1 - x^2)^{-\frac{1}{2}}$.

Proof. Without loss of generality we can assume

$$\mu_0 = \int_{-1}^{1} p(x)\, dx = 1.$$

The requirement that the quadrature formula be exact for $f(x) = 1$ then determines the constant c_n:

$$\int_{-1}^{1} p(x)\, dx = n c_n, \qquad c_n = \frac{1}{n}.$$

Assuming in turn that $f(x) = x$ and $f(x) = x^2$ we obtain the following equations

$$\mu_1 = \int_{-1}^{1} p(x) x\, dx = \frac{1}{n} \sum_{k=1}^{n} x_k^{(n)} = -\frac{1}{n} \beta_n, \qquad n = 1, 2, \ldots$$

$$\mu_2 = \int_{-1}^{1} p(x) x^2\, dx = \frac{1}{n} \sum_{k=1}^{n} [x_k^{(n)}]^2 = \frac{1}{n} \left\{ \left[\sum_{k=1}^{n} x_k^{(n)} \right]^2 - 2 \sum_{j<k} x_j^{(n)} x_k^{(n)} \right\}$$

$$= \frac{1}{n} (\beta_n^2 - 2\gamma_n), \qquad n = 2, 3, \ldots$$

Thus we can find the first two coefficients of $\omega_n(x)$:

$$\beta_n = -n \mu_1, \qquad n = 1, 2, \ldots$$

$$\gamma_1 = 0$$

$$\gamma_n = \frac{1}{2} [\beta_n^2 - n \mu_2] = \frac{n}{2} [n \mu_1^2 - \mu_2], \qquad n = 2, 3, \ldots.$$

---

[3] The requirement that (10.2.1) be exact for $f(x) = x^2$ is only necessary for $n > 1$.

In Section 2.1 we showed that there is a recursion relation between three consecutive polynomials of an orthogonal sequence. If we denote by $P_n(x)$ the orthonormal polynomials for the weight function $p(x)$ then the recursion relation is given by (2.1.10). The polynomial $\omega_n(x)$ differs by only a constant multiple from the corresponding orthonormal polynomial $P_n(x)$ of the same degree. Using the fact that the leading coefficient of $\omega_n(x)$ is unity then the recursion relation for $\omega_n(x)$ can be written in the form

$$x\omega_0(x) = \omega_1(x) + \alpha_0$$

$$x\omega_n(x) = \omega_{n+1}(x) + \alpha_n\omega_n(x) + \lambda_n\omega_{n-1}(x) \qquad n = 1, 2, \ldots$$

Knowing β_n and γ_n we can find the coefficients α_n and λ_n. Indeed, equating the coefficients of x^n on opposite sides of the last equation we find

$$\beta_n = \beta_{n+1} + \alpha_n$$

$$\alpha_n = \beta_n - \beta_{n+1} = -n\mu_1 + (n+1)\mu_1 = \mu_1.$$

All the α_n $(n = 0, 1, \ldots)$ have the same value which for simplicity we denote by α:

$$\alpha_n = \alpha \qquad (n = 0, 1, \ldots).$$

Equating the coefficients of x^{n-1} in the same way we obtain:

$$\gamma_n = \gamma_{n+1} + \alpha_n\beta_n + \lambda_n$$

$$\lambda_n = \gamma_n - \gamma_{n+1} - \alpha_n\beta_n$$

Introducing the quantity σ we can write λ_n as:

$$\lambda_1 = \mu_2 - \mu_1 = \frac{\sigma^2}{2},$$

$$\lambda_n = \frac{1}{2}[\mu_2 - \mu_1] = \frac{\sigma^2}{4}, \qquad n = 2, 3, \ldots.$$

Thus the recursion relation for the polynomials $\omega_n(x)$ is

$$\omega_0(x) = 1, \qquad \omega_1(x) = x - \alpha$$

$$x\omega_n(x) = \omega_{n+1}(x) + \alpha\omega_n(x) + \frac{\sigma^2}{4}\omega_{n-1}(x), \qquad n = 1, 2, \ldots \qquad (10.2.2)$$

We recall now (see Section 2.3) that the Chebyshev polynomial of the first kind $T_n(x) = \cos(n \arccos x) = 2^{n-1}x^n + \cdots$ has the recursion relation

$$T_0(x) = 1, \quad T_1(x) = x$$

$$x T_n(x) = \frac{1}{2} T_{n+1}(x) + \frac{1}{2} T_{n-1}(x).$$

If we reduce the leading coefficient of $T_n(x)$ to unity we obtain the polynomial $T_n^*(x) = 2^{-n+1} T_n(x)$, $T_0^*(x) = T_0(x)$. The recursion relation for $T_n^*(x)$ is

$$T_0^*(x) = 1, \quad T_1^*(x) = x$$

$$x T_n^*(x) = T_{n+1}^*(x) + \frac{1}{4} T_{n-1}^*(x).$$

Finally if the variable x is replaced by $\dfrac{x - \alpha}{\sigma}$ and we introduce the polynomials $T_n^+(x) = \sigma^n T_n^* \left(\dfrac{x - \alpha}{\sigma} \right)$ then for these polynomials we obtain the recursion relation

$$T_0^+(x) = 1, \quad T_1^+(x) = x - \alpha$$

$$(x - \alpha) T_n^+(x) = T_{n+1}^+(x) + \frac{\sigma^2}{4} T_{n-1}^+(x).$$

These coincide with (10.2.2) and because these equations completely determine $\omega_n(x)$ $(n = 0, 1, \dots)$ then

$$\omega_n(x) = T_n^+(x) = \sigma^n T_n^* \left(\frac{x - \alpha}{\sigma} \right) = \frac{\sigma^n}{2^{n-1}} T_n \left(\frac{x - \alpha}{\sigma} \right) \qquad n = 1, 2, \dots .$$

The roots of the polynomial $T_n(x)$ are $\cos \dfrac{2k - 1}{2n} \pi$ $(k = 1, 2, \dots, n)$. They lie inside the segment $[-1, 1]$ and as n increases they become dense in this segment. Hence it follows that the roots of $\omega_n(x)$ lie in the segment $[\alpha - \sigma, \alpha + \sigma]$ and these also become dense in this segment.

On the other hand we showed in Chapter 2 that the roots of polynomials of an arbitrary orthogonal system corresponding to a positive weight function lie inside the segment of orthogonality. From the theory of orthogonal polynomials it is also known that the roots of a sequence of orthogonal polynomials become dense in the segment of orthogonality.[4] Therefore the roots of the polynomials $\omega_n(x)$ belong to $[-1, 1]$ and form a dense set.

---

[4] The more general theorem is known: *If the segment of orthogonality is* $[-1, 1]$ *and if the function* $p(x)$ *is summable and almost everywhere positive there, then the limiting distribution function of the zeros of the orthogonal polynomials coincides with the Chebyshev distribution function* $\mu(x) = \dfrac{1}{\pi} \displaystyle\int_{-1}^{x} \dfrac{dt}{\sqrt{1 - t^2}}$.

We must therefore have $\alpha = 0$ and $\sigma = 1$ and

$$\omega_0(x) = T_0(x) = 1$$

$$\omega_n(x) = 2^{-n+1} T_n(x) \qquad (n = 1, 2, \ldots)$$

The polynomials $T_n(x)$ form an orthogonal system on $[-1, 1]$ with respect to the weight function $(1 - x^2)^{-\frac{1}{2}}$ and to complete the proof of the theorem there only remains to show that for a finite segment of integration and a given weight function the corresponding orthogonal polynomials are unique up to a constant multiple and up to their values on a set of points of measure zero.

Suppose that the $\omega_n(x)$ are orthogonal on $[-1, 1]$ with respect to both $p_1(x)$ and $p_2(x)$. If necessary we can multiply these weight functions by constants so that

$$\int_{-1}^{1} p_1(x)\,dx = \int_{-1}^{1} p_2(x)\,dx = 1.$$

By the orthogonality of $\omega_n(x)$ we must have

$$\int_{-1}^{1} p_1(x)\,\omega_n(x)\,dx = \int_{-1}^{1} p_2(x)\,\omega_n(x)\,dx = 0, \qquad (n = 1, 2, \ldots).$$

Thus the difference $\phi(x) = p_1(x) - p_2(x)$ must satisfy

$$\int_{-1}^{1} \phi(x)\,\omega_n(x)\,dx = 0, \qquad (n = 0, 1, 2, \ldots)$$

which is equivalent to

$$\int_{-1}^{1} \phi(x)\,x^n\,dx = 0 \qquad (n = 0, 1, 2, \ldots).$$

It is known[5] that the system of powers x^n $(n = 0, 1, 2, \ldots)$ is complete in L and thus from the last equation it follows that $\phi(x)$ is equivalent to zero.

10.3. INTEGRALS WITH A CONSTANT WEIGHT FUNCTION

In this section we turn our attention to the much investigated case of a constant weight function. Let us assume that the segment of integration has been transformed into $[-1, 1]$ and consider the quadrature formula

---

[5] See, for example, I. P. Natanson, *Constructive Theory of Functions*, Gostekhizdat, Moscow, Chap. 3, Sec. 1 (Russian).

$$\int_{-1}^{1} f(x)\, dx \approx c_n \sum_{k=1}^{n} f(x_k). \tag{10.3.1}$$

The coefficient c_n and nodes x_k are to be chosen so that the formula is exact for all polynomials of degree $\leq n$. The coefficient c_n is determined from the requirement that (10.3.1) be exact for $f(x) \equiv 1$ and has the value

$$c_n = \frac{2}{n}.$$

Since

$$\int_{-1}^{1} x^k\, dx = \frac{1 - (-1)^{k+1}}{k+1}$$

the system of equations (10.3.1) which the nodes $x_1, \ldots, x_n$ must satisfy is:

$$s_1 = x_1 + x_2 + \cdots + x_n = 0$$

$$s_2 = x_1^2 + x_2^2 + \cdots + x_n^2 = \frac{n}{3}$$

$$s_3 = x_1^3 + x_2^3 + \cdots + x_n^3 = 0 \tag{10.3.2}$$

$$s_4 = x_1^4 + x_2^4 + \cdots + x_n^4 = \frac{n}{5}$$

$$\cdots\cdots\cdots\cdots\cdots\cdots\cdots\cdots\cdots\cdots$$

$$s_n = x_1^n + x_2^n + \cdots + x_n^n = \frac{n}{2}\left[\frac{1 - (-1)^{n+1}}{n+1}\right]$$

The coefficients of the polynomial $\omega(x) = (x - x_1) \cdots (x - x_n)$ must be found from the system of equations (10.1.6) which is in this case:

$$A_1 = 0$$

$$\frac{n}{3} + 2A_2 = 0$$

$$A_3 = 0$$

$$\frac{n}{5} + \frac{n}{3}A_2 + 4A_4 = 0$$

$$A_5 = 0 \tag{10.3.3}$$

$$\frac{n}{7} + \frac{n}{5}A_2 + \frac{n}{3}A_4 + 6A_6 = 0$$

$$A_7 = 0$$

$$\cdots\cdots\cdots\cdots\cdots\cdots\cdots\cdots\cdots\cdots$$

Here all the A_k with odd subscripts are zero and the polynomial $\omega(x)$ has the form

$$\omega(x) = x^n + A_2 x^{n-2} + A_4 x^{n-4} + \cdots .$$

The roots of $\omega(x)$ are the nodes of the formula (10.3.1) and they are symmetrically located on $[-1, 1]$ with respect to the point $x = 0$. If n is odd then one of the nodes coincides with $x = 0$.

It should be noted that if n is an even number $n = 2m$ then the x_k satisfy the equations

$$x_1 + x_2 + \cdots + x_n = 0$$

$$\cdots \cdots \cdots \cdots \cdots \cdots \cdots \cdots \cdots$$

$$x_1^n + x_2^n + \cdots + x_n^n = \frac{n}{n+1} .$$

Since $n + 1 = 2m + 1$ is an odd number and since the x_k are symmetrically located with respect to $x = 0$ then the nodes will also satisfy

$$x_1^{n+1} + x_2^{n+1} + \cdots + x_n^{n+1} = 0.$$

In this case formula (10.3.1) will be exact for one higher degree, that is it will be exact for all polynomials of degree $\leq n + 1$.

We will now construct formula (10.3.1) for low values of n.

For $n = 1$ we have $\omega(x) = x$ and $c_1 = 2$

$$\int_{-1}^{1} f(x)dx \approx 2f(0).$$

For $n = 2$ the coefficient is $c_2 = 1$ and the system of equations for A_1, A_2 is

$$A_1 = 0$$

$$\frac{2}{3} + 2A_2 = 0.$$

Thus

$$\omega(x) = x^2 - \frac{1}{3}$$

$$x_1 = -\frac{\sqrt{3}}{3}, \qquad x_2 = \frac{\sqrt{3}}{3}$$

$$\int_{-1}^{1} f(x)\,dx \approx f\left(\frac{-\sqrt{3}}{3}\right) + f\left(\frac{\sqrt{3}}{3}\right).$$

For $n = 3$ we have $c_3 = \dfrac{2}{3}$ and

$$A_1 = 0$$

$$1 + 2A_2 = 0$$

$$A_3 = 0$$

$$\omega(x) = x^3 - \frac{1}{2}x$$

and the formula is then

$$\int_{-1}^1 f(x)dx \approx \frac{2}{3}\left[f\left(\frac{-\sqrt{2}}{2}\right) + f(0) + f\left(\frac{\sqrt{2}}{2}\right)\right].$$

For $n = 4$ we have $c_4 = \dfrac{1}{2}$ and the following system of equations for the A_k:

$$A_1 = 0$$

$$\frac{4}{3} + 2A_2 = 0$$

$$A_3 = 0$$

$$\frac{4}{5} + \frac{4}{3}A_2 + 4A_4 = 0$$

Thus we obtain

$$A_2 = -\frac{2}{3}, \qquad A_4 = \frac{1}{45}$$

$$\omega(x) = x^4 - \frac{2}{3}x^2 + \frac{1}{45}$$

which has the roots

$$-x_1 = x_4 = \sqrt{\frac{5 + 2\sqrt{5}}{15}}$$

$$-x_2 = x_3 = \sqrt{\frac{5 - 2\sqrt{5}}{15}}.$$

In a similar way we obtain the following polynomials:

$$n = 5, \quad \omega(x) = x^5 - \frac{5}{6} x^3 + \frac{7}{72} x$$

$$n = 6, \quad \omega(x) = x^6 - x^4 + \frac{1}{5} x^2 - \frac{1}{105}$$

$$n = 7, \quad \omega(x) = x^7 - \frac{7}{6} x^5 + \frac{119}{360} x^3 - \frac{149}{6480} x$$

$$n = 9, \quad \omega(x) = x^9 - \frac{3}{2} x^7 + \frac{27}{40} x^5 - \frac{57}{560} x^3 + \frac{53}{22400} x.$$

For $n = 8$ two of the roots of $\omega(x)$ are complex and it is impossible to construct a Chebyshev formula (10.3.1) in this case with real roots. Here we tabulate the decimal values of the nodes in (10.3.1) for[6] $n = 1(1)7, 9$.

$n = 1$		$n = 6$	
0.00000	00000	0.26663	54015
		0.42251	86538
$n = 2$		0.86624	68181
0.57735	02691		
		$n = 7$	
$n = 3$		0.00000	00000
0.00000	00000	0.32391	18105
0.70710	67812	0.52965	67753
		0.88386	17008
$n = 4$			
0.18759	24741	$n = 9$	
0.79465	44723	0.00000	00000
		0.16790	61842
$n = 5$		0.52876	17831
0.00000	00000	0.60101	86554
0.37454	14096	0.91158	93077
0.83249	74870		

We could also calculate the nodes for the Chebyshev formulas for $n > 9$ but in every case it turns out that some of the roots of $\omega(x)$ will be complex and it will be impossible to construct formula (10.3.1) with real nodes. The general question as to the existence of Chebyshev formulas for $n > 9$ with all real nodes remained unanswered until S. N. Bernstein proved that such formulas do not exist. The remainder of this chapter is devoted to a somewhat simplified presentation of his results.

We prove four preliminary lemmas.

---

[6]These values are from the paper by Salzer, *J. Math. Phys.*, Vol. 26, 1947, pp. 191–194.

Lemma 1. *Let the formula*

$$\int_{-1}^{1} f(x)dx \approx \frac{2}{n} \sum_{k=1}^{n} f(x_k) \tag{10.3.4}$$

be exact for all polynomials of degree $\leq 2m - 1$ where $m < n$. Let ξ_m denote the largest root of the mth degree Legendre polynomial $P_m(x)$. Then, assuming that the x_k are enumerated in order of size:

$$x_n > \xi_m.$$

Proof. Consider

$$f(x) = \frac{P_m^2(x)}{x \cdot \xi_m}.$$

The function $P_m(x)/(x - \xi_m)$ is a polynomial of degree $m - 1$ and since $P_m(x)$ is orthogonal on $[-1, 1]$ to all polynomials of lower degree then

$$\int_{-1}^{1} f(x)dx = 0.$$

On the other hand $f(x)$ is a polynomial of degree $2m - 1$ and equation (10.3.4) must be exact for this function. Therefore

$$\sum_{k=1}^{n} f(x_k) = 0.$$

The polynomial $f(x) = P_m^2(x)/(x - \xi_m)$ has m distinct roots and therefore not all terms in the last sum can be zero. Thus this sum must contain positive and negative terms. But $f(x)$ takes on positive values only for $x > \xi_m$. Thus we can find a node x_k for which $x_k > \xi_m$ and hence the largest node must also be greater than ξ_m.

The following arguments are based on comparisons of (10.3.4) with Gauss quadrature formulas with m nodes

$$\int_{-1}^{1} f(x)dx \approx \sum_{i=1}^{m} A_i f(\xi_i)$$

$$\tag{10.3.5}$$

$$P_m(\xi_i) = 0 \qquad (i = 1, \ldots, m) \qquad A_i = \frac{2}{(1 - \xi_i^2)[P_m'(\xi_i)]^2}.$$

Lemma 2. *If formula (10.3.4) is exact for all polynomials of degree $\leq 2m - 1$ where $m < n$ then*

$$A_m > \frac{2}{n}. \tag{10.3.6}$$

Proof. Let

$$f(x) = \left[\frac{P_m(x)}{(x - \xi_m)P'_m(\xi_m)} \right]^2.$$

Then $f(\xi_m) = 1$ and at the other ξ_i, $i < m$, $f(\xi_i) = 0$. Therefore for $f(x)$ the quadrature sum (10.3.5) becomes:

$$A_m f(\xi_m) = A_m.$$

The function $f(x)$ is a polynomial of degree $2m - 2$ and both (10.3.4) and (10.3.5) must be exact for this function. Therefore

$$\frac{2}{n} \sum_{k=1}^{n} f(x_k) = A_m.$$

Because $f(x) \geq 0$ for all x it follows that

$$\frac{2}{n} f(x_n) \leq A_m. \tag{10.3.7}$$

Writing

$$f(x) = [P'_m(\xi_m)]^{-2}(x - \xi_1)^2 \cdots (x - \xi_{m-1})^2$$

we see that, for $x \geq \xi_m$, $f(x)$ is an increasing function of x and since $x_n > \xi_m$ we have $f(x_n) > f(\xi_m) = 1$. Combining this with (10.3.7) proves the lemma.

In order to estimate the coefficient

$$A_m = \frac{2}{(1 - \xi_m^2)[P'_m(\xi_m)]^2}$$

in formula (10.3.5) we will obtain estimates for ξ_m and $P'_m(\xi_m)$.

Lemma 3. *For any value of* m *the largest root* ξ_m *of* $P_m(x)$ *satisfies the inequality*

$$1 - \xi_m < \frac{3}{m(m + 1)}. \tag{10.3.8}$$

Proof. We begin with the differential equation satisfied by $P_m(x)$:

$$\frac{d}{dx} [(1 - x^2)P'_m(x)] + m(m + 1)P_m(x) = 0.$$

Integrating both terms in this equation between the limits ξ_m and 1 we obtain

$$(1 - \xi_m^2)P_m'(\xi_m) = m(m + 1) \int_{\xi_m}^{1} P_m(x)dx.$$

Let us replace the polynomial $P_m(x)$ in this integral by its expansion in terms of powers of $x - \xi_m$

$$P_m(x) = \sum_{i=1}^{m} \frac{(x - \xi_m)^i}{i!} P_m^{(i)}(\xi_m).$$

Carrying out the integration gives

$$(1 - \xi_m^2)P_m'(\xi_m) = m(m + 1) \sum_{i=1}^{m} \frac{(1 - \xi_m)^{i+1}}{(i + 1)!} P_m^{(i)}(\xi_m)$$

Between each pair of adjacent roots ξ_j, ξ_{j+1} of the polynomial $P_m(x)$ there lies a root of $P_m'(x)$. There are $m - 1$ such roots of $P_m'(x)$ and no others. The $m - 2$ roots of the second derivative of $P_m(x)$ lie between adjacent roots of $P_m'(x)$ and so forth. Thus for any i all the roots of $P_m^{(i)}(x)$ lie in the interval $[\xi_1, \xi_m]$ and none of these roots are greater than ξ_m. Therefore $P_m^{(i)}(\xi_m) > 0$ and all terms on the right side of the last equation are positive. For a sufficiently precise estimate for ξ_m we can replace this sum by only its first two terms. Then dividing both sides by $1 - \xi_m$ we obtain the inequality

$$(1 + \xi_m)P_m'(\xi_m) > m(m + 1) \times$$

$$\times \left[\frac{1}{2}(1 - \xi_m)P_m'(\xi_m) + \frac{1}{6}(1 - \xi_m)^2 P_m''(\xi_m) \right].$$

The value of $P_m''(\xi_m)$ is easily found from the equation

$$(1 - x^2)P_m''(x) - 2xP_m'(x) + m(m + 1)P_m(x) = 0$$

by substituting $x = \xi_m$:

$$P_m''(\xi_m) = \frac{2\xi_m}{1 - \xi_m^2} P_m'(\xi_m). \qquad (10.3.9)$$

Substituting this value in the inequality and cancelling the factor $P_m'(\xi_m)$ gives:

$$1 + \xi_m > m(m + 1) \left[\frac{1}{2}(1 - \xi_m) + \frac{1}{3} \frac{\xi_m(1 - \xi_m)}{1 + \xi_m} \right].$$

This inequality is made stronger if in the second term inside the brackets we replace $1 + \xi_m$ by the larger value 2:

$$1 + \xi_m > m(m + 1)\left[\frac{1}{2}(1 - \xi_m) + \frac{1}{6}\xi_m(1 - \xi_m)\right].$$

Setting $\lambda = m(m + 1)$ we can write this equation as

$$\lambda\xi_m^2 + 2(3 + \lambda)\xi_m + 6 - 3\lambda > 0. \qquad (10.3.10)$$

Let us form the equations

$$\lambda z^2 + 2(3 + \lambda)z + 6 - 3\lambda = 0.$$

$$z = \frac{\pm\sqrt{4\lambda^2 + 9} - 3 - \lambda}{\lambda}.$$

If ξ_m satisfies the inequality (10.3.10) then ξ_m must be larger than the positive value of z:

$$\xi_m > \frac{\sqrt{4\lambda^2 + 9} - 3 - \lambda}{\lambda} > \frac{\sqrt{4\lambda^2} - 3 - \lambda}{\lambda}.$$

This gives

$$\xi_m > 1 - \frac{3}{\lambda} = 1 - \frac{3}{m(m + 1)}$$

$$1 - \xi_m < \frac{3}{m(m + 1)}.$$

This proves lemma 3.

Lemma 4. *The value of the derivative* $P'_m(\xi_m)$ *of the Legendre polynomial* $P_m(x)$ *at the largest root* $x = \xi_m$ *satisfies the inequality*

$$P'_m(\xi_m) > \frac{2}{3(1 - \xi_m)}\left[1 - \frac{\Gamma(m + 4)}{288\,\Gamma(m - 2)}(1 - \xi_m)^3\right]. \qquad (10.3.11)$$

Proof. Making use of Taylor's series with two terms and the integral form of the remainder:

$$P_m(x) = P'_m(\xi_m)(x - \xi_m) +$$

$$+ \frac{1}{2}P''_m(\xi_m)(x - \xi_m)^2 + \frac{1}{2}\int_{\xi_m}^{x} P_m^{(3)}(t)(x - t)^2 dt.$$

For $x = 1$, using $P_m(1) = 1$, this becomes:

$$1 = P'_m(\xi_m)(1 - \xi_m) + \frac{1}{2} P''_m(\xi_m)(1 - \xi_m)^2 +$$

$$+ \frac{1}{2} \int_{\xi_m}^1 P_m^{(3)}(t)(1 - t)^2 dt. \quad (10.3.12)$$

Consider $P_m^{(3)}(t)$. In the proof of Lemma 3 we showed that all the roots of $P''_m(x)$ are less than ξ_m. Therefore $P_m^{(3)}(x)$ is a monotonically increasing function on $[\xi_m, 1]$ and its greatest value is achieved for $x = 1$. The value of $P_m^{(3)}(1)$ can be easily found using the differential equation

$$(1 - x^2)P''_m(x) - 2xP'_m(x) + m(m + 1)P_m(x) = 0.$$

Setting here $x = 1$ we find

$$P'_m(1) = \frac{m(m + 1)}{2}.$$

Differentiating gives

$$(1 - x^2)P_m^{(3)}(x) - 4xP''_m(x) + (m + 2)(m - 1)P'_m(x) = 0$$

and again setting $x = 1$ gives

$$P''_m(1) = \frac{(m + 2)(m - 1)}{4} P'_m(1) = \frac{(m + 2)(m + 1)m(m - 1)}{8}.$$

Differentiating once more

$$(1 - x^2)P_m^{(4)}(x) - 6xP_m^{(3)}(x) + (m + 3)(m - 2)P''_m(x) = 0$$

and substituting $x = 1$:

$$P_m^{(3)}(1) = \frac{(m + 3)(m - 2)}{6} P''_m(1) =$$

$$= \frac{(m + 3)(m + 2) \cdots (m - 2)}{48} = \frac{\Gamma(m + 4)}{48\,\Gamma(m - 2)}.$$

Substituting in (10.3.12) for $P''_m(\xi_m)$ its expression (10.3.9) and for $P_m^{(3)}(t)$ its upper bound on $[\xi_m, 1]$ leads to the inequality:

$$P'_m(\xi_m)(1 - \xi_m)\left[1 + \frac{\xi_m}{1 + \xi_m}\right] + \frac{\Gamma(m + 4)}{48\,\Gamma(m - 2)} \frac{(1 - \xi_m)^3}{3!} > 1.$$

This, together with $\dfrac{\xi_m}{1 + \xi_m} < \dfrac{1}{2}$, establishes (10.3.11).

We can now easily find an estimate for $A_m = \dfrac{2}{(1 - \xi_m^2)[P'_m(\xi_m)]^2}$.

Substituting for $P'_m(\xi_m)$ its smaller value from (10.3.11)

$$A_m < \frac{9(1 - \xi_m)}{2(1 + \xi_m)} \left[1 - \frac{\Gamma(m + 4)}{288\,\Gamma(m - 2)}(1 - \xi_m)^3 \right]^{-2}.$$

It will suffice to use a cruder inequality for A_m for $m \geq 6$. As m increases the value of ξ_m also increases and since $\xi_6 = 0.93246..$ we are justified in assuming $1 + \xi_m > 1.93$. We also replace $1 - \xi_m$ by the larger value $\dfrac{3}{m(m + 1)}$. We now estimate the value inside the brackets

$$(m + 3)(m - 2) = m(m + 1) - 6 < m(m + 1)$$

$$(m + 2)(m - 1) = m(m + 1) - 2 < m(m + 1)$$

$$\frac{\Gamma(m + 4)}{\Gamma(m - 2)} = (m + 3)(m + 2)(m + 1)m(m - 1)(m - 2) < m^3(m + 1)^3$$

$$1 - \frac{\Gamma(m + 4)}{288\,\Gamma(m - 2)}(1 - \xi_m)^3 > 1 - \frac{m^3(m + 1)^3}{288} \frac{3^3}{m^3(m + 1)^3} = \frac{29}{32}$$

$$A_m < \frac{27\,(32)^2}{2\,(1.93)\,(29)^2} \frac{1}{m\,(m + 1)} \approx \frac{8.517}{m(m + 1)}. \tag{10.3.13}$$

Theorem 2. *For $n \geq 10$ there is no formula (10.3.4) with all real roots which is exact for all polynomials of degree $\leq n$.*

Proof. Let us consider those values of n for which formula (10.3.4) exists. Let us suppose that n is an odd integer: $n = 2m - 1$. Then $m = \frac{1}{2}(n + 1)$ and $m(m + 1) = \frac{1}{4}(n + 1)(n + 3)$ and A_m must satisfy the inequality $A_m < \dfrac{4\,(8.517)}{(n + 1)(n + 3)}$. By Lemma 2 we must have

$$\frac{4\,(8.517)}{(n + 1)(n + 3)} > \frac{2}{n}$$

or

$$n^2 - (13.034)\,n + 3 < 0$$

$$n < 13.$$

Thus formula (10.3.4) does not exist for $n \geq 13$. But for $n = 11$ it also does not exist because then $m = 6$, $A_6 = 0.173\ldots$, $\dfrac{2}{11} = 0.1818..$

and the inequality $\dfrac{2}{11} < A_6$ is not satisfied.

Suppose now that n is even. Then (10.3.4) must be exact for polynomials of degree $\le n + 1$. Set $n + 1 = 2m - 1$, $m = \dfrac{1}{2}(n + 2)$. By (10.3.13) and (10.3.6) we must have

$$\frac{4\,(8.517)}{(n + 2)\,(n + 4)} > \frac{2}{n}$$

and hence

$$n < 11.$$

This means that for $n > 10$ formula (10.3.4) does not exist. For $n = 10$ it also does not exist because the inequality

$$A_6 = 0.173.. > \frac{2}{10} = 0.2$$

is clearly not valid.

REFERENCES

S. N. Bernstein, "On quadrature formulas of Cotes and Chebyshev," *Dokl. Akad. Nauk SSSR*, Vol. 14, 1937, pp. 323–26; Collected Works, Vol. 2, pp. 200–04 (Russian).

S. N. Bernstein, "On a system of indeterminate equations," Collected Works, Vol. 2, pp. 236–42 (Russian).

P. L. Chebyshev, "On Quadratures," *J. Math. Pures Appl.*, (2) Vol. 19, 1874, pp. 19–34; Collected Works, Vol. 3, Moscow, 1948, p. 49–62.

Ia. L. Geronimus, "On Gauss' and Chebyshev's quadrature formulas," *Dokl. Akad. Nauk SSSR*, Vol. 51, 1946, pp. 655–58 (Russian).

Ia. L. Geronimus, *Theory of Orthogonal Polynomials*, Gostekhizdat, Moscow, 1950.

V. I. Krylov, "Mechanical quadratures with equal coefficients for the integrals $\displaystyle\int_0^\infty e^{-x} f(x)dx$ and $\displaystyle\int_{-\infty}^\infty e^{-x^2} f(x)dx$," *Dokl. Akad. Nauk SSSR*, Vol. 2, 1958, pp. 187–92 (Russian).

R. O. Kuz'min, "On the distribution of roots of polynomials connected with quadratures of Chebyshev," *Izv. Akad. Nauk SSSR, Ser. Mat.*, Vol. 4, 1938, pp. 427–44 (Russian).

K. A. Posse, "Sur les quadratures," *Nouv. Ann. de Math.*, (2) Vol. 14, 1875, pp. 49–62.

H. E. Salzer, "Equally weighted quadrature formulas over semi-infinite and infinite intervals," *J. Math. Phys.*, Vol. 34, 1955, pp. 54–63.

N. Ia. Sonin, "On the approximate evaluation of definite integrals and on the related integral functions," *Warshawskia Universitetskia Izvestia*, Vol. 1, 1887, pp. 1–76 (Russian).

H. S. Wilf, "The possibility of Tschebycheff quadrature on infinite intervals," *Proc. Nat. Acad. Sci.*, Vol. 47, 1961, pp. 209–13.

CHAPTER 11

Increasing the Precision of Quadrature Formulas

11.1. TWO APPROACHES TO THE PROBLEM

Let us consider a certain completely defined quadrature formula

$$\int_a^b p(x) f(x)\, dx \approx \sum_{k=1}^{n} A_k f(x_k) \tag{11.1.1}$$

where the weight function $p(x)$, the coefficients A_k and the nodes x_k are fixed; $f(x)$ is any function for which both sides of (11.1.1) are defined.

We will be interested in the remainder of formula (11.1.1)

$$R(f) = \int_a^b p(x) f(x)\, dx - \sum_{k=1}^{n} A_k f(x_k). \tag{11.1.2}$$

By increasing the precision of the approximate quadrature we mean the addition of some quantity to the quadrature sum $\sum_k A_k f(x_k)$ which will decrease the size of the remainder.

The value of $R(f)$ depends both on the quadrature formula, that is on $p(x)$ and the x_k and A_k, and also on the properties of the integrand $f(x)$. A method for increasing the precision of the formula must also depend on these same factors. It is possible to construct such methods for a given class of quadrature formulas with similar properties or for a class of functions which possess certain common structural properties.

In the remainder of this section we discuss two methods for increasing the precision of quadrature formulas.

1. In most practical applications mechanical quadrature formulas are intended for use with integrands which possess some degree of smoothness. One might expect, for example, that Simpson's formula

$$\int_a^b f(x)\,dx \approx \frac{h}{3}[f_0 + f_n + 2(f_2 + f_4 + \cdots + f_{n-2}) + 4(f_1 + f_3 + \cdots + f_{n-1})]$$

$$h = \frac{b-a}{n}$$

will give reasonably good results if $f(x)$ is continuous on the entire segment $[a, b]$ and on each of the segments $[a, a + 2h]$, $[a + 2h, a + 4h]$, ... it can be approximated reasonably well by a second or third degree polynomial.

Similarly it is to be expected that a quadrature formula of the highest algebraic degree of precision with n nodes will give an approximation which is close to the true value if $f(x)$ can be closely approximated on the entire segment $[a, b]$ by an algebraic polynomial of degree $2n - 1$.

It is not always possible to apply formula (11.1.1) to calculate an improper integral of an unbounded function since one or more values $f(x_k)$ in the quadrature sum may be infinite. But even if it is possible to apply the formula the error might be very large. A large error can also be obtained in integrating a continuous function which has an unbounded derivative or in integrating an analytic function which has singular points close to the interval of integration.

In cases such as these the accuracy of the approximate integration can be appreciably improved if a preliminary transformation can be applied to the integrand which removes or weakens the singularities of $f(x)$. This can be done if the integrand can be split into two parts

$$f(x) = f_1(x) + f_2(x)$$

where $f_1(x)$ is a function which contains "most" of the singularity of $f(x)$ for which the integral $\int_a^b p(x)f_1(x)\,dx$ can be evaluated exactly. The function $f_2(x)$ should be relatively smooth so that the integral $\int_a^b p(x)f_2(x)\,dx$ can be closely approximated by a quadrature formula.

In Section 11.2 we discuss several methods for removing or weakening the singularities of $f(x)$.

2. In most cases quadrature formulas can estimate an integral to any degree of precision provided that a sufficiently large number of nodes are used. The number of nodes which must be used to obtain a desired

accuracy can be determined in principle by employing the methods we have discussed for estimating the remainder $R(f)$. Such estimates, however, are usually intended for a wide class of functions and do not take into account the individual properties of a particular integrand. Therefore, as a rule, these estimates are too large and only serve as a rough estimate for the number of nodes which are necessary.

To decide on the number of nodes to be used in a calculation one usually takes into account not only the estimate for the remainder but also other information such as experience derived from previous calculations, comparisons with similar integrals or a comparison of the results of integrations carried out by different methods. The value of n obtained in this way will often give the desired accuracy but we can not be absolutely certain that it will. We then have the problem of checking the result and if it is not sufficiently accurate of increasing the accuracy.

We will assume that $f(x)$ is sufficiently smooth so that a large value for $R(f)$ can only result from using an insufficiently exact quadrature formula.

To increase the precision of the formula we must find additional terms to add to the right side of (11.1.1) so that the new formula will be more precise than (11.1.1).

It is clear that these new terms must account for the principle part of the remainder $R(f)$. There are many different ways in which the "principle part" of the remainder can be defined and we must determine a simple method to calculate the part which is appropriate to this problem. We discuss two such methods in the last two sections of this chapter.

Suppose that by some method a new term has been found for (11.1.1). If the correction provided by this term improves the accuracy to the desired degree then the computation is completed. If the desired accuracy is not achieved with the first term then the process is repeated and another term is found. It is usually impossible to determine beforehand how many steps will be necessary and therefore we must construct a sequence of principle parts for the remainder (11.1.2) for our initial formula.

11.2. WEAKENING THE SINGULARITY OF THE INTEGRAND

As we pointed out in the previous section we can improve the accuracy of an approximate integration by weakening the singularity of the integrand by splitting it into two parts $f(x) = f_1(x) + f_2(x)$ where $f_1(x)$ contains "most" of the singularity of $f(x)$ such that the integral $\int_a^b p(x) f_1(x)\, dx$ can be calculated exactly and where the integral of the second part $\int_a^b p(x) f_2(x)\, dx$ can be closely approximated by a quadrature formula.

The particular method used will depend on the character of the singularities of $f(x)$ and on the weight function $p(x)$. Let us consider some simple examples of such methods.

1. Suppose we are given the integral

$$\int_a^b (x - x_1)^\alpha \phi(x)\, dx \qquad (11.2.1)$$

where x_1 is a point in or close to the segment $[a, b]$. To be definite let us assume that x_1 belongs to $[a, b]$. We also assume that α is greater than -1 and is not an integer, that $\phi(x)$ is continuous on $[a, b]$, that $\phi(x)$ has derivatives up to a certain order m at x_1 and that $\phi(x_1) \neq 0$.

For $\alpha < 0$ the above integral will be improper. If $\alpha > 0$ then the integrand will not have derivatives of all orders at x_1. Thus quadrature formulas might give a large error for this integrand.

Let us split off from the Taylor series expansion of $\phi(x)$ around the point x_1 the first k terms ($k < m$) and write $f(x)$ as

$$f(x) = (x - x_1)^\alpha \phi(x) = f_1(x) + f_2(x)$$

where

$$f_1(x) = (x - x_1)^\alpha \left[\phi(x_1) + \frac{\phi'(x_1)}{1!}(x - x_1) + \cdots + \frac{\phi^{(k-1)}(x_1)}{(k-1)!}(x - x_1)^{k-1} \right]$$

$$f_2(x) = (x - x_1)^\alpha \times$$
$$\times \left[\phi(x) - \phi(x_1) - \frac{\phi'(x_1)}{1!}(x - x_1) - \cdots - \frac{\phi^{(k-1)}(x_1)}{(k-1)!}(x - x_1)^{k-1} \right]$$

Thus the original integral will also be split into two parts

$$\int_a^b (x - x_1)^\alpha \phi(x)\, dx = \int_a^b f_1(x)\, dx + \int_a^b f_2(x)\, dx.$$

The first of these integrals can be calculated exactly by elementary methods. At x_1 the function $f_2(x)$ is differentiable k more times than the original function. Therefore the integral $\int_a^b f_2(x)\, dx$ can be calculated with greater accuracy than (11.2.1) by a quadrature formula.

As an example consider the integral

$$\int_0^1 \sqrt{1 - x^2} \, dx = \frac{\pi}{4} \approx 0.785398163 \ldots .$$

At the upper limit $x = 1$ the function $\sqrt{1 - x^2}$ has an algebraic singularity. Let us remove the factor $\sqrt{1 - x}$ and expand $\sqrt{1 + x}$ in powers of $x - 1$ taking two terms in the expansion:

$$\sqrt{1 + x} = \sqrt{2} \left(1 - \frac{1 - x}{4} \right) + \left[\sqrt{1 + x} - \sqrt{2} \left(1 - \frac{1 - x}{4} \right) \right].$$

The integral then splits into two integrals the first of which can be integrated exactly:

$$I_1 = \int_0^1 \sqrt{2} \ \sqrt{1 - x} \left(1 - \frac{1 - x}{4} \right) dx = \frac{17\sqrt{2}}{30} \approx 0.801388 \ldots .$$

The second integral

$$I_2 = \int_0^1 \sqrt{1 - x} \left[\sqrt{1 + x} - \sqrt{2} \left(\frac{3}{4} + \frac{1}{4} x \right) \right] dx$$

can be calculated by Simpson's formula (6.3.5) with three nodes:

$$f_2(0) = 1 - \frac{3\sqrt{2}}{4} \approx -0.060660$$

$$4 f_2 \left(\frac{1}{2} \right) = 2\sqrt{3} - \frac{7}{2} \approx -0.035898$$

$$f_2(1) = 0$$

$$I_2 \approx \frac{1}{6} \left[f_2(0) + 4 f_2 \left(\frac{1}{2} \right) + f_2(1) \right] \approx -0.016035$$

$$\int_0^1 \sqrt{1 - x^2} \, dx = I_1 + I_2 \approx 0.785353.$$

This result is exact to four significant figures. Applying Simpson's formula with three and five nodes directly to the original integrand gives 0.637 and 0.744 respectively.

2. A similar transformation can be carried out when the integrand has singularities at several points. Suppose the integral has the form

$$\int_a^b f(x) \, dx = \int_a^b (x - x_1)^{\alpha_1} (x - x_2)^{\alpha_2} \cdots (x - x_n)^{\alpha_n} \phi(x) \, dx. \qquad (11.2.2)$$

We combine all but the first factor

$$(x - x_2)^{\alpha_2} \cdots (x - x_n)^{\alpha_n} \phi(x)$$

and expand this function in a Taylor series in powers of $x - x_1$. Taking the first k terms of this expansion we split the integral as before into two parts

$$f(x) = f_1(x) + [f(x) - f_1(x)]$$

where $f_1(x)$ is a sum of powers and at the point x_1 $f(x) - f_1(x)$ has derivatives of higher order than $f(x)$. In a similar way we can expand around the other points $x_2, \ldots, x_n$ and obtain

$$f(x) = f_k(x) + [f(x) - f_k(x)], \qquad k = 2, \ldots, n.$$

We can then split the original integral into two parts

$$\int_a^b f(x)\, dx = \int_a^b [f_1(x) + f_2(x) + \cdots + f_n(x)]\, dx +$$

$$+ \int_a^b [f(x) - f_1(x) - f_2(x) - \cdots - f_n(x)]\, dx$$

where the first integral is easily calculated exactly. The function in the second integral has higher order derivatives than $f(x)$ and a quadrature formula applied to this integral will give a more accurate result than when applied to (11.2.2).

3. Taylor's formula can be used to weaken the singularity of the integrand any time that the integral has the form

$$\int_a^b \psi(x)\, \phi(x)\, dx$$

where $\psi(x)$ has a singularity at a point x_1 provided that the integrals $\int_a^b \psi(x)(x - x_1)^i dx$ can be calculated exactly and that the function $\phi(x)$ is differentiable several times at the point x_1. An example is the integral

$$\int_a^b (x - x_1)^{\alpha} \ln^p |x - x_1|\, \phi(x)\, dx$$

where α is a real number greater than -1 and p is an integer.

4. Consider the integral

$$\int_a^b \psi[\phi(x)]\, dx \qquad (11.2.3)$$

where $\psi(t)$ has a singularity at $t = 0$ and $\phi(x)$ is a continuously differentiable function which is zero at $x = x_1$ and such that $\phi'(x_1) = A \neq 0$.

To weaken the singularity of the integrand we can split it into two parts

$$\psi[\phi(x)] = \psi[A(x - x_1)] + \{\psi[\phi(x)] - \psi[A(x - x_1)]\}$$

and if the first integral $\int_a^b \psi[A(x - x_1)]\, dx$ can be calculated exactly then a quadrature formula applied to the second integral will give a more exact result than when it is applied to (11.2.3).

As an example consider the integral

$$\int_0^{\pi/2} ln \ sin \ x \ dx = -\frac{\pi}{2} ln \ 2 \approx -1.089045.$$

This integrand has a logarithmic singularity at $x = 0$. We remove from $\sin x$ the first term of its expansion in powers of x and write the integral in the following way:

$$\int_0^{\pi/2} ln \ sin \ x \ dx = \int_0^{\pi/2} ln \ x \ dx + \int_0^{\pi/2} ln \ \frac{sin \ x}{x} \ dx = I_1 + I_2$$

$$I_1 = \int_0^{\pi/2} ln \ x \ dx = \frac{\pi}{2}\left(ln \ \frac{\pi}{2} - 1\right) \approx -0.861451.$$

The function $y(x) = ln \ \dfrac{sin \ x}{x}$ has no singular points in $\left[0, \dfrac{\pi}{2}\right]$. To calculate I_2 we use Simpson's formula with 3 nodes:

$$I_2 \approx \frac{\pi}{12}\left[y(0) + 4y\left(\frac{\pi}{4}\right) + y\left(\frac{\pi}{2}\right)\right] \approx -0.228189.$$

Thus

$$\int_0^{\pi/2} ln \ sin \ x \ dx = I_1 + I_2 \approx -1.089640.$$

11.3. EULER'S METHOD FOR EXPANDING THE REMAINDER

We now consider the problem of increasing the precision of a quadrature formula by removing the principle part of the remainder. The most appropriate way of doing this depends on the properties of the remainder and there are many different methods which may be used.

In this chapter we discuss two of these methods, the first of which is closely related to the Euler-Maclaurin sum formula.

The simplest type of Euler's formula serves to increase the accuracy of the simple one-point formula. It is another form of the method for expanding an arbitrary function in Bernoulli polynomials.

Let $f(x)$ have ν continuous derivatives on the finite segment $[a, b]$. In Chapter 1 we established the representation (1.4.2) which expresses $f(x)$ in terms of Bernoulli polynomials and the periodic functions $B_\nu^*(x)$. This representation can be written in the form

$$
\int_a^b f(t)\,dt = (b-a)f(x) - (b-a)\,B_1\left(\frac{x-a}{b-a}\right)[f(b)-f(a)] -
$$

$$
- \frac{(b-a)^2}{2!}\,B_2\left(\frac{x-a}{b-a}\right)[f'(b)-f'(a)] - \cdots -
$$

$$
- \frac{(b-a)^{\nu-1}}{(\nu-1)!}\,B_{\nu-1}\left(\frac{x-a}{b-a}\right)[f^{(\nu-2)}(b)-f^{(\nu-2)}(a)] +
$$

$$
\tag{11.3.1}
$$

$$
+ \frac{(b-a)^\nu}{\nu!}\int_a^b f^{(\nu)}(t)\left[B_\nu^*\left(\frac{x-t}{b-a}\right) - B_\nu^*\left(\frac{x-a}{b-a}\right)\right]dt.
$$

The first term on the right side of this equation $(b-a)f(x)$ gives an approximate value for the integral $\int_a^b f(t)\,dt$ and is a one-point formula which uses the point x. The approximate equation

$$
\int_a^b f(t)\,dt \approx (b-a)f(x)
$$

will be exact when $f(t)$ is a constant function.

If we adjoin to the term $(b-a)f(x)$ the second term on the right side we obtain

$$
\int_a^b f(t)\,dt \approx (b-a)f(x) - (b-a)\,B_1\left(\frac{x-a}{b-a}\right)[f(b)-f(a)]
$$

which is exact for any linear function. If we add a third term then the resulting equation is exact for any quadratic polynomial and so forth. Adding one term at a time from (11.3.1) increases the algebraic degree of precision of the formula each time by one. We can expect, at least in certain cases, that each new term will increase the accuracy of the approximate integration.

The integral on the right side of (11.3.1) is the remainder term in the final quadrature formula. Below we will investigate this integral further.

Equation (11.3.1) is more valuable than we have indicated for from it we can construct, in principle, a method for increasing the precision of any quadrature formula for use with a constant weight function. Let us consider an arbitrary quadrature formula of the form

$$\int_a^b f(t)\,dt \approx (b-a) \sum_{k=1}^n A_k f(x_k). \tag{11.3.2}$$

Let us assume that this formula is exact if $f(t)$ is a constant, that is $\sum_{k=1}^n A_k = 1$. Then it is obvious that (11.3.2) is a linear combination of n elementary one point formulas

$$\int_a^b f(t)\,dt \approx (b-a) f(x_k) \qquad (k = 1, 2, \ldots, n).$$

Therefore a linear combination with coefficients A_k of n equations (11.3.1) with $x = x_1,\ x = x_2, \ldots,\ x = x_n$ gives a new equation which will increase the accuracy of the formula (11.3.2) to an arbitrarily high degree.

One can see that similar equations can also be constructed for quadrature formulas for the approximate evaluation of an integral $\int_a^b p(t)f(t)\,dt$ with any summable weight function $p(t)$. Such formulas are formally very simple to derive by using the theorem on the expansion of a function in Bernoulli polynomials together with special forms of integral representations for the remainder of the quadrature formulas. But it is not clear in which cases the formulas obtained in this way will actually increase the accuracy of the quadrature formulas and in which cases they will give a worse result.

We will begin our discussion from an intuitive point of view and will derive a method to increase the accuracy which will be very generally applicable. Our discussion will also clarify the conditions under which formulas of Euler's type are to be preferred over other methods.

We assume again that the segment $[a,\ b]$ is finite and that $f(x)$ has a continuous derivative of order $m + s$ on $[a,\ b]$ where m and s are positive integers which will enter into the following discussion.

We will consider the remainder $R(f)$ of the quadrature formula

$$\int_a^b p(x)f(x)\,dx = \sum_{k=1}^n A_k f(x_k) + R(f) \tag{11.3.3}$$

which we assume is exact for all polynomials of degree $\leq m - 1$.

The function $f(x)$ can be represented by the Taylor series:

$$f(x) = \sum_{i=0}^{m-1} \frac{f^{(i)}(a)}{i!} (x-a)^i + \int_a^x f^{(m)}(t) \frac{(x-t)^{m-1}}{(m-1)!} \, dt =$$

$$= P_{m-1}(x) + \int_a^b f^{(m)}(t) E(x-t) \frac{(x-t)^{m-1}}{(m-1)!} \, dt.$$

Now, since $R(P_{m-1}) = 0$, the remainder $R(f)$ will be:

$$R(f) = \int_a^b p(x) \int_a^b f^{(m)}(t) E(x-t) \frac{(x-t)^{m-1}}{(m-1)!} \, dt \, dx -$$

$$- \sum_{k=1}^n A_k \int_a^b f^{(m)}(t) E(x_k - t) \frac{(x_k - t)^{m-1}}{(m-1)!} \, dt.$$

The assumptions of the continuity of $f^{(m)}(x)$, the summability of $p(x)$, and the finiteness of $[a, b]$ allow us to change the order of this double integral. This allows us to construct a representation for $R(f)$ which will be useful for analyzing the remainder and especially for selecting its "principal part":

$$R(f) = \int_a^b f^{(m)}(t) K(t) \, dt \tag{11.3.4}$$

where the kernel $K(t)$ is given by

$$K(t) = \int_t^b p(x) \frac{(x-t)^{m-1}}{(m-1)!} \, dx - \sum_{k=1}^n A_k E(x_k - t) \frac{(x_k - t)^{m-1}}{(m-1)!}. \tag{11.3.5}$$

When $K(t)$ is a "slowly varying" function the part of $K(t)$ which most influences the numerical value of $R(f)$ is the average value of the kernel. The principle part of $R(f)$ can then be separated by writing

$$K(t) = C_0 + [K(t) - C_0] \qquad \text{where } C_0 = (b-a)^{-1} \int_a^b K(t) \, dt.$$

Then

$$R(f) = C_0 \int_a^b f^{(m)}(t) \, dt + \int_a^b f^{(m)} [K(t) - C_0] \, dt =$$

$$= C_0 [f^{(m-1)}(b) - f^{(m-1)}(a)] + \int_a^b f^{(m+1)}(t) L_1(t) \, dt$$

where

$$L_1(t) = \int_a^t [C_0 - K(x)]\, dx.$$

If the new kernel $L_1(t)$ is again a "slowly varying" function we can again separate the principle part from the integral

$$\int_a^b f^{(m+1)}(t) L_1(t)\, dt$$

and so forth.

After performing this operation s times the original quadrature formula (11.3.3) will be transformed into an equation of Euler's form which can be used to increase the accuracy of (11.3.3) provided that the functions $L_0 = K,\ L_1,\ L_2,\ \ldots$ do not have large variation:

$$\int_a^b p(x) f(x)\, dx = \sum_{k=1}^n A_k f(x_k) + C_0 [f^{(m-1)}(b) - f^{(m-1)}(a)] + \cdots +$$

$$+ C_{s-1} [f^{(m+s-2)}(b) - f^{(m+s-2)}(a)] + R_s(f) \qquad (11.3.6)$$

$$C_i = (b - a)^{-1} \int_a^b L_i(t)\, dt,$$

$$L_0(t) = K(t)$$

$$L_{i+1}(t) = \int_a^t [C_i - L_i(x)]\, dx \qquad (11.3.6^*)$$

$$R_s(f) = \int_a^b f^{(m+s)}(t) L_s(t)\, dt.$$

Equations (11.3.6*) give a method for sequentially calculating the C_i and $L_i(t)$. However, we can find a representation for C_i and $L_i(t)$ directly from the kernel $K(t)$. To do this we return to the initial quadrature formula (11.3.3) with the integral representation for the remainder

$$R(f) = \int_a^b f^{(m)}(t) K(t)\, dt.$$

Replacing $f^{(m)}(t)$ by its expansion in terms of Bernoulli polynomials

$$f^{(m)}(t) = (b-a)^{-1} \int_a^b f^{(m)}(x)\,dx +$$

$$+ \sum_{i=1}^{s-1} \frac{(b-a)^{i-1}}{i!} B_i\left(\frac{t-a}{b-a}\right)[f^{(m+i-1)}(b) - f^{(m+i-1)}(a)] -$$

$$- \frac{(b-a)^{s-1}}{s!} \int_a^b f^{(m+s)}(x)\left[B_s^*\left(\frac{t-x}{b-a}\right) - B_s^*\left(\frac{t-a}{b-a}\right)\right]dx$$

and integrating we obtain

$$\int_a^b p(x) f(x)\,dx = \sum_{k=1}^n A_k f(x_k) +$$

$$+ (b-a)^{-1} \int_a^b K(t)\,dt\,[f^{(m-1)}(b) - f^{(m-1)}(a)] +$$

$$+ \sum_{i=1}^{s-1} \frac{(b-a)^{i-1}}{i!} \int_a^b K(t) B_i\left(\frac{t-a}{b-a}\right)dt \times$$

$$\times [f^{(m+i-1)}(b) - f^{(m+i-1)}(a)] - \frac{(b-a)^{s-1}}{s!} \int_a^b K(t) \times$$

$$\times \int_a^b f^{(m+s)}(x)\left[B_s^*\left(\frac{t-x}{b-a}\right) - B_s^*\left(\frac{t-a}{b-a}\right)\right]dt\,dx$$

which must coincide with (11.3.6) for any function $f(x)$ which has a continuous derivative of order $m + s$ on $[a, b]$.

This can happen only when the coefficients of the terms $[f^{(m+i-1)}(b) - f^{(m+i-1)}(a)]$ are equal and when $R_s(f)$ in (11.3.6) coincides with the last term in the previous equation. Thus we have shown that

$$C_i = \frac{(b-a)^{i-1}}{i!} \int_a^b K(t) B_i\left(\frac{t-a}{b-a}\right) dt \qquad (11.3.7)$$

$$L_s(t) = -\frac{(b-a)^{s-1}}{s!} \int_a^b K(x)\left[B_s^*\left(\frac{x-t}{b-a}\right) - B_s^*\left(\frac{x-a}{b-a}\right)\right]dx. \qquad (11.3.8)$$

There is a simple interpretation for the C_i and $L_i(t)$. Comparing (11.3.7) with the integral representation for the remainder $R(f)$ given by (11.3.4) we see that C_i is the remainder when the quadrature formula is applied to a function which has for its mth derivative $\dfrac{(b-a)^{i-1}}{i!} B_i\left(\dfrac{t-a}{b-a}\right)$.

Recalling the rule (1.2.6) for differentiating a Bernoulli polynomial we see that the polynomial

$$\frac{(b-a)^{m+i-1}}{(m+i)!} B_{m+i}\left(\frac{t-a}{b-a}\right)$$

has this property. Thus

$$C_i = \frac{(b-a)^{m+i-1}}{(m+i)!} R\left[B_{m+i}\left(\frac{t-a}{b-a}\right)\right] =$$

$$= \frac{(b-a)^{m+i-1}}{(m+i)!} \left\{\int_a^b p(t) B_{m+i}\left(\frac{t-a}{b-a}\right) dt - \right.$$

$$\left. - \sum_{k=1}^n A_k B_{m+i}\left(\frac{x_k-a}{b-a}\right) \right\}. \qquad (11.3.9)$$

This equation provides a simple method for calculating the C_i.

Similarly we obtain for $L_s(t)$ the expression

$$L_s(t) = -\frac{(b-a)^{m+s-1}}{(m+s)!} R_x\left[B_{m+s}^*\left(\frac{x-t}{b-a}\right) - \right.$$

$$\left. - B_{m+s}^*\left(\frac{x-a}{b-a}\right)\right] \qquad (11.3.10)$$

where R_x indicates the remainder when the quadrature formula is applied with respect to the variable x.

Now we construct some special cases of Euler's formula. We begin by obtaining the Euler-Maclaurin[1] formula for increasing the accuracy of the trapezoidal rule.

Consider the simple trapezoidal formula

$$\int_a^b f(x)dx = \frac{b-a}{2}[f(a)+f(b)] + R(f) \qquad (11.3.11)$$

which is exact for linear polynomials and for which we must take $m = 2$. To construct (11.3.6) we must first compute the coefficients C_i. The easiest method in this case is to use (11.3.9).

---

[1]For other Euler-Maclaurin formulas see J. F. Steffensen, *Interpolation*, Chap. 13.

The polynomials $B_n(x)$, $n = 2, 3, \ldots$, have the property that $B_n(0) = B_n(1)$ so that

$$\int_a^b B_{i+2}\left(\frac{t-a}{b-a}\right) dt = \frac{b-a}{i+3}\left[B_{i+3}(1) - B_{i+3}(0)\right] = 0$$

$$C_i = -\frac{(b-a)^{i+2}}{2(i+2)!}\left[B_{i+2}(0) + B_{i+2}(1)\right] =$$

$$= -\frac{(b-a)^{i+2}}{(i+2)!}\left[\frac{1+(-1)^{i+2}}{2}\right]B_{i+2}.$$

All the odd order Bernoulli numbers, except B_1, are zero so that $C_1 = C_3 = C_5 = \cdots = 0$. The coefficients C_i with even subscript $i = 2j$ are

$$C_{2j} = -\frac{(b-a)^{2j+2}}{(2j+2)!} B_{2j+2}. \tag{11.3.12}$$

The first few C_{2j} are:

$$C_0 = -\frac{(b-a)^2}{12}, \qquad C_2 = \frac{(b-a)^4}{720}, \qquad C_4 = -\frac{(b-a)^6}{30240},$$

$$C_6 = \frac{(b-a)^8}{1209600}, \qquad C_8 = -\frac{(b-a)^{10}}{47900160}.$$

To construct the remainder $R_s(f)$ in (11.3.6) we will calculate the kernel $L_s(t)$ from (11.3.10):

$$L_s(t) = -\frac{(b-a)^{s+1}}{(s+2)!} \left\{ \int_a^b \left[B_{s+2}^*\left(\frac{x-t}{b-a}\right) - B_{s+2}^*\left(\frac{x-a}{b-a}\right)\right] dx - \right.$$

$$- \frac{b-a}{2}\left[B_{s+2}^*\left(\frac{a-t}{b-a}\right) - B_{s+2}^*(0) + \right.$$

$$+ B_{s+2}^*\left(\frac{b-t}{b-a}\right) - B_{s+2}^*(1)\Bigg]\Bigg\}.$$

The period of $B_{s+2}^*\left(\dfrac{x-t}{b-a}\right)$ is $b-a$ so that

$$\int_a^b B_{s+2}^*\left(\frac{x-t}{b-a}\right) dx = \int_a^b B_{s+2}^*\left(\frac{x-a}{b-a}\right) dx$$

and the integral in the expression for $L_s(t)$ is zero. Also

$$B^*_{s+2}\left(\frac{a-t}{b-a}\right) = B^*_{s+2}\left(\frac{b-t}{b-a}\right), \qquad B^*_{s+2}(0) = B^*_{s+2}(1) = B_{s+2}$$

and defining $y^*_k(x) = B^*_k(x) - B_k$ we obtain

$$L_s(t) = \frac{(b-a)^{s+2}}{(s+2)!}\left[B^*_{s+2}\left(\frac{b-t}{b-a}\right) - B_{s+2}\right] =$$

$$= \frac{(b-a)^{s+2}}{(s+2)!} y^*_{s+2}\left(\frac{b-t}{b-a}\right).$$

We can now write equation (11.3.6) for the trapezoidal rule. Since the C_i are zero for all odd i we have

$$\int_a^b f(x)dx = \frac{(b-a)}{2}[f(a) + f(b)] -$$

(11.3.13)

$$- \sum_{k=1}^{\nu-1} \frac{(b-a)^{2k}}{(2k)!} B_{2k}[f^{(2k-1)}(b) - f^{(2k-1)}(a)] + \rho_{2\nu}(f)$$

where the remainder $\rho_{2\nu}(f)$ is either

$$\rho_{2\nu}(f) = \frac{(b-a)^{2\nu-1}}{(2\nu-1)!} \int_a^b f^{(2\nu-1)}(t) y^*_{2\nu-1}\left(\frac{b-t}{b-a}\right)dt$$

or

(11.3.14)

$$\rho_{2\nu}(f) = \frac{(b-a)^{2\nu}}{(2\nu)!} \int_a^b f^{(2\nu)}(t) y^*_{2\nu}\left(\frac{b-t}{b-a}\right)dt$$

depending on whether $f(t)$ has a continuous derivative of order $2\nu - 1$ or 2ν.

In the following discussion we assume that $f(t)$ has a continuous derivative of order 2ν so that $\rho_{2\nu}(f)$ satisfies the second equation of (11.3.14) and will transform this equation into a somewhat simpler form. We make the transformation $t = a + (b - a)u$, $0 \le u \le 1$. Using the relationships

$$y^*_{2\nu}\left(\frac{b-t}{b-a}\right) = y^*_{2\nu}(1-u) = B^*_{2\nu}(1-u) - B_{2\nu} =$$

$$= B_{2\nu}(u) - B_{2\nu} = y_{2\nu}(u)$$

we obtain

$$\rho_{2\nu}(f) = \frac{(b-a)^{2\nu+1}}{(2\nu)!} \int_0^1 f^{(2\nu)}(a + (b-a)u) y_{2\nu}(u)du \qquad (11.3.15)$$

In order to obtain an equation for increasing the precision of the repeated trapezoidal formula (6.3.4) we divide the segment $[a, b]$ into any number n of equal parts of length $h = \dfrac{b - a}{n}$ and apply (11.3.13) to the subsegment $[a + ph, a + (p + 1)h]$:

$$\int_{a+ph}^{a+(p+1)h} f(x)dx = \frac{h}{2}\{f[a + ph] + f[a + (p + 1)h]\} -$$

$$- \sum_{k=1}^{\nu-1} \frac{h^{2k}}{(2k)!} B_{2k}\{f^{(2k-1)}[a + (p + 1)h] -$$

$$- f^{(2k-1)}[a + ph]\} + \rho_{2\nu}^{(p)}(f)$$

$$\rho_{2\nu}^{(p)}(f) = \frac{h^{2\nu+1}}{(2\nu)!} \int_0^1 f^{(2\nu)}(a + ph + hu)y_{2\nu}(u)du.$$

By adding these equations for $p = 0, 1, \ldots, n - 1$ we obtain

$$\int_a^b f(x)dx = T_n - \sum_{k=1}^{\nu-1} \frac{h^{2k}}{(2k)!} B_{2k}[f^{(2k-1)}(b) - f^{(2k-1)}(a)] + \rho_{2\nu}(f) =$$

$$= T_n - \frac{h^2}{12}[f'(b) - f'(a)] + \frac{h^4}{720}[f^{(3)}(b) - f^{(3)}(a)] -$$

$$- \frac{h^6}{30240}[f^{(5)}(b) - f^{(5)}(a)] + \tag{11.3.16}$$

$$+ \frac{h^8}{1209600}[f^{(7)}(b) - f^{(7)}(a)] -$$

$$- \frac{h^{10}}{47900160}[f^{(9)}(b) - f^{(9)}(a)] + \cdots + \rho_{2\nu}(f)$$

where

$$T_n = h\left[\frac{1}{2}f(a) + f(a + h) + \cdots + f(a + (n - 1)h) + \frac{1}{2}f(b)\right]$$

and

$$\rho_{2\nu}(f) = \frac{h^{2\nu+1}}{(2\nu)!} \int_0^1 y_{2\nu}(u) \sum_{p=0}^{n-1} f^{(2\nu)}(a + ph + hu)du =$$

$$= \frac{h^{2\nu+1}}{(2\nu)!} \int_0^1 [B_{2\nu}(u) - B_{2\nu}] \sum_{p=0}^{n-1} f^{(2\nu)}(a + ph + hu)du.$$

Equation (11.3.16) is the well known Euler-Maclaurin sum formula relating the integral $\int_a^b f(x)dx$ to the sum of integrand values at equally spaced points:

$$S_n = f(a) + f(a + h) + \cdots + f(b) = h^{-1} T_n + \frac{1}{2}[f(a) + f(b)].$$

From (11.3.16) we can calculate S_n if we know the value of the integral or the value of the integral in terms of S_n. We are only interested in this second application.

If ν increases without bound then the terms in the summation in (11.3.16) become the infinite series

$$\sum_{k=1}^{\infty} \frac{h^{2k}}{(2k)!} B_{2k} [f^{(2k-1)}(b) - f^{(2k-1)}(a)] \qquad (11.3.17)$$

We recall that, for large integers k, the Bernoulli numbers B_{2k} grow very rapidly and are approximately equal to

$$B_{2k} \approx 2(-1)^{k-1}(2k)!\,(2\pi)^{-2k}$$

Therefore the series (11.3.17) converges for only a very small subclass of the functions we have been considering. In spite of this shortcoming the Euler-Maclaurin formulas are often used because for the first few values of ν the remainder decreases and the first few corrections applied to T_n significantly increase the accuracy of the trapezoidal formula.

We now prove three simple theorems about the remainder $\rho_{2\nu}(f)$.

Theorem 1. *If $f^{(2\nu)}(x)$ is continuous on $[a, b]$ then there exists a point $\xi \in [a, b]$ for which*

$$\rho_{2\nu}(f) = -\frac{nh^{2\nu+1}}{(2\nu)!} B_{2\nu} f^{(2\nu)}(\xi). \qquad (11.3.18)$$

Proof. In Section 1.2 we showed that the function $y_{2\nu}(u)$ does not change sign on the interval $[0, 1]$:

$$(-1)^\nu y_{2\nu}(u) > 0 \qquad \text{for} \qquad 0 < u < 1.$$

Consider the integral

$$I_\nu = \int_0^1 (-1)^\nu y_{2\nu}(u) \sum_{p=0}^{n-1} f^{(2\nu)}(a + ph + hu)\, du.$$

Let m and M denote the smallest and greatest values of $f^{(2\nu)}(x)$ on

$[a, b]$. Then it is clear that

$$(-1)^\nu nm \int_0^1 y_{2\nu}(u)\,du \le I_\nu \le (-1)^\nu nM \int_0^1 y_{2\nu}(u)\,du$$

and since

$$\int_0^1 y_{2\nu}(u)\,du = \int_0^1 [B_{2\nu}(u) - B_{2\nu}]\,du =$$

$$= \frac{1}{2\nu + 1}[B_{2\nu+1}(1) - B_{2\nu+1}(0)] - B_{2\nu} = -B_{2\nu}$$

then

$$I_\nu = (-1)^{\nu+1} nP B_{2\nu}$$

where $m \le P \le M$. From the continuity of $f^{(2\nu)}(x)$ there must be a point $\xi \in [a, b]$ for which $f^{(2\nu)}(\xi) = P$ so that

$$I_\nu = (-1)^{\nu+1} nB_{2\nu} f^{(2\nu)}(\xi).$$

Since $\rho_{2\nu}(f) = (-1)^\nu \dfrac{h^{2\nu+1}}{(2\nu)!} I_\nu$ the theorem is proved.

Theorem 2. *If $f^{(2\nu)}(x)$ is continuous and does not change sign on $[a, b]$ then $\rho_{2\nu}(f)$ can be written in the form*

$$\rho_{2\nu}(f) = -\theta(2 - 2^{-2\nu+1})\frac{h^{2\nu}B_{2\nu}}{(2\nu)!}[f^{(2\nu-1)}(b) - f^{(2\nu-1)}(a)]$$

$$0 < \theta < 1. \qquad (11.3.19)$$

Proof. In Section 1.2 we showed that $(-1)^k y_{2k}(x)$ does not change sign on $[0, 1]$ and that this function increases for $0 \le x \le \dfrac{1}{2}$ and decreases for $\dfrac{1}{2} \le x \le 1$ with its largest value at $x = \dfrac{1}{2}$ where

$$(-1)^k y_{2k}\left(\frac{1}{2}\right) = -(-1)^k(2 - 2^{-2k+1})B_{2k}.$$

Therefore $\rho_{2\nu}(f)$ has the same sign as

$$y_{2\nu}\left(\frac{1}{2}\right)\frac{h^{2\nu+1}}{(2\nu)!}\int_0^1 \sum_{p=0}^{n-1} f^{(2\nu)}(a + ph + hu)\,du$$

and, in absolute value, is less than this quantity. Therefore

$$\rho_{2\nu}(f) = \theta y_{2\nu}\left(\frac{1}{2}\right)\frac{h^{2\nu+1}}{(2\nu)!} \int_0^1 \sum_{p=0}^{n-1} f^{(2\nu)}(a + ph + hu)du, \qquad 0 < \theta < 1.$$

But

$$y_{2k}\left(\frac{1}{2}\right) = -(2 - 2^{-2k+1})B_{2k}$$

and

$$\int_0^1 \sum_{p=0}^{n-1} f^{(2\nu)}(a + ph + hu)du =$$

$$= h^{-1} \sum_{p=0}^{n-1} \{f^{(2\nu-1)}[a + h(p + 1)] - f^{(2\nu-1)}[a + hp]\} =$$

$$= h^{-1}[f^{(2\nu-1)}(b) - f^{(2\nu-1)}(a)]$$

which then establishes (11.3.19).

From this second theorem we see, provided $f^{(2\nu)}(x)$ satisfies the necessary assumptions, that $\rho_{2\nu}(f)$ has the same sign as the first neglected term in (11.3.16) and is smaller, in absolute value, than twice this term.

It turns out that under certain assumptions on $f(x)$ the remainder $\rho_{2\nu}(f)$ in the Euler-Maclaurin formula (11.3.16) has an estimate similar to the estimate for the partial sum of an alternating series.

Theorem 3. *If $f(x)$ has a continuous derivative of order $2\nu + 2$ on $[a, b]$ and for each $x \in [a, b]$ either*

$$f^{(2\nu)}(x) \geq 0 \qquad and \qquad f^{(2\nu+2)}(x) \geq 0$$

or

$$f^{(2\nu)}(x) \leq 0 \qquad and \qquad f^{(2\nu+2)}(x) \leq 0$$

then $\rho_{2\nu}(f)$ has the same sign as

$$-\frac{h^{2\nu}B_{2\nu}}{(2\nu)!}[f^{(2\nu-1)}(b) - f^{(2\nu-1)}(a)]$$

and is less, in absolute value, than this term.

Proof. The remainders $\rho_{2\nu}(f)$ and $\rho_{2\nu+2}(f)$ satisfy the relationship

$$\rho_{2\nu}(f) = -\frac{h^{2\nu}B_{2\nu}}{(2\nu)!}[f^{(2\nu-1)}(b) - f^{(2\nu-1)}(a)] + \rho_{2\nu+2}(f)$$

which can be written as

$$\frac{h^{2\nu+1}}{(2\nu)!} \int_0^1 y_{2\nu}(u) \sum_{p=0}^{n-1} f^{(2\nu)}(a + ph + hu)du +$$

$$+ \frac{h^{2\nu+3}}{(2\nu+2)!} \int_0^1 [-y_{2\nu+2}(u)] \sum_{p=0}^{n-1} f^{(2\nu+2)}(a + ph + hu) du =$$

$$= -\frac{h^{2\nu} B_{2\nu}}{(2\nu)!} [f^{(2\nu-1)}(b) - f^{(2\nu-1)}(a)].$$

In Section 1.2 we showed that $y_{2\nu}(u)$ and $-y_{2\nu+2}(u)$ have the same sign on [0, 1]. If $f^{(2\nu)}(x)$ and $f^{(2\nu+2)}(x)$ also have the same sign then both terms on the left side of the last equation must also have the same sign. Therefore each of these terms must have the same sign as the right side and can not be larger, in absolute value, than this term.

Let us apply the Euler-Maclaurin formula (11.3.16) to approximate the integral

$$\int_0^1 \frac{dx}{1 + x} = ln\ 2.$$

Here $a = 0$, $b = 1$ and we will divide [0, 1] into 10 equal parts so that $n = 10$, $h = 0.1$. In the formula we will use two terms in addition to T_n and thus $\nu = 3$:

$$\int_0^1 \frac{dx}{1 + x} \approx T_n - \frac{h^2}{12} [f'(1) - f'(0)] + \frac{h^4}{720} [f^{(3)}(1) - f^{(3)}(0)].$$

$$T_n = (0.1)\left[\left(\frac{1}{2}\right)\frac{1}{1} + \frac{1}{1.1} + \cdots + \frac{1}{1.9} + \left(\frac{1}{2}\right)\frac{1}{2}\right] = 0.693771403.$$

$$-\frac{h^2}{12} [f'(1) - f'(0)] = -\frac{1}{1200}\left[1 - \frac{1}{4}\right] = -0.000625.$$

$$\frac{h^4}{720} [f^{(3)}(1) - f^{(3)}(0)] = \frac{6 \times 10^{-4}}{720}\left[1 - \frac{1}{24}\right] = 0.000000781.$$

Therefore the approximate value of the integral is

$$0.693771403 - 0.000625 + 0.000000781 = 0.693147184.$$

We now estimate the error in this approximation. We note that the derivatives $f^{(6)}(x) = \frac{6!}{(1 + x)^7}$ and $f^{(8)}(x) = \frac{8!}{(1 + x)^9}$ are both positive throughout $[0, 1]$ and therefore we can apply Theorem 3. The first neglected term in the formula is

$$-\frac{h^6}{30240}[f^{(5)}(1) - f^{(5)}(0)] = -\frac{120 \times 10^{-6}}{30240}\left[1 - \frac{1}{64}\right] < -0.000000004.$$

Therefore

$$0.693147180 < \int_0^1 \frac{dx}{1+x} < 0.693147184.$$

We will now construct an Euler type formula for increasing the accuracy of the repeated Simpson's rule (6.3.10). To do this we first construct equation (11.3.6) for

$$\int_a^{a+2h} f(x)dx = \frac{h}{3}[f(a) + 4f(a+h) + f(a+2h)] + R(f). \qquad (11.3.20)$$

This formula is exact for all cubic polynomials and thus we must take $m = 4$.

We again use (11.3.9) to calculate the coefficients C_i:

$$C_i = \frac{(2h)^{i+3}}{(i+4)!}\left\{\int_a^{a+2h} B_{i+4}\left(\frac{t-a}{2h}\right)dt - \right.$$

$$\left. - \frac{h}{3}\left[B_{i+4}(0) + 4B_{i+4}\left(\frac{1}{2}\right) + B_{i+4}(1)\right]\right\} =$$

$$= \frac{(2h)^{i+4}}{3(i+4)!}(1 - 2^{-i-2})B_{i+4}.$$

Since $B_{2k+1} = 0$ $(k = 1, 2, \ldots)$ then only the C_i with even subscripts will be nonzero.

To find $R_s(f)$ we calculate $L_s(t)$:

$$L_s(t) = -\frac{(2h)^{s+3}}{(s+4)!}\left\{\int_a^{a+2h}\left[B_{s+4}^*\left(\frac{x-t}{2h}\right) - B_{s+4}\left(\frac{x-a}{2h}\right)\right]dx - \right.$$

$$- \frac{h}{3}\left[B_{s+4}^*\left(\frac{a-t}{2h}\right) - B_{s+4}(0) + \right.$$

$$+ 4\left[B_{s+4}^*\left(\frac{a+h-t}{2h}\right) - B_{s+4}\left(\frac{1}{2}\right)\right] +$$

$$\left. + B_{s+4}^*\left(1 + \frac{a-t}{2h}\right) - B_{s+4}(1)\right]\right\}.$$

The integral in this expression is zero and replacing the function $B^*_{s+4}(x)$ by the function $y^*_{s+4}(x) = B^*_{s+4}(x) - B_{s+4}$ we obtain

$$L_s(t) = \frac{(2h)^{s+4}}{3(s+4)!} \left\{ y_{s+4}\left(\frac{2h+a-t}{2h}\right) + \right.$$

$$\left. + 2\left[y^*_{s+4}\left(\frac{h+a-t}{2h}\right) - y_{s+4}\left(\frac{1}{2}\right)\right] \right\}.$$

Substituting $t = a + 2hu$ $(0 \le u \le 1)$ we can write the remainder $R_s(f)$ in the form

$$R_s(f) = \frac{(2h)^{s+5}}{3(s+4)!} \int_0^1 f^{(s+4)}(a+2hu) \times$$

$$\times \left\{ y_{s+4}(1-u) + 2\left[y^*_{s+4}\left(\frac{1}{2}-u\right) - y_{s+4}\left(\frac{1}{2}\right)\right] \right\} du.$$

Thus Simpson's formula (11.3.20) can be written as

$$\int_a^{a+2h} f(x)dx = \frac{h}{3}[f(a) + 4f(a+h) + f(a+2h)] +$$

$$+ \sum_{k=2}^{\nu-1} \frac{(2h)^{2k}}{3(2k)!}(1 - 2^{-2k+2})B_{2k} \times$$

$$\times [f^{(2k-1)}(a+2h) - f^{(2k-1)}(a)] + \rho_{2\nu}(f). \qquad (11.3.20^*)$$

where the remainder $\rho_{2\nu}(f)$ can be written as either

$$\rho_{2\nu}(f) = \frac{(2h)^{2\nu}}{3(2\nu-1)!} \int_0^1 f^{(2\nu-1)}(a+2hu) \times$$

$$\times \left\{ y_{2\nu-1}(1-u) + 2y^*_{2\nu-1}\left(\frac{1}{2}-u\right) \right\} du.$$

or

$$\rho_{2\nu}(f) = \frac{(2h)^{2\nu+1}}{3(2\nu)!} \int_0^1 f^{(2\nu)}(a+2hu) \times$$

$$\times \left\{ y_{2\nu}(u) + 2\left[y^*_{2\nu}\left(\frac{1}{2}-u\right) - y_{2\nu}\left(\frac{1}{2}\right)\right] \right\} du.$$

depending on whether $f(x)$ has a continuous derivative of order $2\nu - 1$ or 2ν. Below we will assume that $f^{(2\nu)}(x)$ exists and is continuous and will, therefore, use the second expression for the remainder.

Let us now consider the repeated Simpson's rule (6.3.10). We divide the segment of integration $[a, b]$ into an even number n of equal sub-segments and apply (11.3.20*) to the double segment $[a + 2hp, a + 2h(p + 1)]$. Writing these equations for $p = 0, 1, \ldots, \frac{n}{2} - 1$ and adding we obtain

$$\int_a^b f(x)dx = U_n + \sum_{k=2}^{\nu-1} \frac{(2h)^{2k}}{3(2k)!} (1 - 2^{-2k+2}) B_{2k} \times$$

$$\times [f^{(2k-1)}(b) - f^{(2k-1)}(a)] + \rho_{2\nu}(f) =$$

$$= U_n - \frac{h^4}{180} [f^{(3)}(b) - f^{(3)}(a)] + \frac{h^6}{1512} [f^{(5)}(b) - f^{(5)}(a)] -$$

$$- \frac{h^8}{14400} [f^{(7)}(b) - f^{(7)}(a)] + \cdots + \rho_{2\nu}(f) \quad (11.3.21)$$

where

$$U_n = \frac{h}{3} \{f(a) + f(b) + 2[f(a + 2h) + \cdots + f(a + (n - 2)h)] +$$

$$+ 4[f(a + h) + \cdots + f(a + (n - 1)h)]\}$$

$$\rho_{2\nu}(f) = \frac{(2h)^{2\nu+1}}{3(2\nu)!} \int_0^1 \left\{ y_{2\nu}(u) + 2\left[y_{2\nu}^*\left(\frac{1}{2} - u\right) - y_{2\nu}\left(\frac{1}{2}\right)\right] \right\} \times$$

$$\times \sum_{p=0}^{\frac{n}{2}-1} f^{(2\nu)}(a + 2ph + 2hu)du. \quad (11.3.22)$$

In order to study the remainder $\rho_{2\nu}(f)$ it will be necessary to investigate the kernel

$$F(u) = y_{2\nu}(u) + 2\left[y_{2\nu}^*\left(\frac{1}{2} - u\right) - y_{2\nu}\left(\frac{1}{2}\right)\right].$$

To do this we need the following Lemma.

Lemma. *For each $\nu \geq 1$ the function*

$$\phi_{2\nu+1}(x) = y_{2\nu+1}(x) - 2y_{2\nu+1}\left(\frac{1}{2} - x\right)$$

has no zeros inside the segment $\left[0, \frac{1}{2}\right]$ and the sign of this function is *given by*

$$(-1)^{\nu}\phi_{2\nu+1}(x) > 0, \qquad 0 < x < \frac{1}{2}.$$

Proof. Assume $\nu \geq 1$. Since $y_{2\nu+1}(0) = y_{2\nu+1}\left(\frac{1}{2}\right) = 0$ then it is clear that $x = 0$ and $x = \frac{1}{2}$ are zeros of $\phi_{2\nu+1}(x)$. Let us suppose that the point $a \left(0 < a < \frac{1}{2}\right)$ was also a zero of $\phi_{2\nu+1}(x)$. Then inside each of the segments $[0, a]$ and $\left[a, \frac{1}{2}\right]$ the derivative $\phi'_{2\nu+1}(x)$ will have at least one zero. Therefore the second derivative $\phi''_{2\nu+1}(x)$ will have at least one zero inside $\left[0, \frac{1}{2}\right]$. But

$$\phi''_{2\nu+1}(x) = (2\nu + 1)(2\nu)\left[y_{2\nu-1}(x) - 2y_{2\nu-1}\left(\frac{1}{2} - x\right)\right] =$$
$$= (2\nu + 1)(2\nu)\phi_{2\nu-1}(x).$$

Thus, from the assumption that $\phi_{2\nu+1}(x)$ has a zero inside $\left[0, \frac{1}{2}\right]$ it follows that $\phi_{2\nu-1}(x)$ also has a zero inside this segment. From this it follows that $\phi_3(x)$ would have a zero inside $\left[0, \frac{1}{2}\right]$. However, we can easily verify that $\phi_3(x) = 3x^2\left(x - \frac{1}{2}\right)$ and this function has no zeros inside $\left[0, \frac{1}{2}\right]$. To determine the sign of $\phi_{2\nu+1}(x)$ it is sufficient to determine the sign of $\phi_{2\nu+1}\left(\frac{1}{4}\right)$:

$$\phi_{2\nu+1}\left(\frac{1}{4}\right) = -y_{2\nu+1}\left(\frac{1}{4}\right)$$

and in Section 1.2 we showed that

$$(-1)^{\nu+1}y_{2\nu+1}(x) > 0 \qquad \text{for} \qquad 0 < x < \frac{1}{2}.$$

Therefore

$$(-1)^{\nu}\phi_{2\nu+1}(x) > 0, \qquad 0 < x < \frac{1}{2}.$$

This proves the lemma.

Let us now consider the function $(-1)^{\nu-1} F(u)$ for $0 \le u \le \dfrac{1}{2}$. We have

$$(-1)^{\nu-1} F'(u) = 2\nu(-1)^{\nu-1} \left[y_{2\nu-1}(u) - 2y_{2\nu-1}\left(\frac{1}{2} - u\right) \right] =$$

$$= 2\nu(-1)^{\nu-1} \phi_{2\nu-1}(u).$$

By the Lemma we see that $(-1)^{\nu-1} F'(u) > 0$ which means that $(-1)^{\nu-1} F(u)$ is a monotonically increasing function for $0 \le u \le \dfrac{1}{2}$. Since $F(0) = 0$ it follows that $(-1)^{\nu-1} F(u) > 0$ for $0 < u \le \dfrac{1}{2}$.

In order to see how $F(u)$ behaves on $\dfrac{1}{2} \le u \le 1$ it will be sufficient to show that $F(u)$ is symmetric with respect to $u = \dfrac{1}{2}$: $F(1 - u) = F(u)$. Indeed

$$F(1 - u) = y_{2\nu}(1 - u) + 2\left[y_{2\nu}^{*}\left(u - \frac{1}{2}\right) - y_{2\nu}\left(\frac{1}{2}\right) \right]$$

$$y_{2\nu}(1 - u) = y_{2\nu}(u)$$

$$y_{2\nu}^{*}\left(u - \frac{1}{2}\right) = y_{2\nu}^{*}\left(u + \frac{1}{2}\right) = y_{2\nu}^{*}\left(\frac{1}{2} - u\right)$$

from which we see that $F(1 - u) = F(u)$.

Thus it follows that $(-1)^{\nu-1} F(u)$ is a positive monotonically decreasing function on $\dfrac{1}{2} \le u < 1$ and that this function has a relative maximum at $u = \dfrac{1}{2}$:

$$\max_{[0,\,1]} (-1)^{\nu-1} F(u) = -(-1)^{\nu-1} y_{2\nu}\left(\frac{1}{2}\right) =$$

$$= (-1)^{\nu-1} 2(1 - 2^{-2\nu}) B_{2\nu} \qquad (11.3.23)$$

These properties of the kernel $F(u)$ permit us to prove three theorems about the remainder $\rho_{2\nu}(f)$ of (11.3.22) analogous to the theorems about the remainder of (11.3.16).

We omit the proofs of these theorems because the proofs exactly follow the proofs of the preceding theorems.

Theorem 4. *If $f(x)$ has a continuous derivative of order 2ν on $[a, b]$ then there exists a point $\xi \in [a, b]$ for which the remainder of (11.3.21) satisfies*

$$\rho_{2\nu}(f) = \frac{nh(2h)^{2\nu}}{3(2\nu)!}(1 - 2^{-2\nu+2})B_{2\nu}f^{(2\nu)}(\xi). \qquad (11.3.24)$$

Comparing this representation for the remainder of (11.3.21) with the representation (11.3.18) for the remainder of the Euler-Maclaurin formula (11.3.16) we see that if $f^{(2\nu)}(x)$ does not change sign on $[a, b]$ then the remainders of these two quadrature formulas have opposite signs. Hence we have the following useful rule:

If the derivative $f^{(2\nu)}(x)$ does not change sign on $[a, b]$ then the exact value of $\displaystyle\int_a^b f(x)dx$ lies between the approximate values obtained from (11.3.16) and (11.3.21) by neglecting the remainder terms $\rho_{2\nu}(f)$ in these equations.

Theorem 5. *If the derivative $f^{(2\nu)}(x)$ does not change sign on $[a, b]$ then the remainder $\rho_{2\nu}(f)$ of formula (11.3.21) can be written in the form*

$$\rho_{2\nu}(f) = 2\theta\frac{(2h)^{2\nu}}{3(2\nu)!}(1 - 2^{-2\nu})B_{2\nu}[f^{(2\nu-1)}(b) - f^{(2\nu-1)}(a)]$$

$$0 < \theta < 1. \qquad (11.3.25)$$

From (11.3.25) we see that the remainder $\rho_{2\nu}(f)$ has the same sign as the first neglected term of (11.3.21).

Theorem 6. *If the function $f(x)$ has a continuous derivative of order $2\nu + 2$ and for all x in $[a, b]$ either*

$$f^{(2\nu)}(x) \geq 0 \qquad and \qquad f^{(2\nu+2)}(x) \geq 0$$

or

$$f^{(2\nu)}(x) \leq 0 \qquad and \qquad f^{(2\nu+2)}(x) \leq 0$$

then the remainder $\rho_{2\nu}(f)$ of (11.3.21) has the same sign as the first neglected term

$$\frac{(2h)^{2\nu}}{3(2\nu)!}(1 - 2^{-2\nu+2})B_{2\nu}[f^{(2\nu-1)}(b) - f^{(2\nu-1)}(a)]$$

and is not greater, in absolute value, than this term.

We now give the series in (11.3.6) for increasing the precision of certain other special quadrature formulas.

1. The Newton-Cotes formula with 4 nodes:

$$\int_a^{a+3h} f(x)\,dx \approx \frac{3h}{8}\left[f(a) + 3f(a+h) + 3f(a+2h) + f(a+3h)\right] +$$

$$+ \sum_{k=2} \frac{(3h)^{2k}}{8(2k)!}(1 - 3^{-2k+2})\,B_{2k}\,[f^{(2k-1)}(a+3h) - f^{(2k-1)}(a)] =$$

$$= \frac{3h}{8}\left[f(a) + 3f(a+h) + 3f(a+2h) + f(a+3h)\right] -$$

$$- \frac{h^4}{80}\,[f^{(3)}(a+3h) - f^{(3)}(a)] + \frac{h^6}{336}\,[f^{(5)}(a+3h) - f^{(5)}(a)] -$$

$$- \frac{13h^8}{19200}\,[f^{(7)}(a+3h) - f^{(7)}(a)] + \dots . \tag{11.3.26}$$

2. The quadrature formula of the highest degree of precision $2n - 1$ for the segment $[-1, 1]$ and the weight function $(1-x)^\alpha(1+x)^\beta(\alpha, \beta > -1)$; the nodes are the zeros of the Jacobi polynomial $P_n^{(\alpha,\beta)}(x)$:

$$\int_{-1}^1 (1-x)^\alpha(1+x)^\beta f(x)\,dx \approx \sum_{k=1}^n A_k f(x_k) +$$

$$+ C_0[f^{(2n-1)}(1) - f^{(2n-1)}(-1)] + C_1[f^{(2n)}(1) - f^{(2n)}(-1)] + \dots$$

$$C_0 = \frac{2^{\alpha+\beta+2n}\,n!\,\Gamma(\alpha+\beta+n+1)\,\Gamma(\alpha+n+1)\,\Gamma(\beta+n+1)}{(2n)!(\alpha+\beta+2n+1)\,[\Gamma(\alpha+\beta+2n+1)]^2}$$

$$C_1 = \frac{\beta-\alpha}{\alpha+\beta+2n}\left[\frac{\alpha+\beta}{\alpha+\beta+2n+2} + 2n\right] \times$$

$$\times \frac{n!\,2^{\alpha+\beta+2n}\,\Gamma(\alpha+\beta+n+1)\,\Gamma(\alpha+n+1)\,\Gamma(\beta+n+1)}{(2n+1)!\,\Gamma(\alpha+\beta+2n+1)\,\Gamma(\alpha+\beta+2n+2)}.$$

For the special ultraspherical case, $\alpha = \beta$, the C_i with odd subscripts are zero:

$$C_0 = \frac{2^{2\alpha}n!\,\Gamma(2\alpha+n+1)}{(2n)!(2\alpha+2n+1)}\left[\frac{2^n\,\Gamma(\alpha+n+1)}{\Gamma(2\alpha+2n+1)}\right]^2$$

$$C_2 = \frac{2^{2\alpha}n!\,\Gamma(2\alpha+n+1)}{(2n+2)!}\left[\frac{2^n\,\Gamma(\alpha+n+1)}{\Gamma(2\alpha+2n+1)}\right]^2 \times$$

$$\times \left[\frac{2n^2 + 2(2\alpha+1)n + 2\alpha - 1}{(2\alpha+2n-1)(2\alpha+2n+1)(2\alpha+2n+3)} + \frac{n(n-1)}{(2\alpha+2n-1)(2\alpha+2n+1)} -\right.$$

$$\left. - \frac{(n+1)(2n+1)}{3(2\alpha+2n+1)}\right].$$

3. For the Gauss formulas $\alpha = \beta = 0$ and the nodes are the zeros of the nth degree Legendre polynomial:

$$\int_{-1}^{1} f(x)dx \approx \sum_{k=1}^{n} A_k f(x_k) + \frac{1}{(2n+1)!}\left[\frac{2^n (n!)^2}{(2n)!}\right]^2 \times$$

$$\times [f^{(2n-1)}(1) - f^{(2n-1)}(-1)] +$$

$$+ \frac{1}{(2n+2)!}\left[\frac{2^n (n!)^2}{(2n)!}\right]^2 \left[\frac{-n(4n^2+5n-2)}{3(2n-1)(2n+3)}\right] \times$$

$$\times [f^{(2n+1)}(1) - f^{(2n+1)}(-1)] + \cdots.$$

To apply formulas of Euler's type it is necessary to find the values of the derivative of the integrand at the ends of the segment $[a, b]$ and in many cases this may be difficult to do. We can construct other formulas for increasing the precision of quadrature formulas in which the corrective terms are expressed in terms of differences or values of the integrand in place of its derivatives. There can be a wide variety of such formulas since the derivatives can be replaced by finite differences in many different ways. As an illustration we show one example of how this may be done.

Suppose we wish to calculate derivatives of $f(x)$ at the point a. To do this we interpolate on $f(x)$ using its values at certain points. The form of the interpolating polynomial depends on the choice of the nodes and we will assume that we can only use the values of $f(x)$ at the equally spaced points $a + kh$ $(k = 0, 1, \ldots)$ which belong to the segment $[a, b]$.

We will use Newton's representation of the interpolating polynomial with the nodes $x_0 = a$, $x_1 = a + h$, $x_2 = a + 2h$, $\ldots$:

$$f(x) = f(a) + (x - a)f(a, a + h) +$$

$$+ (x - a)(x - a - h)f(a, a + h, a + 2h) + \cdots + r(x)$$

where $r(x)$ is the remainder of the interpolation. For equally spaced nodes the divided differences are easily expressed in terms of differences:

$$f(a, a + h, \ldots, a + kh) = h^k k! \Delta^k f_0, \qquad f_0 = f(a).$$

Making the change of variable $x = a + ht$ we obtain the well known Newton-Gregory formula which is useful for interpolating near the beginning of a table

$$f(a + ht) = f_0 + \frac{t}{1!}\Delta f_0 + \frac{t(t-1)}{2!}\Delta^2 f_0 +$$

$$+ \frac{t(t-1)(t-2)}{3!}\Delta^3 f_0 + \cdots.$$

Taking derivatives of both sides of this equation and setting $t = 0$ gives

$$hf'(a) = \Delta f_0 - \frac{1}{2}\Delta^2 f_0 + \frac{1}{3}\Delta^3 f_0 - \frac{1}{4}\Delta^4 f_0 + \frac{1}{5}\Delta^5 f_0 - \cdots$$

$$h^2 f''(a) = \Delta^2 f_0 - \Delta^3 f_0 + \frac{11}{12}\Delta^4 f_0 - \frac{5}{6}\Delta^5 f_0 + \cdots$$

$$h^3 f^{(3)}(a) = \Delta^3 f_0 - \frac{3}{2}\Delta^4 f_0 + \frac{7}{4}\Delta^5 f_0 - \cdots$$

$$h^4 f^{(4)}(a) = \Delta^4 f_0 - 2\Delta^5 f_0 + \cdots$$

$$h^5 f^{(5)}(a) = \Delta^5 f_0 - \cdots$$

. . . .

In a similar way we can find the values of the derivatives $f^{(k)}(b)$ $(k = 1, 2, \ldots)$ by interpolating on $f(x)$ close to $b = a + nh$. We use the same representation for the interpolating polynomial and set $x_0 = a + nh$, $x_1 = a + (n - 1)h$, $x_2 = a + (n - 2)h$, $\ldots$:

$$f(x) = f(a + nh) + (x - a - nh)f(a + nh, a + (n - 1)h) +$$
$$+ (x - a - nh)(x - a - (n - 1)h)f(a + nh, a + (n - 1)h, a + (n - 2)h) +$$
$$+ \cdots + r(x)$$

or

$$f(a + th) = f_n + \frac{t}{1!}\Delta f_{n-1} + \frac{t(t + 1)}{2!}\Delta^2 f_{n-2} +$$
$$+ \frac{t(t + 1)(t + 2)}{3!}\Delta^3 f_{n-3} + \cdots.$$

Thus we obtain

$$hf'(b) = \Delta f_{n-1} + \frac{1}{2}\Delta^2 f_{n-2} + \frac{1}{3}\Delta^3 f_{n-3} + \frac{1}{4}\Delta^4 f_{n-4} + \frac{1}{5}\Delta^5 f_{n-5} + \cdots$$

$$h^2 f''(b) = \Delta^2 f_{n-2} + \Delta^3 f_{n-3} + \frac{11}{12}\Delta^4 f_{n-4} + \frac{5}{6}\Delta^5 f_{n-5} + \cdots$$

$$h^3 f^{(3)}(b) = \Delta^3 f_{n-3} + \frac{3}{2}\Delta^4 f_{n-4} + \frac{7}{4}\Delta^5 f_{n-5} + \cdots$$

$$h^4 f^{(4)}(b) = \Delta^4 f_{n-4} + 2\Delta^5 f_{n-5} + \cdots$$

$$h^5 f^{(5)}(b) = \Delta^5 f_{n-5} + \cdots$$

. . . .

Suppose we wish to replace the derivatives in the Euler-Maclaurin formula (11.3.16) by finite differences. If we substitute the above expressions for the derivatives we obtain the Gregory formula:

$$
\int_a^{a+nh} f(x)dx = T_n - \frac{h}{12}(\Delta f_{n-1} - \Delta f_0) - \frac{h}{24}(\Delta^2 f_{n-2} + \Delta^2 f_0) -
$$

$$
- \frac{19h}{720}(\Delta^3 f_{n-3} - \Delta^3 f_0) - \frac{3h}{160}(\Delta^4 f_{n-4} + \Delta^4 f_0) -
$$

$$
- \frac{863h}{60480}(\Delta^5 f_{n-5} - \Delta^5 f_0) - \frac{275h}{24192}(\Delta^6 f_{n-6} + \Delta^6 f_0) -
$$

$$
- \cdots - C_k h[\Delta^k f_{n-k} + (-1)^k \Delta^k f_0] + R_1(f) \qquad (11.3.27)
$$

where it can be shown that

$$
C_k = \frac{(-1)^k}{(k+1)!} \int_0^1 x(x-1) \cdots (x-k) \, dx.
$$

11.4. INCREASING THE PRECISION WHEN THE INTEGRAL REPRESENTATION OF THE REMAINDER CONTAINS A SHORT PRINCIPLE SUBINTERVAL

As in the preceding section we will consider a mechanical quadrature formula for an arbitrary weight function

$$
\int_a^b p(x)f(x)\,dx = \sum_{k=1}^n A_k f(x_k) + R(f). \qquad (11.4.1)
$$

If the algebraic degree of precision of (11.4.1) is $m-1$ and if $f(x)$ has a continuous derivative of order m on $[a, b]$ then, as shown in the preceding section, $R(f)$ can in many cases be written in the form

$$
R(f) = \int_a^b f^{(m)}(x)K(x)\,dx \qquad (11.4.2)
$$

where the kernel $K(x)$ is independent of $f(x)$.

Let us assume that in $[a, b]$ there exists a subinterval $[\alpha, \beta]$ outside of which $K(x)$ has a neglectably small value or that $K(x)$ rapidly becomes small away from $[\alpha, \beta]$. Then the value of the integral (11.4.2) will be mainly due to its value on $[\alpha, \beta]$. In addition let $f^{(m)}(x)$ have "small variation" on $[\alpha, \beta]$ or, what is essentially the same, let $[\alpha, \beta]$ have a relatively small length. In order to remove the principle part of the remainder $R(f)$ let us assume that it suffices to write $f^{(m)}(x)$ as the sum of two terms

$$f^{(m)}(x) = f^{(m)}(\alpha_0) + [f^{(m)}(x) - f^{(m)}(\alpha_0)]$$

where α_0 is a point in the subsegment $[\alpha, \beta]$:

$$R(f) = f^{(m)}(\alpha_0) \int_a^b K(x)\, dx + \int_a^b [f^{(m)}(x) - f^{(m)}(\alpha_0)] K(x)\, dx.$$

The choice of the point α_0 is still arbitrary. When the kernel does not change sign it is natural to take α_0 as the point of the x-axis about which $K(x)$ is concentrated:

$$\alpha_0 = \frac{\displaystyle\int_a^b x K(x)\, dx}{\displaystyle\int_a^b K(x)\, dx}.$$

Let us suppose that $f(x)$ has a continuous derivative of order $m + 2s$. We transform the last expression for $R(f)$ by expanding $f^{(m)}(x) - f^{(m)}(\alpha_0)$ in a Taylor series with two terms

$$f^{(m)}(x) - f^{(m)}(\alpha_0) = f^{(m+1)}(\alpha_0)(x - \alpha_0) + \int_a^x f^{(m+2)}(t)(x - t)\, dt =$$

$$= f^{(m+1)}(\alpha_0)(x - \alpha_0) + \int_a^b f^{(m+2)}(t) \times$$

$$\times [E(x - t) - E(\alpha_0 - t)](x - t)\, dt.$$

We substitute this expression for $f^{(m)}(x) - f^{(m)}(\alpha_0)$ into the expression for $R(f)$ and integrate. By our choice of α_0, $\displaystyle\int_a^b K(t)(t - \alpha_0)\, dt = 0$ and

$$R(f) = C_0 f^{(m)}(\alpha_0) + \int_a^b f^{(m+2)}(t) K_1(t)\, dt$$

$$C_0 = \int_a^b K(x)\, dx, \qquad K_1(t) = \int_a^b K(x)[E(x - t) - E(\alpha_0 - t)](x - t)\, dx.$$

If we perform this transformation s times we obtain the following formula which is sometimes useful for sequentially increasing the precision of a quadrature formula:

$$\int_a^b p(x) f(x) \, dx = \sum_{k=1}^n A_k f(x_k) + C_0 f^{(m)}(a_0) +$$
$$+ C_1 f^{(m+2)}(a_1) + \cdots +$$
$$+ C_{s-1} f^{(m+2s-2)}(a_{s-1}) +$$
$$+ \int_a^b f^{(m+2s)}(x) K_s(x) \, dx \qquad (11.4.3)$$

$$K_0(x) = K(x), \qquad K_{i+1}(x) = \int_a^b K_i(t) [E(t-x) - E(a_i - x)] (t-x) \, dt$$

$$C_i = \int_a^b K_i(x) \, dx, \quad a_i = C_i^{-1} \int_a^b x K_i(x) \, dx.$$

The above expression for $K_{i+1}(x)$ can be written as

$$K_{i+1}(x) = \begin{cases} \displaystyle\int_a^x K_i(t)(x-t) \, dt & a \le x < a_i \\[2ex] \displaystyle\int_x^b K_i(t)(t-x) \, dt & a_i < x \le b. \end{cases}$$

Thus we see that if $K_i(x)$ does not change sign on $a \le x \le b$ then $K_{i+1}(x)$ also does not change sign on this interval and $K_{i+1}(x)$ has the same sign as $K_i(x)$. In particular if the initial kernel $K(x)$ of the remainder (11.4.2) is positive throughout $[a, b]$ then all the kernels $K_i(x)$, $i = 1, 2, \ldots$, will also be positive throughout $[a, b]$.

Let us now consider the quadrature formula with n nodes with the highest algebraic degree of precision $2n - 1$ for the weight function $(1 - x)^p (1 + x)^q$, $p, q > -1$:

$$\int_{-1}^1 (1-x)^p (1+x)^q f(x) \, dx = \sum_{k=1}^n A_k f(x_k) + R(f). \qquad (11.4.4)$$

The nodes of this formula are the zeros of the Jacobi polynomial $P_n^{(p,q)}(x)$. We will assume that these nodes are innumerated in increasing order: $-1 < x_1 < \cdots < x_n < 1$.

First we show that when p or q is large there is a principle subinterval $[\alpha, \beta]$ for the integral representation of $R(f)$. We will show this by constructing an electrostatic analogy for the roots of $P_n^{(p,q)}(x)$.

It is known that $P_n^{(p,q)}(x)$ satisfies the differential equation[2]

$$\frac{d^2}{dx^2} P_n^{(p,q)}(x) + \left(\frac{p+1}{x-1} + \frac{q+1}{x+1}\right) \frac{d}{dx} P_n^{(p,q)}(x) +$$

$$+ \frac{n(p+q+n+1)}{1-x^2} P_n^{(p,q)}(x) = 0.$$

Substituting $x = x_k$ makes the third term on the left side vanish and since $\frac{d}{dx} P_n^{(p,q)}(x_k) \neq 0$ we can divide by this term to obtain

$$\sum_{i \neq k} \frac{2}{x_k - x_i} + \frac{p+1}{x_k - 1} + \frac{q+1}{x_k + 1} = 0 \quad (k = 1, 2, \ldots, n). \qquad (11.4.5)$$

This equation has a simple physical interpretation.

Consider a planar electrostatic field in which particles with like charges are repelled with a force proportional to their charge and inversely proportional to the distance between them. If two particles with charges m_1 and m_2 lie on the x-axis at the points x_1 and x_2 then the force which one particle exerts on the other is

$$\frac{\lambda m_1 m_2}{x_2 - x_1}.$$

Let particles with charges of $p + 1$ and $q + 1$ be fixed at the respective points $x = +1$ and $x = -1$. In addition we place n particles of charge 2 inside the segment $[-1, 1]$ and assume that these are only free to move along the x-axis. Let x_k $(k = 1, 2, \ldots, n)$ denote the coordinates of the free particles. If the free particles are at equilibrium then the force on each free particle is zero. Thus for the particle at x_k

$$\sum_{i \neq k} \frac{4\lambda}{x_k - x_i} + \frac{2\lambda(p+1)}{x_k - 1} + \frac{2\lambda(q+1)}{x_k + 1} = 0 \quad (k = 1, 2, \ldots, n).$$

Thus the equations of equilibrium differ from (11.4.5) by only the multiple 2λ and the position of these particles will coincide with the zeros of the Jacobi polynomial $P_n^{(p,q)}(x)$.

When the charges $p + 1$ and $q + 1$ of the fixed particles are large they will strongly repel the free particles and will force then to concentrate in a "small" subinterval so that $[x_1, x_n]$ will have a relatively small length. When p is significantly larger than q the interval $[x_1, x_n]$ will be close to -1; conversely if q is significantly larger than p the interval $[x_1, x_n]$ will be close to $+1$.

---

[2] See G. Szegö, *Orthogonal Polynomials*, Amer. Math. Soc. Colloq. Publ., 1939, p. 59.

The remainder $R(f)$ of the quadrature formula (11.4.4) has a representation of the form (11.4.2). In general the kernel is given by (11.3.5) which in the present case is

$$K(x) = \int_x^1 (1-t)^p (1+t)^q \frac{(t-x)^{2n-1}}{(2n-1)!} \, dt - \sum_{k=1}^n A_k E(x_k - x) \frac{(x_k - x)^{2n-1}}{(2n-1)!}.$$

In particular for a point x outside $[x_1, x_n]$:

$$K(x) = \int_{-1}^x (1-t)^p (1+t)^q \frac{(x-t)^{2n-1}}{(2n-1)!} \, dt \qquad \text{for} \qquad -1 \le x \le x_1$$

$$K(x) = \int_x^1 (1-t)^p (1+t)^q \frac{(t-x)^{2n-1}}{(2n-1)!} \, dt \qquad \text{for} \qquad x_n \le x \le 1.$$

Consider, for example, the case $x_n \le x \le 1$. The factor $1 + t$ lies between the limits $1 + x_n \le 1 + t \le 2$. Therefore $K(x)$ is greater than

$$(1 + x_n)^q \int_x^1 (1-t)^p \frac{(t-x)^{2n-1}}{(2n-1)!} \, dt =$$

$$= (1 + x_n)^q \frac{(1-x)^{p+2n}}{(2n-1)!} \int_0^1 (1-u)^p u^{2n-1} du =$$

$$= (1 + x_n)^q C (1-x)^{p+2n}$$

and less than

$$2^q \int_x^1 (1-t)^p \frac{(t-x)^{2n-1}}{(2n-1)!} \, dt = 2^q C (1-x)^{p+2n}.$$

As x increases from x_n up to 1 the kernel $K(x)$ approaches zero as $(1-x)^{p+2n}$. If p is large $K(x)$ will approach zero very rapidly. If p is not large but if q is large then x_n will be close to unity and $(1-x)^{p+2n}$ will again be a small number.

From this discussion we can expect that the method outlined above can be used to increase the accuracy of formula (11.4.4) when p or q is large.

Formula (11.4.3) for the quadrature formula (11.4.4) is:

$$\int_{-1}^1 (1-x)^p (1+x)^q f(x) dx = \sum_{k=1}^n A_k f(x_k) + C_0 f^{(2n)}(a_0) +$$

$$+ C_1 f^{(2n+2)}(a_1) + \cdots + C_{s-1} f^{(2n+2s-2)}(a_{s-1}) +$$

$$+ \int_{-1}^1 f^{(2n+2s)}(x) K_s(x) dx. \tag{11.4.6}$$

The coefficient $C_0 = \int_{-1}^{1} K(x)\,dx$ of the first corrective term is the remainder when the quadrature formula is applied to a function for which $f^{(2n)}(x) \equiv 1$. We can take this function to be

$$f(x) = \frac{1}{(2n)!\,a_n^2}\,[P_n^{(p,q)}(x)]^2$$

where a_n is the leading coefficient of $P_n^{(p,q)}(x)$. Since $P_n^{(p,q)}(x_k) = 0$ then

$$C_0 = R(f) = \frac{1}{(2n)!\,a_n^2} \left\{ \int_{-1}^{1} (1-x)^p(1+x)^q[P_n^{(p,q)}(x)]^2 dx - \right.$$

$$\left. - \sum_{k=1}^{n} A_k[P_n^{(p,q)}(x_k)]^2 \right\} =$$

$$= \frac{1}{(2n)!\,a_n^2} \int_{-1}^{1} (1-x)^p(1+x)^q[P_n^{(p,q)}(x)]^2\,dx.$$

The value of a_n is given by (2.2.2) and the value of the integral is (2.2.5) and thus

$$C_0 = \frac{2^{p+q+2n+1}n!\,\Gamma(p+n+1)\Gamma(q+n+1)\Gamma(p+q+n+1)}{(2n)!\,\Gamma(p+q+2n+1)\Gamma(p+q+2n+2)}.$$

We now calculate α_0. We note that the integral in the expression $\alpha_0 = C_0^{-1} \int_{-1}^{1} xK(x)\,dx$ is the remainder when formula (11.4.4) is applied to a function $f(x)$ which has a derivative of order $2n$ equal to x. Writing

$$P_n^{(p,q)}(x) = a_n x^n + b_n x^{n-1} + \cdots$$

we see that we can take $f(x)$ to be

$$f(x) = \frac{1}{(2n+1)!\,a_n^2}\left[x - \frac{2b_n}{a_n}\right][P_n^{(p,q)}(x)]^2.$$

Since $f(x_k) = 0$ we have

$$\int_{-1}^{1} xK(x)\,dx = R(f) =$$

$$= \frac{1}{(2n + 1)!a_n^2} \left\{ \int_{-1}^{1} (1 - x)^p (1 + x)^q x [P_n^{(p,q)}(x)]^2 dx - \right.$$

$$\left. - \frac{2b_n}{a_n} \int_{-1}^{1} (1 - x)^p (1 + x)^q [P_n^{(p,q)}(x)]^2 dx \right\}. \quad (11.4.7)$$

We have found that the second integral inside the brackets is $(2n)!a_n^2 C_0$. The coefficients a_n and b_n of the Jacobi polynomial are known to have the values[3]

$$a_n = \frac{\Gamma(p + q + 2n + 1)}{2^n n! \Gamma(p + q + n + 1)} \qquad b_n = \frac{n(p - q)\Gamma(p + q + 2n)}{2^n n! \Gamma(p + q + n + 1)}.$$

Therefore

$$\frac{b_n}{a_n} = \frac{n(p - q)}{p + q + 2n}.$$

We now calculate the first integral inside the brackets in (11.4.7). The following recursion relation is known for Jacobi polynomials[4]

$$(p + q + 2n)(p + q + 2n + 1)(p + q + 2n + 2)x P_n^{(p,q)}(x) =$$

$$= 2(n + 1)(p + q + n + 1)(p + q + 2n)P_{n+1}^{(p,q)}(x) +$$

$$+ (q^2 - p^2)(p + q + 2n + 1)P_n^{(p,q)}(x) +$$

$$+ 2(p + n)(q + n)(p + q + 2n + 2)P_{n-1}^{(p,q)}(x).$$

We multiply both sides of this equation by $(1 - x)^p (1 + x)^q P_n^{(p,q)}(x)$ and integrate over $[-1, 1]$ and thus obtain

$$\int_{-1}^{1} (1 - x)^p (1 + x)^q x [P_n^{(p,q)}(x)]^2 dx =$$

$$= \frac{q^2 - p^2}{(p + q + 2n)(p + q + 2n + 2)} \int_{-1}^{1} (1 - x)^p (1 + x)^q [P_n^{(p,q)}(x)]^2 dx =$$

$$= \frac{(q^2 - p^2)(2n)!a_n C_0}{(p + q + 2n)(p + q + 2n + 2)}.$$

[3]See D. Jackson, *Fourier Series and Orthogonal Polynomials*, Math. Assoc. Amer., 1941, p. 171.

[4]See G. Szego, *Orthogonal Polynomials*, Amer. Math. Soc. Colloq. Publ., 1939, p. 70.

Thus we obtain

$$\int_{-1}^{1} x K(x)\, dx =$$

$$= \frac{q - p}{2n + 1}\left[\frac{p + q}{(p + q + 2n)(p + q + 2n + 2)} + \frac{2n}{p + q + 2n}\right] C_0.$$

Rewriting this expression gives

$$\alpha_0 = \frac{q - p}{2n + 1}\left[\frac{n}{p + q + 2n} + \frac{n + 1}{p + q + 2n + 2}\right].$$

In the special ultraspherical case $p = q$ and the α_k $(k = 0, 1, \ldots)$ will be zero and formula (11.4.6) will be

$$\int_{-1}^{1} (1 - x)^p f(x)\, dx = \sum_{k=1}^{n} A_k f(x_k) +$$

$$+ \frac{2^{2p + 2n + 1} n\,![\Gamma(p + n + 1)]^2 \Gamma(2p + n + 1)}{(2n)!\,\Gamma(2p + 2n + 1)\Gamma(2p + 2n + 2)} \times$$

$$\times \left\{ f^{(2n)}(0) + \frac{1}{2(2n + 1)(2n + 2)} \times\right.$$

$$\left.\times \left[\frac{(n + 1)(n + 2)}{2p + 2n + 3} + \frac{n(n - 1)}{2p + 2n - 1}\right] f^{(2n+2)}(0) + \cdots\right\}.$$

Suppose we wish to approximate the integral

$$\int_{-1}^{1} (1 - x^2)^2 e^x\, dx = 8e - 56e^{-1} \approx 1.145006.$$

Here $p = q = 2$ and let us take $n = 1$ so that the formula has the form

$$\int_{-1}^{1} (1 - x^2)^2 f(x)\, dx = A_1 f(x_1) + \frac{8}{105}\left[f^{(2)}(0) + \frac{1}{36} f^{(4)}(0) + \cdots\right].$$

Since the weight function is symmetric about the origin $x_1 = 0$ and

$$A_1 = \int_{-1}^{1} (1 - x^2)^2\, dx = \frac{16}{15} \approx 1.066667.$$

The formula with only the first term

$$A_1 f(x_1) = \frac{16}{15}\,(1) \approx 1.066667$$

gives a very poor result. The first two terms give

$$A_1 f(x_1) + \frac{8}{105} f^{(2)}(0) = \frac{16}{15} + \frac{8}{105} = \frac{8}{7} \approx 1.142857$$

and three terms give

$$A_1 f(x_1) + \frac{8}{105} f^{(2)}(0) + \frac{2}{945} f^{(4)}(0) = \frac{8}{7} + \frac{2}{945} = \frac{1082}{945} \approx 1.144974$$

which differs from the exact value in only the sixth significant figure.

The method of removing from the integral representation of the remainder several successive "principal parts" which we have discussed in this section in connection with increasing the accuracy of mechanical quadrature formulas is closely related to a problem in the constructive theory of functions which is usually called the problem of interpolation by derivatives of successive orders or the problem of Abel-Goncharov.[5]

Let $f(x)$ be a function with $n + 1$ derivatives defined on $[\alpha, \beta]$ and consider the $n + 1$ points $\xi_0, \xi_1, \ldots, \xi_n$. We wish to find a polynomial $P(x)$ of degree $\leq n$ which satisfies the conditions

$$P^{(i)}(\xi_i) = f^{(i)}(\xi_i) \qquad (i = 0, 1, \ldots, n). \tag{11.4.8}$$

It is easy to find an explicit expression for $P(x)$. From the last condition (11.4.8) we can take

$$P^{(n)}(x) = f^{(n)}(\xi_n).$$

Integrating this equation between the limits ξ_{n-1} and x and using the second from the last condition (11.4.8):

$$P^{(n-1)}(x) = f^{(n-1)}(\xi_{n-1}) + f^{(n)}(\xi_n) \int_{\xi_{n-1}}^{x} dt_n.$$

Continuing in this way we obtain after n steps

$$P(x) = f(\xi_0) + f'(\xi_1) \int_{\xi_0}^{x} dt_1 + f''(\xi_2) \int_{\xi_0}^{x} \int_{\xi_1}^{t_1} dt_2 \, dt_1 + \cdots +$$

$$+ f^{(n)}(\xi_n) \int_{\xi_0}^{x} \int_{\xi_1}^{t_1} \cdots \int_{\xi_{n-1}}^{t_{n-1}} dt_n \cdots dt_2 \, dt_1. \tag{11.4.9}$$

Introducing the notation

---

[5] See V. L. Goncharov, *The Theory of Interpolation and Approximation of Functions,* Moscow, 1954, pp. 84–87 (Russian) and M. A. Evgrafov, *The Interpolation Problem of Abel-Goncharov,* Moscow, 1954 (Russian) which contains a bibliography on this subject.

$$L_0(x) = 1, \qquad L_i(x) = \int_{\xi_0}^{x} \int_{\xi_1}^{t_1} \cdots \int_{\xi_{i-1}}^{t_{i-1}} dt_i \cdots dt_2 \, dt_1$$

we can write $P(x)$ in the form

$$P(x) = \sum_{i=0}^{m} f^{(i)}(\xi_i) L_i(x). \tag{11.4.10}$$

Consider the remainder of the interpolation

$$r(x) = f(x) - P(x).$$

Under certain assumptions on the function $f(x)$ we can construct another representation for the remainder which is better suited for studying and estimating $r(x)$.

Let the point x and the nodes ξ_k $(k = 0, 1, \ldots, n)$ belong to the segment $[\alpha, \beta]$. If $f(x)$ has a continuous derivative of order $n + 1$ on $[\alpha, \beta]$ then the remainder of the interpolation $r(x)$ can be represented in the form:

$$r(x) = \int_{\xi_0}^{x} \int_{\xi_1}^{t_1} \cdots \int_{\xi_n}^{t_n} f^{(n+1)}(t_{n+1}) \, dt_{n+1} \cdots dt_2 \, dt_1. \tag{11.4.11}$$

The validity of this representation follows from the fact that at the nodes ξ_i the remainder $r(x)$ must satisfy

$$r(\xi_0) = 0, \qquad r'(\xi_1) = 0, \qquad \ldots, \qquad r^{(n)}(\xi_n) = 0$$

and in addition that

$$r^{(n+1)}(x) = f^{(n+1)}(x).$$

We now return to the expression for the remainder of formula (11.4.1):

$$R(f) = \int_{a}^{b} f^{(m)}(x) K(x) \, dx. \tag{11.4.12}$$

In order to remove the principle part of the remainder $R(f)$ suppose we have selected, by some means, a point ξ_0 so that we can split $f^{(m)}(x)$ into two parts

$$f^{(m)}(x) = f^{(m)}(\xi_0) + [f^{(m)}(x) - f^{(m)}(\xi_0)] =$$

$$= f^{(m)}(\xi_0) + \int_{\xi_0}^{x} f^{(m+1)}(t) \, dt =$$

$$= f^{(m)}(\xi_0) + \int_{a}^{b} f^{(m+1)}(t) [E(x - t) - E(\xi_0 - t)] \, dt.$$

Previously we denoted the selected point by α_0 and choose it so that the remainder was concentrated around it. Now we will not say how ξ_0 is selected and assume that it is arbitrary.

Substituting the above expression for $f^{(m)}(x)$ into (11.4.12) we obtain

$$R(f) = D_0 f^{(m)}(\xi_0) + \int_a^b f^{(m+1)}(x) N_1(x)\, dx$$

$$D_0 = \int_a^b K(x)\, dx, \qquad N_1(x) = \int_a^b K(t)[E(t-x) - E(\xi_0 - x)]\, dt.$$

In order to remove the second principle part from $R(f)$ let us select a second point ξ_1 and expand $f^{(m+1)}(x)$ into two parts

$$f^{(m+1)}(x) = f^{(m+1)}(\xi_1) + [f^{(m+1)}(x) - f^{(m+1)}(\xi_1)]$$

and so forth. After carrying out this transformation s times we obtain

$$R(f) = D_0 f^{(m)}(\xi_0) + D_1 f^{(m+1)}(\xi_1) + \cdots + D_{s-1} f^{(m+s-1)}(\xi_{s-1}) +$$
$$+ \int_a^b f^{(m+s)}(x) N_s(x)\, dx \qquad (11.4.13)$$

$$N_0(x) = K(x), \qquad N_{i+1}(x) = \int_a^b N_i(t)[E(t-x) - E(\xi_i - x)]\, dt$$

$$D_i = \int_a^b N_i(x)\, dx.$$

This expansion for the remainder is clearly similar to the expansion for $R(f)$ in equation (11.4.3). The only difference between the two expansion is that the points $\alpha_0, \alpha_1, \ldots$ were chosen in a definite way and the points $\xi_0, \xi_1, \ldots$ are arbitrary. But if we select the ξ_i so that $\xi_0 = \alpha_0$, $\xi_2 = \alpha_1, \ldots$, then it is clear that the expansion (11.4.13) for $R(f)$ will coincide with the expansion (11.4.3).

Equation (11.4.13) can be obtained in another way which is closely connected with the above mentioned problem of Abel-Goncharov. Taking the nodes $\xi_0, \xi_1, \ldots, \xi_{s-1}$ we interpolate on $f^{(m)}(x)$ by a sequence of its derivatives

$$f^{(m)}(x) = \sum_{i=0}^{s-1} f^{(m+i)}(\xi_i) L_i(x) + r(x). \qquad (11.4.14)$$

If we substitute this expansion for $f^{(m)}(x)$ into (11.4.12) we obtain the representation

$$R(f) = \sum_{i=0}^{s-1} f^{(m+i)}(\xi_i) \int_a^b K(x) L_i(x) \, dx +$$

$$+ \int_a^b K(x) r(x) \, dx. \qquad (11.4.15)$$

This representation must clearly coincide with (11.4.13) for any function which has a continuous derivative of order $m + s$. Therefore

$$D_i = \int_a^b K(x) L_i(x) \, dx$$

$$\int_a^b f^{(m+s)}(x) N_s(x) \, dx = \int_a^b K(x) r(x) \, dx. \qquad (11.4.16)$$

This relationship between interpolation on $f^{(m)}(x)$ and the expansion of the remainder of quadrature formulas in "principal parts" is useful in the theory of quadrature formulas in the following way.

If $f(x)$ has continuous derivatives of all orders then in (11.4.13) we can increase s without limit. Then the sum $\sum_{i=0}^{s-1} D_i f^{(m+i)}(\xi_i)$ can be replaced by the infinite series

$$R(f) \approx D_0 f^{(m)}(\xi_0) + D_1 f^{(m+1)}(\xi_1) + \cdots. \qquad (11.4.17)$$

This series will converge to $R(f)$ if and only if

$$\lim_{s \to \infty} \int_a^b f^{(m+s)}(x) N_s(x) \, dx = 0.$$

From (11.4.16) this is equivalent to

$$\lim_{s \to \infty} \int_a^b K(x) r(x) \, dx = 0.$$

Thus the possibility of expanding the remainder $R(f)$ in a series (11.4.16) of "principal parts" is related to the convergence of the Abel-Goncharov interpolation (11.4.14) for the function $f^{(m)}(x)$.

In particular if $[a, b]$ is finite and if $r(x)$ converges to zero uniformly with respect to x then the expansion

$$R(f) = D_0 f^{(m)}(\xi_0) + D_1 f^{(m+1)}(\xi_1) + \cdots$$

is certainly possible.

For a discussion of the conditions under which the Abel-Goncharov interpolation converges the reader is referred to the book by M. A. Evgrafov.

REFERENCES

M. A. Evgrafov, *The Interpolation Problem of Abel-Goncharov*, Moscow, 1954, (Russian).

L. V. Kantorovich, "On approximate calculation of certain types of definite integrals and other applications of the method of removal of singularities," *Mat. Sbornik*, Vol. 41, 1934, pp. 235–45 (Russian).

A. N. Krylov, *Lectures on Approximate Computations*, Moscow-Leningrad, 1950 (Russian).

V. I. Krylov, "Increasing the accuracy of mechanical quadratures. Formulas of Euler's form," *Dokl. Akad. Nauk SSSR*, Vol. 96, 1954, pp. 429–32 (Russian).

V. I. Krylov, "Increasing the accuracy of mechanical quadratures when the main part of the integration is over a small interval in the integral representation of the remainder of quadrature," *Dokl. Akad. Nauk SSSR*, Vol. 101, 1955, pp. 989–91 (Russian).

V. N. Smirnov, "Increase in the precision of formulas of mechanical quadrature of Chebyshev-Hermite type. Increase in the precision of quadratures Chebyshev-Laguerre," *Sbornik Trudov Kuibyshevskovo Indust. Inst.*, No. 5, 1955 (Russian).

V. N. Smirnov, "On quadrature formulas of Gauss," *Sbornik Trudov Kuibyshevskovo Indust. Inst.*, No. 6, 1956 (Russian).

J. F. Steffensen, *Interpolation*, Williams and Wilkins, Baltimore, 1927.

CHAPTER 12

Convergence of the Quadrature Process

12.1. INTRODUCTION

In this chapter we consider a sequence of quadrature formulas with n nodes $(n = 1, 2, 3, \ldots)$. Such a sequence is defined by two triangular matrices: a matrix of nodes

$$
X = \begin{bmatrix}
x_1^{(1)} \\
x_1^{(2)} & x_2^{(2)} \\
\cdots\cdots\cdots\cdots\cdots \\
x_1^{(n)} & x_2^{(n)} \cdots x_n^{(n)} \\
\cdots\cdots\cdots\cdots\cdots
\end{bmatrix}
\tag{12.1.1}
$$

and a matrix of coefficients,

$$
A = \begin{bmatrix}
A_1^{(1)} \\
A_1^{(2)} & A_2^{(2)} \\
\cdots\cdots\cdots\cdots\cdots \\
A_1^{(n)} & A_2^{(n)} \cdots A_n^{(n)} \\
\cdots\cdots\cdots\cdots\cdots
\end{bmatrix}.
\tag{12.1.2}
$$

Consider the quadrature formula which corresponds to the nth row of these matrices:

$$
\int_a^b p(x) f(x) \, dx = \sum_{k=1}^{n} A_k^{(n)} f(x_k^{(n)}) + R_n(f) = Q_n(f) + R_n(f).
\tag{12.1.3}
$$

We will say that the sequence of quadrature formulas, defined by the

242

matrices X and A, converges if

$$\lim_{n\to\infty} Q_n(f) = \lim_{n\to\infty} \sum_{k=1}^{n} A_k^{(n)} f(x_k^{(n)}) = \int_a^b p(x) f(x) \, dx. \tag{12.1.4}$$

Whether or not the process converges depends on the properties of the integrand $f(x)$ and also on the properties of the quadrature formulas. A study of the convergence of the process consists of studying what relationship between the integrand $f(x)$ and the matrices X and A will lead to a convergent process.

There are two basic problems:

1. Given the matrices X and A determine for what class of functions F equation (12.1.4) will hold.

2. Given a class of functions F determine the properties that the matrices X and A must have to assure convergence of the process for all $f(x) \, \epsilon \, F$.

In the rest of this chapter we discuss the solutions to these problems for certain particular cases of practical importance in the theory of quadrature formulas.

We limit our discussion to finite segments of integration and will not be concerned with the harder problem of convergence of quadrature formulas for integrals with infinite limits.

12.2. CONVERGENCE OF INTERPOLATORY QUADRATURE FORMULAS FOR ANALYTIC FUNCTIONS

In order to simplify the proofs in this section and to make them more general we will write the integral $\int_a^b p(x) f(x) \, dx$ as a Stieltjes integral. Suppose that we are given a certain function $\sigma(x)$ of bounded variation on the segment $[a, b]$ and consider the integral $\int_a^b f(x) \, d\sigma(x)$. Let us take n points $x_k^{(n)}$ $(k = 1, \ldots, n)$ on the segment $[a, b]$ and construct the interpolatory quadrature formula

$$\int_a^b f(x) \, d\sigma(x) = \sum_{k=1}^{n} A_k^{(n)} f(x_k^{(n)}) + R_n(f) \tag{12.2.1}$$

$$\omega_n(x) = \prod_{k=1}^{n} (x - x_k^{(n)}), \quad A_k^{(n)} = \int_a^b \frac{\omega_n(x)}{(x - x_k^{(n)}) \, \omega_n'(x_k^{(n)})} \, d\sigma(x).$$

A sequence of such formulas is completely defined by the matrix of their nodes (12.1.1).

It is remarkable that we can give an effective and simple criterion to decide whether or not the interpolatory quadrature process converges for analytic functions. Such a criterion can be formulated by means of a function which can be interpreted as the limiting distribution function of the nodes $x_k^{(n)}$ $(k = 1, \ldots, n)$ as $n \longrightarrow \infty$.

The nodes $x_k^{(n)}$ are assumed to lie on the segment $[a, b]$ and a distribution function for these nodes will be defined on this segment.

Consider a unit mass of arbitrary form distributed on the segment $[a, b]$. If x is any point of this segment then for the value of $\mu(x)$ at the point x we take the mass which lies to the left of x. In particular, since there is no mass to the left of a $\mu(a) = 0$ and $\mu(b) = 1$.

Thus it is clear that the function $\mu(x)$ must have the following properties:

1. $\mu(a) = 0$;

2. $\mu(x)$ is a monotone nondecreasing function of x which is continuous from the left at each point inside $[a, b]$;

3. $\mu(b) = 1$.

These properties follow from the definition and each function possessing these properties will be called a distribution function for the segment $[a, b]$.

Let us be given a sequence of distribution functions $\mu_n(x), n = 1, 2, \ldots$. We will say that this sequence converges fundamentally to a function $\mu(x)$ if $\mu_n(x) \longrightarrow \mu(x)$ at each point of continuity[1] of $\mu(x)$.

We now consider the nth row of the matrix X

$$x_1^{(n)}, x_2^{(n)}, \ldots, x_n^{(n)}$$

and assume that the nodes $x_k^{(n)}$ are enumerated in increasing order. We assign to each of these nodes a mass of $\frac{1}{n}$. To this row of the matrix X there then corresponds a distribution function $\mu_n(x)$.

If there exists a function $\mu(x)$, which possesses the above three properties, to which the sequence $\mu_n(x)$ converges fundamentally:

$$\mu_n(x) \underset{\text{fund.}}{\longrightarrow} \mu(x) \text{ as } n \longrightarrow \infty$$

then we call $\mu(x)$ the limiting distribution function for the matrix X.

We will only be concerned with cases for which $\mu(x)$ exists.[2]

Let $r_n(x)$ denote the remainder of the interpolation for $f(x)$ using its values at the nodes $x_k^{(n)}$ $(k = 1, \ldots, n)$:

---

[1] Note that at the end points we have $\mu_n(a) = \mu(a) = 0$ and $\mu_n(b) = \mu(b) = 1$ and thus the sequence will always converge at the ends of the segment.

[2] From known theorems concerning distribution functions we could consider X to be an arbitrary matrix.

$$r_n(x) = f(x) - \sum_{k=1}^{n} \frac{\omega_n(x)}{(x - x_k^{(n)})\omega_n'(x_k^{(n)})} f(x_k^{(n)}) = f(x) - L_n(x).$$

The remainder $R_n(f)$ of the quadrature (12.2.1) is the integral of $r_n(x)$:

$$R_n(f) = \int_a^b r_n(x)\, d\sigma(x).$$

The convergence of the quadrature process is closely related to the convergence of the interpolation and in particular if $r_n(x) \longrightarrow 0$ uniformly with respect to $x \in [a, b]$ as $n \longrightarrow \infty$ then $R_n(f) \longrightarrow 0$, that is the quadrature process will also converge. To investigate the convergence of the quadrature process (12.2.1) we will first study convergence of the interpolation.

We assume that $f(z)$ is an analytic function of z and that it is holomorphic in a certain simply connected region B of the complex plane and that the segment $[a, b]$ of the real line is contained in the interior of B. Let l denote a simple closed rectifiable curve which is contained in B and which encloses $[a, b]$.

By (3.2.11) the remainder of the interpolation can be represented as the contour integral

$$r_n(x) = \frac{1}{2\pi i} \int_l \frac{\omega_n(x) f(z)}{\omega_n(z)(z - x)}\, dz \tag{12.2.2}$$

where x is any point inside l. Let $\mu(x)$ be the limiting distribution function of the nodes of the matrix X. The logarithmic potential

$$u(z) = \int_a^b \ln \frac{1}{|z - t|}\, d\mu(t) \tag{12.2.3}$$

is very useful for studying $r_n(x)$ as $n \longrightarrow \infty$. The function $u(z)$ is harmonic and is holomorphic everywhere in the complex plane except at the point at infinity and on the segment $[a, b]$. As z approaches the point at infinity $u(z)$ approaches $-\infty$.

Consider the curve

$$u(z) = C.$$

If C is a large negative number then this curve encloses the segment $[a, b]$ and will be "close" to a circle with a large radius. We call this curve l_C and denote by B_C the part of the plane which lies inside it. As C increases in the positive direction B_C will become smaller. We denote by λ the least upper bound of the values of C for which $[a, b]$ lies inside B_C. For each $C < \lambda$ the curve l_C will enclose $[a, b]$.

We will denote by χ the open region of the z plane for which $u(z) < \lambda$. The complement of χ will be denoted by β.

Theorem 1. *If $f(z)$ is an analytic function holomorphic in a certain domain D which contains β in its interior then*

$$r_n(x) \longrightarrow 0 \quad \text{as} \quad n \longrightarrow \infty$$

uniformly with respect to $x \epsilon \beta$.

Proof. Since β lies in the interior of D there exists a number $C' < \lambda$ for which $B_{C'} \cup l_{C'}$ also lies inside D.

Take an arbitrary number C'' between C' and $\lambda: C' < C'' < \lambda$. The curve $l_{C''}$ is enclosed by $l_{C'}$ and contains β and thus $[a, b]$ in its interior.

We take $l_{C'}$ as the contour of integration in the integral representation of $r_n(x)$. We also assume that x lies on $l_{C''}$.

Let M denote the largest value of $|f(z)|$ on $l_{C'}$ and δ the distance between $l_{C'}$ and $l_{C''}$. The following estimate is valid:

$$\left| r_n(x) \right| \leq \frac{M}{2\pi\delta} \int_{l_{C'}} \frac{|\omega_n(x)|}{|\omega_n(z)|}\, ds.$$

Consider $\left| \omega_n(z) \right|^{-1}$:

$$\left| \omega_n(z) \right|^{-1} = \exp \sum_{k=1}^{n} ln \frac{1}{\left| z - x_k^{(n)} \right|}.$$

As above, we assign to each node $x_k^{(n)}$ $(k = 1, \ldots, n)$ a mass of $\frac{1}{n}$ and introduce the corresponding distribution function $\mu_n(x)$. Clearly

$$\int_a^b ln \frac{1}{\left| z - t \right|}\, d\mu_n(t) = \frac{1}{n} \sum_{k=1}^{n} ln \frac{1}{\left| z - x_k^{(n)} \right|}$$

and therefore

$$\left| \omega_n(z) \right|^{-1} = \exp n \int_a^b ln \frac{1}{\left| z - t \right|}\, d\mu_n(t).$$

As $n \longrightarrow \infty$ $\mu_n(t)$ converges fundamentally to the limiting distribution function of the nodes $\mu(t)$. The point z lies outside $[a, b]$ and $ln \frac{1}{\left| z - t \right|}$ is a continuous function of $t \epsilon [a, b]$. According to the theorem on passage to the limit for Stieltjes integrals, which is often called Helly's second theorem,[3] we have

$$\int_a^b \ln \frac{1}{|z-t|} \, d\mu_n(t) \longrightarrow \int_a^b \ln \frac{1}{|z-t|} \, d\mu(t) = C' \qquad \text{as } n \longrightarrow \infty.$$

Here z plays the role of a parameter and the convergence will be uniform with respect to $z \, \epsilon \, l_{C'}$. This can be shown by the usual proof of Helly's theorem. Thus there exists a number n' such that for $n > n'$ and for any $z \, \epsilon \, l_{C'}$ we have

$$C' - \frac{1}{3}(C'' - C') < \int_a^b \ln \frac{1}{|z-t|} \, d\mu_n(t) < C' + \frac{1}{3}(C'' - C').$$

Similarly for $x \, \epsilon \, l_{C''}$ we have

$$\int_a^b \ln \frac{1}{|x-t|} \, d\mu_n(t) \longrightarrow \int_a^b \ln \frac{1}{|x-t|} \, d\mu(t) = C'' \qquad \text{as } n \longrightarrow \infty$$

uniformly with respect to x.

Therefore there exists an n'' such that for $n > n''$ and for each $x \, \epsilon \, l_{C''}$

$$C'' - \frac{1}{3}(C'' - C') < \int_a^b \ln \frac{1}{|x-t|} \, d\mu_n(t) < C'' + \frac{1}{3}(C'' - C').$$

Taking $n_0 = \max(n', n'')$ we can assert that for $n > n_0$ and any $z \, \epsilon \, l_{C'}$, $x \, \epsilon \, l_{C''}$ we have

$$\int_a^b \ln \frac{1}{|z-t|} \, d\mu_n(t) - \int_a^b \ln \frac{1}{|x-t|} \, d\mu_n(t) <$$

$$< \left[C' + \frac{1}{3}(C'' - C') \right] - \left[C'' - \frac{1}{3}(C'' - C') \right] = -\frac{1}{3}(C'' - C').$$

Hence we obtain the estimate

$$\frac{|\omega_n(x)|}{|\omega_n(z)|} < \exp \left[-\frac{n}{3}(C'' - C') \right].$$

---

[3] V. I. Glivenko, *The Stieltjes Integral*, Moscow, 1936, Sec. 14 (Russian). Also see I. P. Natanson, *Theory of Functions of a Real Variable*, Ungar, New York, 1955, Chap. 8, Sec. 7, where Helly's theorem is proved with slightly different assumptions than $\mu_n \underset{\text{fund.}}{\longrightarrow} \mu$. But it is easy to see that with slight modifications this proof also holds for the case $\mu_n \underset{\text{fund.}}{\longrightarrow} \mu$.

This gives

$$|r_n(x)| < \frac{Ms}{2\pi\delta} \exp\left[-\frac{n}{3}(C'' - C')\right], \quad n > n_0, \; x \in l_{C''} \quad (12.2.3^*)$$

where s is the length of $l_{C'}$.

From (12.2.3*) it follows that as $n \longrightarrow \infty$, $r_n(x) \longrightarrow 0$ uniformly on $l_{C''}$. Since $r_n(x)$ is an analytic function which is holomorphic in $l_{C''} \cup B_{C''}$ then $r_n(x)$ will converge uniformly to zero in the entire region $l_{C''} \cup B_{C''}$. In particular this will be true on β which lies inside $l_{C''}$. This completes the proof of Theorem 1.

Theorem 1 immediately leads to the following theorem on convergence of the interpolatory quadrature process.

Theorem 2. *Let* [a, b] *be a finite segment. If* f(x) *is an analytic function which is holomorphic in a certain region containing the set* β *in its interior then for any function* σ(x) *the interpolatory quadrature process defined by* (12.2.1) *converges:*

$$R_n(f) = \int_a^b r_n(x) \, d\sigma(x) \longrightarrow 0 \qquad \text{as } n \longrightarrow \infty.$$

We now discuss the case when the limit function $\mu(x)$ corresponds to a uniform distribution of a unit mass on [a, b]:

$$\mu(x) = \frac{(x-a)}{(b-a)} \qquad a \leq x \leq b. \tag{12.2.4}$$

This case corresponds to a sequence of quadrature formulas with equally spaced nodes, one instance of which are the Newton-Cotes formulas.

For simplicity of notation we assume that the segment [a, b] has been transformed into [0,1]:

$$\mu(x) = x.$$

In this case the logarithmic potential (12.2.3) is

$$u(z) = \int_0^1 \ln \frac{1}{|z-t|} \, dt.$$

Since

$$\int_0^1 \ln |z-t| \, dt = \operatorname{Re} \int_0^1 \ln(z-t) \, dt = \operatorname{Re}\{(1-z)\ln(1-z) + z \ln z - 1\}$$

then

$$u(z) = \operatorname{Re}\{1 - z \ln z - (1-z)\ln(1-z)\} =$$

$$= 1 - x \ln \sqrt{x^2 + y^2} - (1-x)\ln\sqrt{(1-x)^2 + y^2} + y \operatorname{Arctan} \frac{y}{x - x^2 - y^2}.$$

Curves for $u(z) = C$ are depicted in Fig. 8. The set β consists of the curve which passes through the ends of the segment [0, 1] and the region inside this curve.

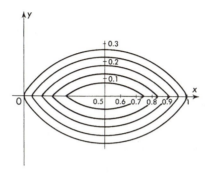

Figure 8.

The greatest horizontal dimension of β is unity and the greatest vertical dimension is about 0.5 and is obtained by the section of β by the line $x = 0.5$.

From this discussion we see that we can guarantee convergence of the Newton-Cotes quadrature process with equally-spaced nodes only when $f(z)$ is an analytic function which is holomorphic in a sufficiently wide region about the segment [0, 1] which contains the indicated region β in its interior.

We consider two more questions on the theory of convergence of interpolatory quadrature formulas which, in a certain sense, are extreme cases of Theorem 2.

It can be expected that if $f(z)$ is a function which is holomorphic in a very wide region about the segment $[a, b]$ then the quadrature process will certainly converge for any function $\sigma(x)$ and for any choice of nodes $x_k^{(n)}$ $(k = 1, \ldots, n)$. We now determine the smallest region in which $f(z)$ must be holomorphic in order that the quadrature process will converge for any $\sigma(x)$ and any choice of $x_k^{(n)}$.

First of all we study the related interpolation problem.

Construct a circle of radius $b - a$ around each of the points a and b and let χ denote the closed region inside the union of these two circles.

Theorem 3. *If $f(z)$ is holomorphic in the region χ then for any choice of nodes on $[a, b]$ the corresponding interpolation process converges*

$$L_n(x) \longrightarrow f(x) \quad \text{as} \quad n \longrightarrow \infty$$

uniformly with respect to all $x \in [a, b]$.

The domain of holomorphy χ is the smallest which will guarantee convergence of the interpolation for any matrix of nodes $x_k^{(n)}$.

Proof. Let x and t be two arbitrary points of the segment $[a, b]$ and z be an arbitrary point in the complex plane. If z is not in χ then $\left|\dfrac{x-t}{z-t}\right| < 1$. If z belongs to χ then we can find points x and t in $[a, b]$ for which $\left|\dfrac{x-t}{z-t}\right| \geq 1$.

We now prove the first part of the theorem. Since $f(z)$ is assumed holomorphic in the closed region χ then it will also be holomorphic in some larger region. Therefore there exists a closed curve l which encloses χ so that $f(z)$ is holomorphic on this curve and in its interior. The remainder of the interpolation has the representation

$$r_n(x) = \frac{1}{2\pi i} \int_l \frac{\omega_n(x)}{\omega_n(z)} \frac{f(z)}{z-x} \, dz.$$

Since x and the $x_k^{(n)}$ lie on $[a, b]$ and z lies on l we have

$$\left| \frac{x - x_k^{(n)}}{z - x_k^{(n)}} \right| < 1.$$

Then there exists a number $q < 1$ which is independent of x and the $x_k^{(n)}$ and z for which

$$\left| \frac{x - x_k^{(n)}}{z - x_k^{(n)}} \right| \leq q < 1.$$

Thus for each $x \, \epsilon \, [a, b]$ and each $z \, \epsilon \, l$

$$\left| \frac{\omega_n(x)}{\omega_n(z)} \right| \leq q^n$$

and then

$$\left| r_n(x) \right| \leq \frac{q^n}{2\pi} \max_{z \, \epsilon \, l} \int_l \frac{|f(z)|}{|z-x|} \, ds = A_0 q^n.$$

Hence it follows that $r_n(x) \longrightarrow 0$ uniformly with respect to x.

We now show that the region χ can not be made smaller. To do this it suffices to show that for any $a \, \epsilon \, \chi$ we can always find a function $f(z)$ which is holomorphic everywhere in χ except at a and also a system of nodes $x_k^{(n)}$ for which the interpolation process for $f(z)$ will diverge at a certain point $x \, \epsilon \, [a, b]$.

Suppose a is any point of χ. We can assume that a does not lie on $[a, b]$. Consider the function $f(z) = \dfrac{1}{z - a}$. This function is holomorphic on the entire plane except at $z = a$. For the line l in the integral representation for $r_n(x)$ we take the boundary Γ of χ together with a small circle γ drawn around a and connected to Γ by a cut.

The integrals along the sides of the cut cancel so we obtain

$$r_n(x) = r_n\left(\frac{1}{z - a}; x\right) = \frac{\omega_n(x)}{2\pi i} \int_{\Gamma + \gamma} \frac{dz}{\omega_n(z)\,(z - a)\,(z - x)}.$$

The integral over Γ is zero since at infinity the integrand has a zero of multiplicity greater than two. The integral over γ in a clockwise direction is the residue of the integrand at a multiplied by $-2\pi i$

$$r_n\left(\frac{1}{z - a}; x\right) = \frac{\omega_n(x)}{\omega_n(a)\,(x - a)}.$$

Since $a \,\epsilon\, \chi$ there exists points x and t of $[a, b]$ for which $\left|\dfrac{x - t}{a - t}\right| \geq 1$. Let us fix the value of x and take all the nodes of the interpolation to coincide with t. The interpolation must coincide with the value of $f(z) = \dfrac{1}{z - a}$ and its derivatives up to order $n - 1$ at the point t. The interpolating polynomial will be the truncated Taylor series for $\dfrac{1}{z - a}$ close to t. Then

$$\omega_n(z) = (z - t)^n$$

$$r_n\left(\frac{1}{z - a}; x\right) = \left(\frac{x - t}{a - t}\right)^n \frac{1}{x - a}.$$

Since $\left|\dfrac{x - t}{a - t}\right| \geq 1$ the remainder will not approach zero as $n \longrightarrow \infty$ and the interpolation for $\dfrac{1}{z - a}$ will not converge at x.

The above example is a divergent Hermite interpolation process which uses a single node of multiplicity n. But it is clear that if the $x_k^{(n)}$ ($k = 1, \ldots, n$) are taken very close to t and if the $x_k^{(n)}$ approach t sufficiently fast as $n \longrightarrow \infty$ then we can construct an example of a divergent interpolation process with distinct nodes. This completes the proof of the second part of the theorem.

From this result it is not difficult to prove the following theorem.

Theorem 4. *If $f(z)$ is holomorphic in the region χ then the interpolatory quadrature process defined by (12.2.1) is convergent*

$$R_n(f) \longrightarrow 0 \quad as \quad n \longrightarrow \infty$$

for any nodes $x_k^{(n)}$ and any function $\sigma(x)$ of bounded variation on $[a, b]$.

The region of holomorphy χ is the smallest which will guarantee convergence of the quadrature process (12.2.1) for arbitrary $x_k^{(n)}$ and $\sigma(x)$.

Proof. If $f(z)$ is holomorphic in χ then for arbitrary nodes $x_k^{(n)}$ the remainder of the interpolation approaches zero uniformly with respect to x as $n \longrightarrow \infty$. Therefore

$$R_n(f) = \int_a^b r_n(x)\, d\sigma(x)$$

also approaches zero as $n \longrightarrow \infty$ for any $\sigma(x)$ of bounded variation.

To prove the second part of the theorem it suffices to show that if we remove a single point from χ then we can find functions $f(z)$ and $\sigma(x)$ and nodes $x_k^{(n)}$ for which $R_n(f)$ will not approach zero.

Let $\sigma(x)$ be a function which is piece-wise constant with a unit jump at x. Then

$$R_n(f) = \int_a^b r_n(x)\, d\sigma(x) = r_n(x).$$

Convergence of the quadrature process for this $\sigma(x)$ is equivalent to convergence of the interpolation at x. But from the last theorem we see that for any point $\alpha \in \chi$ we can find, for the function $f(z) = \dfrac{1}{z - a}$, a point x and nodes $x_k^{(n)}$ of $[a, b]$ for which $r_n(x)$ diverges. This completes the proof.

In the remainder of this section we will assume that the segment of integration $[a, b]$ has been transformed into $[-1, 1]$. We will call the function

$$\mu(x) = \frac{1}{\pi} \int_{-1}^{x} \frac{dt}{\sqrt{1 - t^2}} \tag{12.2.5}$$

the Chebyshev distribution function. Let us suppose that the matrix X has a limiting distribution function for its nodes and that this function is (12.2.5). This will happen, for example, when the $x_k^{(n)}$ ($k = 1, \ldots, n$; $n = 1, 2, \ldots$) are the roots of the sequence of orthogonal polynomials $P_n(x)$ which are orthogonal on $[-1, 1]$ with respect to an arbitrary summable almost everywhere positive weight function $p(x)$. Such nodes correspond to integration formulas of the highest algebraic degree of precision.

Consider the logarithmic potential

$$u(z) = \frac{1}{\pi} \int_{-1}^{1} ln \; \frac{1}{|z - t|} \; \frac{dt}{\sqrt{1 - t^2}}. \tag{12.2.6}$$

This function is the real part of

$$F(z) = \frac{1}{\pi} \int_{-1}^{1} ln \; \frac{1}{z - t} \; \frac{dt}{\sqrt{1 - t^2}}. \tag{12.2.7}$$

In the complex plane z we draw a cut along the real axis from the point 1 to $-\infty$ and choose that branch of the logarithm for which $arg\,(z - t) = 0$ for real z greater than t

$$F'(z) = -\frac{1}{\pi} \int_{-1}^{1} \frac{dt}{(z - t)\sqrt{1 - t^2}}.$$

This integral is easily seen to be

$$F'(z) = -\frac{1}{\sqrt{z^2 - 1}}$$

where we choose that branch of the root which has a positive value for $z > 1$

$$F(z) = ln \; \frac{K}{z + \sqrt{z^2 - 1}}.$$

The constant K is found from the condition that for large z (12.2.7) can be represented as

$$F(z) = ln \frac{1}{z} + \frac{a_1}{z} + \frac{a_2}{z^2} + \cdots.$$

This gives $K = 2$ and thus

$$F(z) = ln \; \frac{2}{z + \sqrt{z^2 - 1}}$$

$$u(z) = ln \; \frac{2}{\left| z + \sqrt{z^2 - 1} \right|}. \tag{12.2.8}$$

The curve

$$u(z) = C$$

for $C < ln\,2$ is a closed curve enclosing the segment $[-1, 1]$. For $C = ln\,2$ the curve coincides with the segment $[-1, 1]$. The set β consists only of the interval of integration $[-1, 1]$. Thus we have:

Theorem 5. *If the matrix X has for its limiting distribution function the Chebyshev function* (12.2.5) *then*

1. *The corresponding interpolation process converges uniformly with respect to* $x \in [-1, 1]$ *for each function which is analytic on* $[-1, 1]$;

2. *The quadrature process defined by* (12.2.1) *on* $[-1, 1]$ *converges for any function* $f(x)$ *which is analytic on* $[-1, 1]$ *for an arbitrary function* $\sigma(x)$ *which has bounded variation on* $[-1, 1]$.

It is interesting to note that the converse to this theorem is also true. We will now prove the converse for the interpolation process.

Theorem[4] 6. *If the matrix X has the property that the interpolation process converges for all points of the segment* $[-1, 1]$ *for each analytic function on* $[-1, 1]$ *then X has a limiting distribution function which is the Chebyshev function* (12.2.5).

To prove this theorem it will be necessary to become acquainted with certain properties of the logarithmic potential. Consider the sequence of distribution functions which correspond to the rows of the matrix X: $\mu_1(x)$, $\mu_2(x), \ldots$. By Helly's theorem[5] we can always select from this sequence a subsequence which converges fundamentally to a certain distribution function which we denote by $\mu(x)$. In the following we assume that the index n runs through the integers for which $\mu_n(x) \longrightarrow \mu(x)$ fundamentally.

The theorem will be proved if we establish that

$$\mu(x) = \frac{1}{\pi} \int_{-1}^{x} \frac{dt}{\sqrt{1 - t^2}}.$$

For x in $[-1, 1]$ the integral

$$u(x) = \int_{-1}^{1} ln \frac{1}{|x - t|} d\mu(t) \tag{12.2.9}$$

is improper and we will define it as follows. We define the function $ln_N x$ by

$$ln_N x = \begin{cases} ln \ x & \text{for } ln \ x \leq N \\ N & \text{for } ln \ x > N. \end{cases}$$

Then $ln_N \dfrac{1}{|x - t|}$ is bounded and continuous for $t \in [-1, 1]$. The integral

[4] L. Kalmár, "Az interpolációról," *Mathematikai es physikai lapok*, Vol. 32, 1926, p. 120, where a more general theorem is proved.

[5] V. I. Glivenko, *The Stieltjes Integral*, Moscow, 1936, Sec. 13; or I. P. Natanson, *Theory of Functions of a Real Variable*, Ungar, New York, 1955, Chap. 8, Sec. 4.

$\int_{-1}^{1} ln_N \dfrac{1}{|x-t|} d\mu(t)$ is a nondecreasing function of N and we set

$$u(x) = \int_{-1}^{1} ln \frac{1}{|x-t|} d\mu(t) = \lim_{N\to\infty} \int_{-1}^{1} ln_N \frac{1}{|x-t|} d\mu(t).$$

Lemma 1. *If $\mu(t)$ has a derivative at the point $x \in [-1, 1]$ then the integral (12.2.9) is finite at this point.*

Proof. Let x belong to the interior of $[-1, 1]$. For large N we take the interval $x - \delta \le t \le x + \delta$ close to x where $\delta = e^{-N}$. Then

$$u(x) = \int_{-1}^{1} ln \frac{1}{|x-t|} d\mu(t) =$$

$$= \lim_{N\to\infty} \left\{ \int_{-1}^{x-\delta} ln\frac{1}{x-t} d\mu + \int_{x+\delta}^{1} ln\frac{1}{t-x} d\mu + N[\mu(x+\delta) - \mu(x-\delta)] \right\}.$$

After integrating the term in brackets by parts and using $ln\,\delta = -N$, $\mu(1) = 1$, $\mu(-1) = 0$:

$$u(x) = \lim_{N\to\infty} \left\{ ln\frac{1}{1-x} + \int_{-1}^{x-\delta} \frac{\mu(t)}{x-t} dt + \int_{x+\delta}^{1} \frac{\mu(t)}{x-t} dt \right\} =$$

$$= ln\frac{1}{1-x} + \text{princ. value} \int_{-1}^{1} \frac{\mu(t)}{x-t} dt.$$

Since $\mu(t)$ has a derivative at x then princ. value $\int_{-1}^{1} \dfrac{\mu(t)}{x-t} dt$ exists and is finite.

In a similar way we can verify the assertion of the lemma when x is an end point of $[-1, 1]$.

Since the derivative $\mu'(t)$ exists almost everywhere the integral (12.2.9) is finite almost everywhere. This completes the proof of Lemma 1.

Consider the logarithmic potential $u(z) = \int_{-1}^{1} ln \dfrac{1}{|z-t|} d\mu(t)$. Let x be a point of $[-1, 1]$ and assume that the point $z = x + iy$ approaches x in the vertical direction.

Lemma 2. *For any $x \in [-1, 1]$*

$$\lim_{y\to 0} \int_{-1}^{1} ln \frac{1}{|z-t|} d\mu(t) = \int_{-1}^{1} ln \frac{1}{|x-t|} d\mu(t) \qquad (12.2.10)$$

regardless of whether the integral on the right side is finite or infinite.

Proof. Consider a small segment $x - \epsilon \leq t \leq x + \epsilon$ close to x and let E_ϵ be the part of $[-1, 1]$ which remains after we remove the segment $[x - \epsilon, x + \epsilon]$ from $[-1, 1]$. For small values of y we have $|z - t| < 1$ for each $t \epsilon [x - \epsilon, x + \epsilon]$. Then $ln \dfrac{1}{|z - t|} > 0$ and since $\mu(t)$ is nondecreasing then we must have

$$\int_{E_\epsilon} ln \frac{1}{|z - t|} d\mu(t) \leq \int_{-1}^{1} ln \frac{1}{|z - t|} d\mu(t).$$

Passing to the limit as $y \to 0$ in the integral over E_ϵ and using $ln \dfrac{1}{|z - t|} < ln \dfrac{1}{|x - t|}$ we obtain

$$\int_{E_\epsilon} ln \frac{1}{|x - t|} d\mu(t) \leq \liminf_{y \to 0} \int_{-1}^{1} ln \frac{1}{|z - t|} d\mu(t) \leq$$

$$\leq \limsup_{y \to 0} \int_{-1}^{1} ln \frac{1}{|z - t|} d\mu(t) \leq \int_{-1}^{1} ln \frac{1}{|x - t|} d\mu(t).$$

But as $\epsilon \to 0$

$$\int_{E_\epsilon} ln \frac{1}{|x - t|} d\mu(t) \to \int_{-1}^{1} ln \frac{1}{|x - t|} d\mu(t)$$

independent of whether this last integral is finite or infinite. Therefore $\liminf\limits_{y \to 0} \int_{-1}^{1}$ and $\limsup\limits_{y \to 0} \int_{-1}^{1}$ must both coincide with $\int_{-1}^{1} ln \dfrac{1}{|x - t|} d\mu(t)$. This proves lemma 2.

We now study $u_n(x) = \displaystyle\int_{-1}^{1} ln \frac{1}{|x - t|} d\mu_n(t)$. If $x = x_k^{(n)}$ we define $u_n(x) = \infty$.

Lemma 3. *Almost everywhere on* $-1 \leq x \leq 1$

$$\liminf_{n \to \infty} \int_{-1}^{1} ln \frac{1}{|x - t|} d\mu_n(t) = \int_{-1}^{1} ln \frac{1}{|x - t|} d\mu(t). \quad (12.2.11)$$

Proof. By the definition of $ln_N x$ we have

$$\int_{-1}^{1} ln \frac{1}{|x - t|} d\mu_n(t) \leq \int_{-1}^{1} ln_N \frac{1}{|x - t|} d\mu_n(t).$$

Passing to the limit as $n \longrightarrow \infty$ gives

$$\liminf_{n \to \infty} \int_{-1}^{1} ln \frac{1}{|x-t|} \, d\mu_n(t) \geq \int_{-1}^{1} ln_N \frac{1}{|x-t|} \, d\mu(t)$$

and since this inequality is true for each N

$$\liminf_{n \to \infty} \int_{-1}^{1} ln \frac{1}{|x-t|} \, d\mu_n(t) \geq \int_{-1}^{1} ln \frac{1}{|x-t|} \, d\mu(t). \quad (12.2.12)$$

We now find an upper bound for the left side of (12.2.12). The function $\int_{-1}^{1} ln \frac{1}{|x-t|} \, d\mu_n(t)$ is bounded from below by $ln \frac{1}{2}$ and for any $\alpha, \beta \epsilon$ $[-1, 1]$ we have by Fatou's theorem:

$$\int_{\alpha}^{\beta} \liminf_{n \to \infty} \int_{-1}^{1} ln \frac{1}{|x-t|} \, d\mu_n(t) \, dx \leq \liminf_{n \to \infty} \int_{\alpha}^{\beta} \int_{-1}^{1} ln \frac{1}{|x-t|} \, d\mu_n(t) \, dx =$$

$$= \liminf_{n \to \infty} \int_{-1}^{1} \int_{\alpha}^{\beta} ln \frac{1}{|x-t|} \, dx d\mu_n(t).$$

In this last integral we can pass to the limit under the integral sign because $\int_{\alpha}^{\beta} ln \frac{1}{|x-t|} \, dx$ is a continuous function of t:

$$\int_{\alpha}^{\beta} \liminf_{n \to \infty} \int_{-1}^{1} ln \frac{1}{|x-t|} \, d\mu_n(t) \, dx \leq \int_{-1}^{1} \int_{\alpha}^{\beta} ln \frac{1}{|x-t|} \, dx d\mu(t).$$

Since $ln \frac{1}{|x-t|} \geq ln \frac{1}{2}$ from Fubini's theorem we can change the order of integration on the right side of this inequality

$$\int_{\alpha}^{\beta} \liminf_{n \to \infty} \int_{-1}^{1} ln \frac{1}{|x-t|} \, d\mu_n(t) \, dx \leq \int_{\alpha}^{\beta} \int_{-1}^{1} ln \frac{1}{|x-t|} \, d\mu(t) \, dx.$$

This inequality is valid for each $\alpha, \beta \epsilon [-1, 1]$ and thus we have almost everywhere on $-1 \leq x \leq 1$

$$\liminf_{n \to \infty} \int_{-1}^{1} ln \frac{1}{|x-t|} \, d\mu_n(t) \leq \int_{-1}^{1} ln \frac{1}{|x-t|} \, d\mu(t). \quad (12.2.12^*)$$

Combining (12.2.12) and (12.2.12*), completes the proof of Lemma 3.

Let E denote the set of points of $[-1, 1]$ for which (12.2.11) holds and for which $\int_{-1}^{1} ln \frac{1}{|x - t|} d\mu(t)$ is finite. The set E differs from $[-1, 1]$ by only a set of measure zero.

Lemma 4. *If the matrix X has the property that for the above indicated values of n the remainder of the interpolation $r_n(x)$ strives to zero for any $x \in [-1, 1]$ and for any function of the form $f(x) = \frac{1}{x - a}$ where a lies outside $[-1, 1]$ then the potential (12.2.9) is a constant on E.*

Proof. Suppose the converse is true. Then we can show that there exists a function $f(x) = \frac{1}{x - a}$ for which the remainder will not tend to zero.

Let x_1 and x_2 be two points of E for which (12.2.9) has different values. We can assume that

$$u(x_1) < u(x_2).$$

Let $\delta = u(x_2) - u(x_1)$ and take $\epsilon < \delta$. Consider a straight line passing through x_2 parallel to the imaginary axis. As z approaches x_2 along this line $u(z)$ will approach $u(x_2)$ by Lemma 2. As z approaches x_2 we also note that $ln \frac{1}{|z - t|}$ increases for each t. Since $\mu(t)$ is a nondecreasing function then $u(z)$ will also increase as z approaches x_2.

Thus on this line there exists a point $z_2 \neq x_2$ for which

$$u(x_2) - \tfrac{1}{3}\epsilon < u(z_2) < u(x_2).$$

We fix z_2 and construct the function

$$f(x) = \frac{1}{x - z_2}.$$

The remainder of the interpolation for this function is

$$r_n(x) = r_n\left(\frac{1}{x - z_2}; x\right) = \frac{\omega_n(x)}{\omega_n(z_2)(x - z_2)}$$

so that at $x = x_1$

$$r_n(x_1) = \frac{\omega_n(x_1)}{\omega_n(z_2)(x_1 - z_2)}$$

$$\left| r_n(x_1) \right| = \frac{1}{\left| x_1 - z_2 \right|} \frac{\left| \omega_n(x_1) \right|}{\left| \omega_n(z_2) \right|} =$$

$$= \exp n \left[\int_{-1}^{1} \ln \frac{1}{\left| z_2 - t \right|} d\mu_n(t) - \int_{-1}^{1} \ln \frac{1}{\left| x_1 - t \right|} d\mu_n(t) \right] =$$

$$= \frac{1}{\left| x_1 - z_2 \right|} \exp n \left[u_n(z_2) - u_n(x_1) \right].$$

By Lemma 3 there exists an infinite sequence of values of n for which

$$\left| u_n(x_1) - u(x_1) \right| < \tfrac{1}{3} \epsilon.$$

Therefore there exists such a sequence of numbers n for which

$$u_n(z_2) - u_n(x_1) = u(z_2) - u(x_1) - \left[u(z_2) - u_n(z_2) \right] - \left[u_n(x_1) - u(x_1) \right] >$$

$$> \delta - \tfrac{1}{3}\epsilon - \tfrac{1}{3}\epsilon - \tfrac{1}{3}\epsilon = \delta - \epsilon > 0.$$

For these values of n

$$\left| r_n(x_1) \right| > \frac{1}{\left| x_1 - z_2 \right|} \exp n(\delta - \epsilon)$$

and the interpolation thus diverges at x_1 for $f(x) = \dfrac{1}{x - z_2}$. This proves Lemma 4.

In the statement of Theorem 6 we assumed that the interpolation converges on $[-1, 1]$ for each function which is analytic on $[-1, 1]$. Then this will be true for a function of the form $f(x) = \dfrac{1}{x - a}$ for $a \,\overline{\epsilon}\, [-1, 1]$ and we have seen that $u(x)$ is then a constant on E and therefore is a constant almost everywhere on $[-1, 1]$.

In order to complete the proof of Theorem 6 we still have to prove the following lemma.

Lemma 5. *If the logarithmic potential* (12.2.9) *is almost everywhere constant on* $[-1, 1]$ *then* $\mu(t)$ *is the Chebyshev distribution function*

$$\mu(t) = \frac{1}{\pi} \int_{-1}^{t} \frac{dx}{\sqrt{1 - x^2}}.$$

Proof. Consider the potential

$$u(z) = \int_{-1}^{1} \ln \frac{1}{\left| z - t \right|} d\mu(t) \qquad (12.2.13)$$

defined in the z plane. In order to use results from the theory of the
Poisson integral and the theory of trigonometric series we pass from the
z plane to a circle.

In the z plane we make a cut along the segment $[-1, 1]$ and distinguish
the two sides of the cut. We transform (12.2.13) into an integral along
the contour λ consisting of both sides of the cut. To do this it is sufficient to represent (12.2.13) in the form

$$u(z) = \frac{1}{2} \int_{-1}^{1} ln \ \frac{1}{|z - t|} \ d\mu(t) - \frac{1}{2} \int_{1}^{-1} ln \frac{1}{|z - t|} \ d\mu(t)$$

and introduce the function $\nu(t)$ defined on λ by

$$\nu(t) = \begin{cases} \frac{1}{2} \mu(t) & \text{on the top of the cut} \\ 1 - \frac{1}{2}\mu(t) & \text{on the bottom of the cut.} \end{cases} \qquad (12.2.14)$$

Then

$$u(z) = \int_{\lambda} \ ln \ \frac{1}{|z - t|} \ d\nu(t). \qquad (12.2.15)$$

The integration is carried out along the top of the cut from -1 to 1 and in
the opposite direction along the bottom.

In the plane $\zeta = \rho e^{i\phi}$ consider the circle $|\zeta| \leq 1$. This circle is trans-
formed onto the z plane with the cut along $[-1, 1]$ by

$$z = \frac{1}{2}(\zeta + \zeta^{-1}).$$

The point $\tau = e^{i\psi}$ of the circumference corresponds to the point $t = \frac{1}{2}(\tau + \tau^{-1}) = \frac{1}{2}(e^{i\psi} + e^{-i\psi}) = \cos \psi$. As ψ varies from $-\pi$ to π we pass around
the contour λ in the above indicated direction. The function $\nu(t)$, de-
fined on λ, corresponds to the function

$$\nu(t) = \nu(\cos \psi) = F(\psi) \qquad -\pi \leq \psi \leq \pi$$

of the polar angle ψ.

The contour integral (12.2.15) corresponds to the following integral
over the circumference of the circle in the ζ plane:

$$u(z) = \int_{-\pi}^{\pi} ln \left| \frac{2\zeta}{1 - 2\zeta \cos \psi + \zeta^2} \right| dF(\psi) =$$

$$= ln \ 2|\zeta| + \int_{-\pi}^{\pi} ln \ \frac{1}{|\zeta - e^{i\psi}| \ |\zeta - e^{-i\psi}|} \ dF(\psi) =$$

$$= ln \ 2|\zeta| + I(\zeta).$$

The integral $I(\zeta)$ splits into the sum of two logarithmic potentials which are harmonic in the circle $|\zeta| < 1$:

$$I(\zeta) = \int_{-\pi}^{\pi} ln \frac{1}{|\zeta - e^{i\psi}|} \, dF(\psi) + \int_{-\pi}^{\pi} ln \frac{1}{|\zeta - e^{-i\psi}|} \, dF(\psi) = I_1(\zeta) + I_2(\zeta).$$

Because of the similarity of $I_1(\zeta)$ and $I_2(\zeta)$ it will suffice to only study $I_1(\zeta)$. We can see that $I_1(\zeta)$ can be represented as a Poisson-Lebesgue integral.[6] Let E be any measurable set on $[-\pi, \pi]$ with measure $mE \leq \delta$:

$$\int_E I_1(\rho e^{i\phi}) \, d\phi = \int_E \int_{-\pi}^{\pi} ln \frac{1}{|\rho e^{i\phi} - e^{i\psi}|} \, dF(\psi) \, d\phi$$

$$= \int_{-\pi}^{\pi} \left[\int_E ln \frac{1}{|\rho e^{i\phi} - e^{i\psi}|} \, d\phi \right] dF(\psi).$$

Here it was possible to change the order of integration by Fubini's theorem since $ln \dfrac{1}{|\zeta - e^{i\psi}|}$ is bounded from below by $ln \dfrac{1}{2}$. The inside integral has the upper bound

$$\left| \int_E ln \frac{1}{|\rho e^{i\phi} - e^{i\psi}|} \, d\phi \right| \leq \int_E ln \frac{1}{|\sin(\phi - \psi)|} \, d\phi \leq \int_{-\delta/2}^{\delta/2} ln \frac{1}{|\sin x|} \, dx.$$

---

[6] The function $v(\zeta)$ can be represented as a Poisson-Lebesgue integral if it is harmonic in the circle $|\zeta| < 1$ and if there exists on $[-\pi, \pi]$ a summable function $f(\psi)$ for which

$$v(\zeta) = \frac{1}{2\pi} \int_{-\pi}^{\pi} f(\psi) \frac{1 - \rho^2}{1 - 2\rho \cos(\psi - \phi) + \rho^2} \, d\psi.$$

The following theorem is known: A necessary and sufficient condition that a function $v(\zeta)$ which is harmonic in the circle $|\zeta| < 1$ can be represented as a Poisson-Lebesgue integral is that the family of functions $F_\rho(a) = \int_0^a v(\rho e^{i\phi}) \, d\phi$ be uniformly absolutely continuous in a, that is for all $\rho < 1$ and each $\epsilon > 0$ there exists a number $\delta(\epsilon) > 0$ such that for each set E of measure $mE < \delta(\epsilon)$ we have

$$\left| \int_E v(\rho e^{i\phi}) \, d\phi \right| < \epsilon.$$

It is also known that as the point ζ approaches a point $\psi = \psi_0$ of the circumference by any path not tangent to the circle then for almost all values of ψ_0, $v(\zeta) \longrightarrow f(\psi_0)$. See, for example, I. I. Privalov, *Boundary Properties of Analytic Functions*, Gostekhizdat, Moscow, 1950, Chap. 1, Sec. 3 (Russian).

Thus for each $\epsilon > 0$ we can find a $\delta(\epsilon)$ for which

$$\left| \int_E I_1 \left(\rho e^{i\phi} \right) d\phi \right| < \epsilon \qquad \rho < 1.$$

Hence it follows that $I_1(\zeta)$ can be represented as a Poisson-Lebesgue integral.

Therefore $I_1(\zeta)$ and also $I(\zeta)$ can be represented as a Poisson-Lebesgue integral.

As the point $z = x + iy$ approaches the segment $[-1, 1]$ along a line parallel to the imaginary axis, $u(z)$ tends to a constant for almost all x. When we transform to the circle $|\zeta| < 1$ the indicated line transforms into a line which is orthogonal to the circumference $|\zeta| = 1$. As we approach the boundary along this curve $u(z)$ strives almost everywhere to a constant value and since $ln\, 2 |\zeta|$ tends to $ln\, 2$ then on the circumference $I(\zeta)$ will in the limit be almost everywhere constant. Since $I(\zeta)$ can be represented as a Poisson-Lebesgue integral $I(\zeta)$ is a constant everywhere in the circle. But since $I(0) = 0$ then

$$I(\zeta) = 0$$

everywhere in the circle.

It is easy to see that the functions $ln \dfrac{1}{\left| \zeta - e^{i\psi} \right|}$ and $ln \dfrac{1}{\left| \zeta - e^{-i\psi} \right|}$, $\zeta = \rho e^{i\phi}$, have the following expansions in powers of ρ for $\rho < 1$:

$$ln \frac{1}{\left| \zeta - e^{i\psi} \right|} = \sum_{k=1}^{\infty} \frac{1}{k} \rho^k \cos k(\phi - \psi)$$

$$ln \frac{1}{\left| \zeta - e^{-i\psi} \right|} = \sum_{k=1}^{\infty} \frac{1}{k} \rho^k \cos k(\phi + \psi).$$

Therefore

$$I(\zeta) = \int_{-\pi}^{\pi} \sum_{k=1}^{\infty} \frac{1}{k} \rho^k \left[\cos k(\phi - \psi) + \cos k(\phi + \psi) \right] dF(\psi) =$$

$$= 2 \sum_{k=1}^{\infty} \frac{1}{k} \rho^k \cos k\phi \int_{-\pi}^{\pi} \cos k\psi \, dF(\psi) = 0.$$

Hence

$$\int_{-\pi}^{\pi} \cos k\psi \, dF(\psi) = F(\psi) \cos k\psi \Big|_{-\pi}^{\pi} + k \int_{-\pi}^{\pi} F(\psi) \sin k\psi \, d\psi =$$

$$= (-1)^k \left[F(\pi) - F(-\pi) \right] + k \pi b_k = 0.$$

Here b_k is the coefficient of sin $k\psi$ in the Fourier expansion of $F(\psi)$. From $F(\pi) = 1$ and $F(-\pi) = 0$ we have

$$b_k = -\frac{(-1)^k}{k\pi}.$$

From the definitions of $\nu(t)$ and $F(\psi)$ we see that

$$F(\psi) = \begin{cases} \frac{1}{2}\mu(\cos\psi) & -\pi \le \psi \le 0 \\ 1 - \frac{1}{2}\mu(\cos\psi) & 0 \le \psi \le \pi. \end{cases}$$

The even part of $F(\psi)$ is

$$\tfrac{1}{2}[F(\psi) + F(-\psi)] = \tfrac{1}{2}$$

and hence the coefficients a_k of cos $k\psi$ in the Fourier expansion of $F(\psi)$ are

$$a_0 = \tfrac{1}{2}, \quad a_k = 0, \quad k = 1, 2, \ldots.$$

Thus we obtain

$$F(\psi) = \frac{1}{2} - \sum_{k=1}^{\infty} \frac{(-1)^k}{k\pi} \sin k\psi = \frac{1}{2} + \frac{1}{2\pi}\psi.$$

If we return to the z plane we have $t = \cos\psi$ from which follows

$$\mu(t) = 1 - \frac{1}{\pi}\operatorname{Arc}\cos t = \frac{1}{\pi}\int_{-1}^{t}\frac{dx}{\sqrt{1-x^2}}.$$

This proves Lemma 5 and completes the proof of Theorem 6.

From this result it is not difficult to establish the corresponding theorem for quadrature formulas.

Theorem 7. *If the interpolatory quadrature process defined by (12.2.1) for the segment [−1, 1] converges for each function $\sigma(x)$ of bounded variation and for any analytic function $f(x)$ on [−1, 1] then the matrix of nodes X has a limiting distribution function which is the Chebyshev function (12.2.5).*

Proof. Consider the remainder of the quadrature

$$R_n(f) = \int_{-1}^{1} r_n(x)\, d\sigma(x)$$

where $r_n(x)$ is the remainder of the interpolation. Take an arbitrary point x on [−1, 1]. As the function $\sigma(x)$ we take a piece-wise constant function which has a unit jump at x. For such a $\sigma(x)$

$$R_n(f) = r_n(x)$$

and the convergence of the quadrature process is equivalent to convergence of the interpolation. Then the proof is completed by using Theorem 6.

12.3. CONVERGENCE OF THE GENERAL QUADRATURE PROCESS

In this section we study the general quadrature process (12.1.3) defined by the matrix of nodes (12.1.1) and the matrix of coefficients (12.1.2). The weight function $p(x)$ can be any summable function. We assume that we are given a certain class F of functions f. We wish to determine what conditions X and A must satisfy in order that the quadrature process will converge for each $f \epsilon F$. This problem has been studied for many classes F. We consider here only the simplest and most important of these results.

In the remainder of this section we assume that the segment of integration is finite.

Theorem 8. *In order that the quadrature process* (12.1.3) *converge for each continuous function* f *on* $[a, b]$ *the following two conditions are necessary and sufficient:*

1. *The process converge for each polynomial;*
2. *There exists a number* K *for which* [7]

$$\sum_{k=1}^{n} \left| A_k^{(n)} \right| \le K \qquad (12.3.1)$$

for $n = 1, 2, \ldots$.

Proof. If in the class of continuous functions on $[a, b]$ we define a norm by $\|f\| = \max\limits_{[a, b]} |f(x)|$ then this class can be considered as the

Banach space C. The quadrature sum $Q_n(f) = \sum\limits_{k=1}^{n} A_k^{(n)} f(x_k^{(n)})$ and the

integral $I(f) = \int_a^b p(x) f(x)\, dx$ are two linear functionals defined on C.

The values $Q_n(f)$ and $I(f)$ belong to the set of real numbers which is also a Banach space.

---

[7] The sufficiency of this condition was proved by V. A. Steklov, the necessity by G. Polya.

We can then apply Theorem 1 of Section 4.3 which gives conditions for the convergence of a sequence of linear operators. A necessary and sufficient condition that such a sequence converge is that 1) it converge on a set of elements dense in the space where the operators are defined and 2) that the norms of the operators have a common bound.

From the theorem of Weierstrass it is known that we can uniformly approximate each continuous function on $[a, b]$ by means of polynomials and thus the class of polynomials is a set of functions which is dense in C. This establishes the first condition of the theorem.

The norm of the functional $Q_n(f)$ is

$$\|Q_n\| = \sup_{|f| \leq 1} \left| \sum_{k=1}^{n} A_k^{(n)} f(x_k^{(n)}) \right| = \sum_{k=1}^{n} |A_k^{(n)}|.$$

Thus (12.3.1) is the condition that the functionals have a common bound. This completes the proof.

The following two theorems are simple corollaries to Theorem 8.

Theorem 9. *If all the coefficients $A_k^{(n)}$ are nonnegative then in order that the quadrature process converge for each continuous function it is necessary and sufficient that it converge for each polynomial.*

Proof. The necessity of the condition is obvious. If the process converges for each polynomial then for $f(x) \equiv 1$

$$Q_n(1) \longrightarrow \int_a^b p(x)\,dx \qquad \text{as } n \longrightarrow \infty.$$

Therefore the values of $Q_n(1)$, $n = 1, 2, \ldots$, are bounded:

$$Q_n(1) \leq K.$$

But

$$\sum_{k=1}^{n} |A_k^{(n)}| = \sum_{k=1}^{n} A_k^{(n)} = Q_n(1) \leq K$$

and thus by Theorem 8 the quadrature process converges for each continuous function.

Theorem 10. *For an interpolatory quadrature process to converge for any continuous function it is necessary and sufficient that*

$$\sum_{k=1}^{n} |A_k^{(n)}| \leq K < \infty.$$

The second condition of Theorem 8 coincides with the condition of Theorem 10. The first condition of Theorem 8 is fulfilled since if $f(x)$ is a polynomial of degree m then for any $n > m$, $Q_n(f) = \int_a^b p(x) f(x)\, dx.$
This establishes the theorem.

We now discuss conditions for convergence of the quadrature process in classes of differentiable functions.

As above we enumerate the nodes in increasing order and introduce the piece-wise constant functions $F_{n,0}(x)$ for the nodes and coefficients

$$F_{n,0}(x) = \sum_{k=1}^{n} A_k^{(n)} E\left(x - x_k^{(n)}\right).$$

We also consider the primitive functions of any order r of the functions $F_{n,0}(x)$ defined by the initial conditions $F_{n,r}^{(j)}(a) = 0$ $(j = 0, 1, \ldots, r-1)$:

$$F_{n,r}(x) = \int_a^x F_{n,0}(t)\, \frac{(x-t)^{r-1}}{(r-1)!}\, dt =$$

$$= \sum_{k=1}^{n} A_k^{(n)} E\left(x - x_k^{(n)}\right) \frac{\left(x - x_k^{(n)}\right)^r}{r!}. \qquad (12.3.2)$$

Theorem 11. *In order that the quadrature process* (12.1.3) *converge as* $n \longrightarrow \infty$ *for each function* $f \in C_r[a, b]$ *it is necessary and sufficient that the following conditions be fulfilled:*

1. *The process converge for each polynomial;*
2. *The total variation of the primitive functions* $F_{n,r}(x)$ *of order* r *have a common bound for* $n = 1, 2, \ldots$:

$$\operatorname*{Var}_{[a,b]} F_{n,r}(t) \leq M.$$

Proof. If $f \in C_r[a, b]$, $r \geq 1$, then expanding f in a Taylor series about the point b we obtain the representation

$$f(x) = \sum_{i=0}^{r-1} \frac{f^{(i)}(b)}{i!} (x-b)^i + \int_b^x f^{(r)}(t)\, \frac{(x-t)^{r-1}}{(r-1)!}\, dt =$$

$$= \sum_{i=0}^{r-1} \frac{f^{(i)}(b)}{i!} (x-b)^i + (-1)^r \int_a^b f^{(r)}(t) E(t-x) \frac{(t-x)^{r-1}}{(r-1)!}\, dt.$$

Conversely, for any numbers $f^{(i)}(b)$ and any continuous function $f^{(r)}(t)$ on $[a, b]$ the function $f(x)$ defined by this equation belongs to $C_r[a, b]$.

The remainder $R_n(f)$ is

$$R_n(f) = \sum_{i=0}^{r-1} \frac{f^{(i)}(b)}{i!} R_n[(x-b)^i] +$$

$$+ (-1)^r \int_a^b f^{(r)}(t) \left[\int_a^b p(x) E(t-x) \frac{(t-x)^{r-1}}{(r-1)!} dx - \right.$$

$$\left. - \sum_{k=1}^n A_k^{(n)} E(t-x_k^{(n)}) \frac{(t-x_k^{(n)})^{r-1}}{(r-1)!} \right] dt =$$

$$= \sum_{i=0}^{r-1} \frac{f^{(i)}(b)}{i!} R_n[(x-b)^i] + \qquad (12.3.3)$$

$$+ (-1)^r \int_a^b f^{(r)}(t) \left[\int_a^t p(x) \frac{(t-x)^{r-1}}{(r-1)!} dx - F_{n,r-1}(t) \right] dt.$$

Because the parameters $f^{(i)}(b)$ $(i = 0, 1, \ldots, r-1)$ and $f^{(r)}(t)$ are independent, convergence of the quadrature process is equivalent to

$$R_n[(x-b)^i] \longrightarrow 0 \qquad (i = 0, 1, \ldots, r-1) \qquad (12.3.4)$$

$$R_n^*(f^{(r)}) \longrightarrow 0 \qquad\qquad (12.3.5)$$

where

$$R_n^*(f^{(r)}) = \int_a^b f^{(r)}(t) \left[\int_a^t p(x) \frac{(x-t)^{r-1}}{(r-1)!} dx - F_{n,r-1}(t) \right] dt.$$

Condition (12.3.4) means that the quadrature process must converge for each polynomial of degree $\leq r-1$.

Condition (12.3.5) must be satisfied for any continuous function $f^{(r)}(t)$. Introducing the norm $\|f^{(r)}\| = \max_t |f^{(r)}(t)|$ for the class of functions $f^{(r)}(t)$ this class becomes the Banach space C. By Theorem 1 of Section 4.3 condition (12.3.5) is equivalent to the two requirements:

1. The functional $R_n^*(f^{(r)})$ must tend to zero on a set of elements dense in C. For this set we can take the set of polynomials. But the requirement that $R_n^*(f^{(r)}) \longrightarrow 0$ as $n \longrightarrow \infty$ when $f^{(r)}(t)$ is a polynomial together with (12.3.4) is the same as the condition that the quadrature process converge for polynomials.

2. The norm of the functionals R_n^* $(n = 1, 2, \ldots)$ must have a common bound:

$$\|R_n^*\| = \int_a^b \left| \int_a^t p(x) \frac{(t-x)^{r-1}}{(r-1)!} dx - F_{n,r-1}(t) \right| dt \leq L \qquad (n = 1, 2, \ldots)$$

Since $\int_a^b \left| \int_a^t p(x) \frac{(t-x)^{r-1}}{(r-1)!} \, dx \right| dt$ is independent of n then the bounded-ness of $\| R_n^* \|$ is equivalent to

$$\int_a^b | F_{n,r-1}(t) | \, dt \leq M \qquad (n = 1, 2, \ldots).$$

Since $\dfrac{d}{dt} F_{n,r}(t) = F_{n,r-1}(t)$ this last inequality is equivalent to

$$\operatorname*{Var}_{[a,b]} F_{n,r}(t) \leq M \qquad (n = 1, 2, \ldots).$$

It can also be shown that the above discussion is also valid for $r = 0$. This proves Theorem 11.

We mention a particular case of this theorem for the class of functions with a continuous derivative on $[a, b]$, that is the case $r = 1$. The function $F_{n,0}(t)$ is the piece-wise constant function which has the values:

$$F_{n,0}(t) = \sum_{k=1}^{n} A_k^{(n)} E(t - x_k^{(n)}) = \begin{cases} 0 & \text{for } a \leq t < x_1^{(n)} \\ A_1^{(n)} & \text{for } x_1^{(n)} < t < x_2^{(n)} \\ A_1^{(n)} + A_2^{(n)} & \text{for } x_2^{(n)} < t < x_3^{(n)} \\ \cdots\cdots\cdots\cdots \\ A_1^{(n)} + \cdots + A_n^{(n)} & \text{for } x_n^{(n)} < t \leq b. \end{cases}$$

Hence

$$\operatorname*{Var}_{[a,b]} F_{n,1}(t) = \int_a^b | F_{n,0}(t) | \, dt = | A_1^{(n)} | (x_2^{(n)} - x_1^{(n)}) +$$

$$+ | A_1^{(n)} + A_2^{(n)} | (x_3^{(n)} - x_2^{(n)}) + \cdots +$$

$$+ | A_1^{(n)} + \cdots + A_n^{(n)} | (b - x_n^{(n)}).$$

Therefore we have:

Theorem 12. *In order that the quadrature process (12.1.9) converge for any function with a continuous derivative the following conditions are necessary and sufficient:*

1. *The process converge for each polynomial;*
2. *There exists a number M for which*

$$| A_1^{(n)} | (x_2^{(n)} - x_1^{(n)}) + | A_1^{(n)} + A_2^{(n)} | (x_3^{(n)} - x_2^{(n)}) + \qquad (12.3.6)$$

$$+ \cdots + | A_1^{(n)} + \cdots + A_n^{(n)} | (b - x_n^{(n)}) \leq M$$

for $n = 1, 2, \ldots$.

We will say that f belongs to the class $A_r[a, b]$ if $f^{(r)}$ is an absolutely continuous function.

If $f \epsilon A_r[a, b]$ then we can expand it in a Taylor series:

$$f(x) = \sum_{i=0}^{r} \frac{f^{(i)}(b)}{i!}(x-b)^i + \int_b^x f^{(r+1)}(t) \frac{(x-t)^r}{r!} dt =$$
$$= \sum_{i=0}^{r} \frac{f^{(i)}(b)}{i!}(x-b)^i + (-1)^{r+1} \int_a^b f^{(r+1)}(t) E(t-x) \frac{(t-x)^r}{r!} dt. \qquad (12.3.7)$$

Here $f^{(i)}(b)$ $(i = 0, 1, \ldots, r)$ are arbitrary numbers and $f^{(r+1)}(t)$ is an arbitrary summable function on $[a, b]$.

Theorem 13. *The following conditions are necessary and sufficient for the quadrature process to converge for each $f \epsilon A_r[a, b]$:*

1. The process converge for each polynomial;

2. The primitive functions $F_{n,r}(t)$ of order r for $F_{n,0}(t)$ have a common bound

$$|F_{n,r}(t)| \leq M, \quad a \leq x \leq b, \quad n = 1, 2, \ldots. \qquad (12.3.8)$$

Proof. If $f \epsilon A_r[a, b]$ then from (12.3.7) the remainder $R_n(f)$ can be expressed as

$$R_n(f) = \sum_{i=0}^{r} \frac{f^{(i)}(b)}{i!} R_n[(x-b)^i] +$$
$$+ (-1)^{r+1} \int_a^b f^{(r+1)}(t) \left[\int_a^t p(x) \frac{(t-x)^r}{r!} dx - F_{n,r}(t) \right] dt.$$

Thus the convergence of the quadrature process in $A_r[a, b]$ is equivalent to

$$R_n[(x-b)^i] \longrightarrow 0 \quad \text{as } n \longrightarrow \infty \quad (i = 0, 1, \ldots, r) \qquad (12.3.9)$$

and

$$R_n^*(f^{(n+1)}) \longrightarrow 0 \quad \text{as } n \longrightarrow \infty \qquad (12.3.10)$$

$$R_n^*(f^{(n+1)}) = \int_a^b f^{(r+1)}(t) \left[\int_a^t p(x) \frac{(t-x)^r}{r!} dx - F_{n,r}(t) \right] dt.$$

The rest of the argument is very similar to the argument used in proving Theorem 8. We introduce the norm

$$\|f^{(r+1)}\| = \int_a^b |f^{(r+1)}(t)|\, dt.$$

Thus the space of functions $f^{(r+1)}$ coincides with the Banach space L and we can apply Theorem 1 of Section 4.3 to obtain a condition that $R_n^*(f^{(r+1)}) \longrightarrow 0$. The set of polynomials is dense in L. The requirement that R_n^* converge for each polynomial together with (12.3.9) is the same as the requirement that the quadrature process converge for each polynomial. By (4.2.6) the norm of $R_n^*(f^{(r+1)})$ is

$$\|R_n^*\| = (b-a) \max_t \left| \int_a^t p(x) \frac{(t-x)^r}{r!}\, dx - F_{n,r}(t) \right|.$$

The integral $\int_a^t p(x) \dfrac{(t-x)^r}{r!}\, dx$ is independent of n and thus the condition that $\|R_n^*\|$ be bounded for $n = 1, 2, \ldots$ is equivalent to the condition that

$$|F_{n,r}(t)| \le M, \quad a \le t \le b, \quad n = 1, 2, \ldots.$$

This completes the proof.

We now mention the particular case $r = 0$ for which $A_0[a, b]$ is the class of absolutely continuous functions on $[a, b]$. The function $F_{n,0}(t)$ is the piece-wise constant function which has the values

$$0, \quad A_1^{(n)}, \quad A_1^{(n)} + A_2^{(n)}, \quad \ldots, A_1^{(n)} + \cdots + A_n^{(n)}$$

on the segments

$$[a, x_1^{(n)}], \quad [x_1^{(n)}, x_2^{(n)}], \quad \ldots, \quad [x_n^{(n)}, b]$$

respectively. Thus we obtain as a corollary to the last theorem:

Theorem[8] 14. *The following conditions are necessary and sufficient for the quadrature process* (12.1.3) *to converge for each absolutely continuous function f on $[a, b]$:*

1. *The process converge for each polynomial;*
2. *The partial sums of the quadrature coefficients*

$$A_1^{(n)}, \quad A_1^{(n)} + A_2^{(n)}, \quad \ldots, \quad A_1^{(n)} + \cdots + A_n^{(n)}, \quad n = 1, 2, \ldots$$

have a common bound:

$$\left| \sum_{k=1}^i A_k^{(n)} \right| \le M < \infty, \quad i = 1, 2, \ldots, n, \quad n = 1, 2, \ldots. \quad (12.3.11)$$

---

[8]This theorem was first proved, in a slightly different form, by S. M. Lozinskii, *Izv. Akad. Nauk SSSR. Ser. Mat.*, Vol. 4, 1940, pp. 113–26.

We now study convergence in one more class of functions. We will say that f belongs to the class $V_r[a, b]$ if $f^{(r)}$ is a function of bounded variation on $[a, b]$. The characteristic representation of a function in this class can also be obtained from the Taylor series:

$$f(x) = \sum_{i=0}^{r} \frac{f^{(i)}(b)}{i!}(x - b)^i + \int_b^x \frac{(x - t)^r}{r!} \, df^{(r)}(t) =$$

$$= \sum_{i=0}^{r} \frac{f^{(i)}(b)}{i!}(x - b)^i + (-1)^{(r+1)} \int_a^b E(t - x)\frac{(t - x)^r}{r!} df^{(r)}(t). \tag{12.3.12}$$

The parameters $f^{(i)}(b)$ are any numbers and $f^{(r)}(t)$ is any function of bounded variation on $[a, b]$.

Theorem 15. *In order that the quadrature process converge for each $f \in V_r[a, b]$ for $r \geq 1$ it is necessary and sufficient that:*

1. *The process converge for all polynomials of degree $\leq r$;*

2. *The primitive functions $F_{n,r}(x)$ of order r for $F_{n,0}(x)$ have a common bound*

$$|F_{n,r}(x)| \leq M < \infty, \quad a \leq x \leq b, \quad n = 1, 2, \ldots; \tag{12.3.13}$$

3. *For all $t \in [a, b]$*

$$F_{n,r}(t) \longrightarrow \int_a^t p(x)\frac{(t - x)^r}{r!} \, dx \text{ as } n \longrightarrow \infty.$$

Proof. If $f \in V_r[a, b]$ then using (12.3.12) the remainder $R_n(f)$ can be represented in the form:

$$R_n(f) = \sum_{i=0}^{r} \frac{f^{(i)}(b)}{i!} R_n[(x - b)^i] +$$

$$+ (-1)^{r+1} \int_a^b \left[\int_a^t p(x)\frac{(t - x)^r}{r!} \, dx - F_{n,r}(t)\right] df^{(r)}(t).$$

Since the parameters $f^{(i)}(b)$ $(i = 0, 1, \ldots, r)$ and $f^{(r)}(t)$ are independent then the condition that the quadrature process converge for all functions of $V_r[a, b]$ is equivalent to

$$\lim_{n \to \infty} R_n[(x - b)^i] = 0, \quad i = 0, 1, \ldots, r \tag{12.3.14}$$

and

$$\lim_{n \to \infty} R_n^*(f^{(r)}) = 0 \tag{12.3.15}$$

272 Approximate Calculation of Definite Integrals

where

$$R_n^*(f^{(r)}) = \int_a^b \left[\int_a^t p(x) \frac{(t-x)^r}{r!} dx - F_{n,r}(t) \right] df^{(r)}(t).$$

The first of these conditions means that the process must converge for all polynomials of degree $\leq r$. The functional R_n^* is defined on the linear space of functions of bounded variation. Without loss of generality we may assume that $f^{(r)}(a) = 0$. Then as a norm we take $\| f^{(r)} \| = \underset{[a,b]}{\text{Var}} f^{(r)}$. The set of functions then becomes the Banach space V. If $R_n^*(f^{(r)}) \longrightarrow 0$ as $n \longrightarrow \infty$ for each $f^{(r)}$ of V then by Theorem 1 of Section 4.3 the norms of the functionals R_n^* must have a common bound

$$\| R_n^* \| \leq N \qquad n = 1, 2, \ldots. \tag{12.3.16}$$

But

$$\| R_n^* \| = \max_t \left| \int_a^t p(x) \frac{(t-x)^r}{r!} dx - F_{n,r}(t) \right|$$

and since the integral in this expression is independent of n, condition (12.3.16) is equivalent to the second condition of the theorem.

To show the necessity of the third condition let x be an arbitrary point of $[a, b]$ and take $f^{(r)}$ to be a piece-wise constant function with a jump of unity at x. Then

$$R_n^*(f^{(r)}) = \int_a^x p(u) \frac{(x-u)^r}{r!} du - F_{n,r}(x).$$

Such a function $f^{(r)}$ determines the function f up to a polynomial of degree $r - 1$. If the quadrature process converges for this function then

$$R_n^*(f^{(r)}) \longrightarrow 0 \qquad \text{as } n \longrightarrow \infty.$$

This proves the necessity of the third condition.

We must still prove the sufficiency of all three conditions. The condition (12.3.14) is equivalent to the first condition of the theorem. There remains to be shown that the second and third conditions imply (12.3.15). But these conditions imply that

$$\Phi_n(t) = \int_a^t p(x) \frac{(t-x)^r}{r!} dx - F_{n,r}(t)$$

is bounded in absolute value by a certain number for all $t \, \epsilon \, [a, b]$ and all $n = 1, 2, \ldots$ and that for all $t \, \epsilon \, [a, b]$, $\Phi_n(t) \longrightarrow 0$ as $n \longrightarrow \infty$. If we transform the Stieltjes integral in R_n^* into a Lebesgue integral we can see that (12.3.15) will be satisfied.[9]

REFERENCES

Ia. L. Geronimus, *Theory of Orthogonal Polynomials*, Gostekhizdat, Moscow, 1950, Chap. 2, Sections 27, 28 (Russian).

V. I. Krylov, "On determining the smallest region of holomorphy in which convergence of the Hermite interpolation process is assured for any system of nodes," *Dokl. Akad. Nauk SSSR*, Vol. 78, 1951, pp. 857–59 (Russian).

V. I. Krylov, "Convergence of mechanical quadratures in classes of functions of different orders of differentiability," *Dokl. Akad. Nauk SSSR*, Vol. 101, 1955, pp. 801–02 (Russian).

R. O. Kuz′min, "On the theory of mechanical quadrature," *Izv. Leningrad. Polytehn. In-Ta. Otd. Estest. Mat.*, Vol. 32, 1931 (Russian).

S. M. Lozinskiĭ, "On formulas of mechanical quadrature," *Izv. Akad. Nauk SSSR, Ser. Mat..*, Vol. 4, 1940, pp. 113–26 (Russian).

Sh. E. Mikeladze, "Numerical integration," *Uspehi Mat. Nauk*, Vol. 3, 1948, pp. 3–88 (Russian).

I. P. Natanson, *Constructive Theory of Functions*, Part 3, Chap. 5, Gostekhizdat, Moscow, 1949 (Russian).

G. Polya, "Über die Konvergenz von Quadraturverfahren," *Math. Z.*, Vol. 37, 1933, pp. 264–86.

V. A. Steklov, "On the approximate calculation of definite integrals with the aid of formulas of mechanical quadrature," *Izv. Akad. Nauk SSSR*, (6) Vol. 10, 1916, pp. 169–86 (Russian).

V. A. Steklov, "Sur l'approximation des fonctions à l'aide des polynomes de Tchebycheff et sur les quadratures," *Izv. Akad. Nauk SSSR*, (6) Vol. 11, 1917, p. 187–218; pp. 535–66; p. 687–718.

V. A. Steklov, "Remarques sur les quadratures," *Izv. Akad. Nauk SSSR*, (6) Vol. 12, 1918, pp. 99–118.

---

[9]The following theorem is known: If the functions $f_n(x)$ are measurable on $[a, b]$ and if $|f_n(x)| \leq N < \infty$ for all n and if $f_n(x) \longrightarrow f(x)$ almost everywhere on $[a, b]$ then

$$\int_a^b f_n(x)\, dx \longrightarrow \int_a^b f(x)\, dx.$$

Part Three

APPROXIMATE
CALCULATION
OF INDEFINITE
INTEGRALS

CHAPTER 13

Introduction

13.1. PRELIMINARY REMARKS

The problem of calculating an integral with variable limits has been studied considerably less than the problem of calculating a definite integral which we discussed in Part 2.

We mention here several examples of integrals with variable limits which occur in applications. We consider cases in which only one of the limits of integration is variable and the other is fixed.

The simplest integral of this kind occurs in the problem of finding a primitive function. If we are given a function $f(x)$ which is continuous on the segment $[x_0, X]$ then any primitive of this function can be represented by the following formula:

$$y(x) = y_0 + \int_{x_0}^{x} f(t)\,dt \qquad x \in [x_0, X] \qquad (13.1.1)$$

and thus calculating $y(x)$ is equivalent to finding the value of the integral $\int_{x_0}^{x} f(t)\,dt.$

A more complicated example is the following integral which occurs in many applied problems:

$$y(x) = \int_{a}^{x} K(x - t)f(t)\,dt. \qquad (13.1.2)$$

Here $K(x - t)$ can be considered as a weight function whose value on $a \leq t \leq x$ depends only on the distance $x - t$ from the upper limit x; $f(x)$ is an arbitrary function of a certain class.

277

Another example is the Volterra integral equation

$$f(x) = \phi(x) + \int_a^x K(x, t) f(t) dt$$

and certain other problems which involve the integral

$$y(x) = \int_a^x K(x, t) f(t) dt \tag{13.1.3}$$

where the weight function $K(x, t)$ is an arbitrary function of x and t.

Methods for calculating the above integrals must take into account the properties of the weight function. For example, a computational scheme constructed for (13.1.3) can also be applied, in principle, to the calculation of the more special integral (13.1.1). Such a method, however, cannot be expected to be the very best for (13.1.1) since, for example, we might be able to use to advantage the fact that the weight function in (13.1.1) does not change sign. Thus we should develop separate methods for each of the above integrals.

In this book we will be exclusively concerned with the problem of calculating the integral (13.1.1).

Suppose it is necessary to calculate the value of (13.1.1) for a given set of values of the argument x: x_k ($k = 0, 1, 2, \ldots$). We assume that the calculations have been carried up to step n and that we have constructed[1] the following table of values of $y(x_n) = y_n$. We wish to find y_{n+1}. To do this we can use any of the previously calculated values y_k, $k \leq n$, and any values of $f(t)$ which are available for use.

x	y
x_0	y_0
x_1	y_1
...	...
x_n	y_n
x_{n+1}	

If $f(t)$ is given by a table of its values at the nodes x_k we will be restricted in our possible choice of values of $f(t)$ and any computational method will belong to the field of discrete analysis. One possible solution to the problem in this case is presented in Chapter 14.

For the present we assume that to compute y_{n+1} we may use values of $f(t)$ at any points we wish and we assume only that the number of these points is fixed. In this case the points may be selected to reduce the

---

[1] We do not consider, in this book, the problem of constructing the values of $y(x)$ near the beginning or near the end of the table; we only consider the problem of continuing the table.

error in computing y_{n+1}. As in the problem of computing a definite integral it is often desirable to construct formulas of the highest algebraic degree of precision. Formulas of this type will be discussed in Chapters 15 and 16.

The construction of quadrature formulas of the highest algebraic degree of precision for definite integrals is related to the problem of calculating an integral to within a certain precision with the smallest number of integrand values and thus with the least amount of work. In indefinite integration an additional way to reduce the computational work is to use each value of $f(t)$ to calculate not just one value of $y(x)$ but for many steps in the computation.

In Chapters 15 and 16 we discuss this problem of constructing methods which use values of $f(x_k)$ and y_k for calculating several values of $y(x)$. We discuss two methods in detail and do not attempt to treat all aspects of the problem.

The problem of calculating an indefinite integral has another special feature. One usually calculates $y(x)$ for a large number of values of x by the repeated application of some particular method. Each step produces an approximate value for $y(x)$. As a rule, the error will accumulate and increase from step to step. The rate of growth of the error depends on the computational method and for some methods the error can grow very rapidly and in only a few steps produce an undesirably large error.

We can illustrate these remarks with a simple example of a method which gives good accuracy for a small number of steps but which is totally unsuitable when the number of steps is large.

In order to compute $y(x_{n+1})$ suppose we desire to use the two preceding values of $y(x)$ and also the values of its derivative $y'(x) = f(x)$ at these points: $y(x_n)$, $y(x_{n-1})$, $f(x_n)$, $f(x_{n-1})$. Then it is natural to construct an interpolating polynomial using these values of the function and its derivative. This will be the Hermite interpolating polynomial with the two double nodes x_n and x_{n-1}. As can be verified from (3.3.8) this polynomial will be

$$y(x_{n+1}) = -4y(x_n) + 5y(x_{n-1}) + h[4f(x_n) + 2f(x_{n-1})] + r_n(x).$$

If we neglect the remainder $r_n(x)$ we obtain the approximate formula

$$y_{n+1} = -4y_n + 5y_{n-1} + h(4f_n + f_{n-1}) \qquad (13.1.4)$$

which is exact for all algebraic polynomials of degree ≤ 3. To use this formula we must know the first two values of $y(x)$: y_0 and y_1. Let us use (13.1.4) to evaluate the integral

$$y(x) = \int_0^x e^t \, dt = e^x - 1$$

on the segment [0, 1]. At first we take $h = 0.2$ and assume that $y(0) = 0$ and $y(0.2) \approx 0.22140$ are known and from these values calculate the following table which gives the approximate values of $y(x)$ together with the errors in these values.

x	$f(x)$	y_{approx}	$y - y_{approx}$
0.0	1.00000	0.00000	
0.2	1.22140	0.22140	
0.4	1.49182	0.49152	+ 0.00030
0.6	1.82212	0.82294	− 0.00082
0.8	2.22554	1.22026	+ 0.00528
1.0	2.71828	1.74294	− 0.02466

This table shows that the error grows very rapidly as we go farther down the table. The number of significant figures in the calculation shows that the large error is not due to rounding but to other causes.

We can easily see that the rapid rate of growth of the error is not due to the large interval size and that it can not be corrected by decreasing h. In fact let us try to obtain a more exact value of the integral by decreasing the step size to $h = 0.1$.

Here again we assume that we know the first two values of $y(x)$: $y(0) = 0$, $y(0.1) \approx 0.10517$. The new table is as follows.

x	$f(x)$	y_{approx}	$y - y_{approx}$
0.0	1.00000	0.00000	
0.1	1.10517	0.10517	
0.2	1.22140	0.22139	+ 0.00001
0.3	1.34986	0.34988	− 0.00002
0.4	1.49182	0.49165	+ 0.00017
0.5	1.64872	0.64950	− 0.00078
0.6	1.82212	0.81810	+ 0.00402
0.7	2.01375	1.03610	− 0.02235
0.8	2.22554	1.11602	+ 0.10952
0.9	2.45960	2.01039	− 0.55079
1.0	2.71828	− 1.03251	+ 2.75079

This smaller interval size gives a smaller error for only the single value $y(0.4)$. The error grows so rapidly that at the end of the table the error exceeds the size of the function.

It is easy to see that the rapid rate of growth of the error in this example depends entirely on the unsuitable form of the computational method. To calculate the integral in

$$y_{n+1} = y_n + \int_{x_n}^{x_{n+1}} f(t)\, dt$$

let us use the simple trapezoidal formula

$$y_{n+1} = y_n + \frac{h}{2}(f_n + f_{n+1}) \qquad (13.1.5)$$

which is exact when $f(x)$ is any linear function. The algebraic degree of precision of this formula is thus less than that of (13.1.4) and one might expect the values of y_n obtained using (13.1.5) to be less exact than those obtained using (13.1.4). The table below shows that this is indeed true at the beginning of the table. However, the error grows at a much slower rate and the value of $y(1.0)$ is much more exact than that obtained in the previous case.

x	y_{approx}	$y - y_{\text{approx}}$
0.0	0.00000	
0.1	0.10526	-0.00009
0.2	0.22159	-0.00019
0.3	0.35015	-0.00029
0.4	0.49223	-0.00041
0.5	0.64926	-0.00054
0.6	0.82280	-0.00068
0.7	1.01459	-0.00084
0.8	1.22656	-0.00102
0.9	1.46082	-0.00122
1.0	1.71971	-0.00143

Thus it is clear that (13.1.5) is the better of the two formulas for a large number of intervals.

13.2. THE ERROR OF THE COMPUTATION

We denote the exact value of the function

$$y(x) = y_0 + \int_{x_0}^{x} f(t)\,dt$$

at the nodes x_k by $y(x_k)$ $(k = 0, 1, \ldots)$. The approximate values of $y(x_k)$ which are calculated by some computational method we will denote by y_k.

To calculate y_{n+1} let us assume that we use several preceding values of $y(x)$, y_n, y_{n-1}, $\ldots$, y_{n-p}, and $m = m(n)$ values of $f(x)$ at the points $\xi_{n,j}$ $(j = 1, \ldots, m)$. Thus we assume that the computational formula has the following form[2]:

---

[2] The coefficients $A_{n,i}$ and $B_{n,j}$ in this equation may depend on n so that the computational formula would be changed at each step. The step size h may also change from step to step. Equation (13.2.1) is an equation in finite differences for the y_k and in our discussion it is only necessary that this equation has a certain fixed order $p + 1$.

$$y_{n+1} = \sum_{i=0}^{p} A_{n,i} y_{n-i} + \sum_{j=1}^{m} B_{n,j} f(\xi_{n,j}). \qquad (13.2.1)$$

If in this equation we substitute the exact values $y(x_k)$ in place of the approximate values y_k then the equation will be an approximation and it is necessary to add an auxiliary term in order to make it exact

$$y(x_{n+1}) = \sum_{i=0}^{p} A_{n,i} y(x_{n-i}) + \sum_{j=1}^{m} B_{n,j} f(\xi_{n,j}) + r_n. \qquad (13.2.2)$$

We will call r_n the error in formula (13.2.1).

In the form in which (13.2.1) is written we have assumed that the computation is carried out using exact (unrounded) numbers. This, however, will happen only very rarely. This formula must be modified to indicate the method used for rounding. If the operation of rounding is indicated by enclosing the quantity to be rounded in curly brackets then the computational formula is more exactly written as

$$y_{n+1} = \left\{ \sum_{i=0}^{p} A_{n,i} y_{n-i} + \sum_{j=1}^{m} B_{n,j} f(\xi_{n,j}) \right\}_n \qquad (13.2.3)$$

where the subscript n outside the brackets indicates that the rule for rounding can be changed at each step.

To use (13.2.3) we must know the initial values $y_0, y_1, \ldots, y_p$ and we assume that these are given. We will now construct a difference equation for the error

$$\epsilon_k = y(x_k) - y_k.$$

If we denote by $-\alpha_n$ the rounding error which we indicated by brackets in (13.2.3) then (13.2.3) becomes

$$y_{n+1} = \sum_{i=0}^{p} A_{n,i} y_{n-i} + \sum_{j=1}^{m} B_{n,j} f(\xi_{n,j}) - \alpha_n. \qquad (13.2.4)$$

Subtracting (13.2.4) from (13.2.2) gives

$$\epsilon_{n+1} = \sum_{i=0}^{p} A_{n,i} \epsilon_{n-i} + r_n + \alpha_n. \qquad (13.2.5)$$

If the initial values of the error ϵ_k ($k = 0, 1, \ldots, p$) corresponding to the approximate values y_k ($k = 0, 1, \ldots, p$) formed at the start of the computation of the table are known then all following values of ϵ_k ($k > p$) can be sequentially found from equation (13.2.5).

The errors ϵ_n $(n > p)$ depend first of all on the values $\epsilon_0, \ldots, \epsilon_p$, secondly on the rounding errors α_k $(k < n)$ and finally on the errors of the formula (13.2.1) r_k $(k < n)$.

To analyze the error it will be useful to determine how each of the above three factors separately affect ϵ_n. To do this we will write ϵ_n as a sum of three terms which correspond to the errors from each of the three sources:

$$\epsilon_n = E_n + E_n' + E_n''. \tag{13.2.6}$$

Here E_n is the solution of the homogeneous equation

$$E_{n+1} = \sum_{i=0}^{p} A_{n,i} E_{n-i} \tag{13.2.7}$$

subject to the initial conditions

$$E_k = \epsilon_k \qquad k = 0, 1, \ldots, p. \tag{13.2.8}$$

The term E_n' satisfies the nonhomogeneous equation

$$E_{n+1}' = \sum_{i=0}^{p} A_{n,i} E_{n-i}' + \alpha_n \tag{13.2.9}$$

and has the initial conditions

$$E_k' = 0, \qquad k = 0, 1, \ldots, p. \tag{13.2.10}$$

The term E_n'' is the solution of the nonhomogeneous equation

$$E_{n+1}'' = \sum_{i=0}^{p} A_{n,i} E_{n-i}'' + r_n \tag{13.2.11}$$

also with the initial conditions

$$E_k'' = 0, \qquad k = 0, 1, \ldots, p. \tag{13.2.12}$$

Here E_n is the part of the error ϵ_n due to the errors $\epsilon_0, \ldots, \epsilon_p$ in the initial values, E_n' is the part of ϵ_n due to rounding, and E_n'' is the part of ϵ_n due to the error r_n in formula (13.2.1).

A simple expression for E_n in terms of ϵ_k $(k \leq p)$ which will suffice for our purposes can be constructed in the following way. Denote by E_n^i the solution of the homogeneous equation (13.2.7) which satisfies the conditions

$$E_k^i = \begin{cases} 0 & k \neq i \\ & \\ 1 & k = i \end{cases} \qquad i, k = 0, 1, \ldots, p.$$

Then clearly

$$E_n = E_n^0 \epsilon_0 + E_n^1 \epsilon_1 + \cdots + E_n^p \epsilon_p. \qquad (13.2.13)$$

Hence we can easily obtain an estimate for E_n. We will seldom know the exact values of the errors ϵ_k $(k \leq p)$ but we will know that their absolute values do not exceed a certain number ϵ:

$$|\epsilon_k| \leq \epsilon \qquad k \leq p. \qquad (13.2.14)$$

If we assume that the initial errors ϵ_k can have arbitrary values subject to (13.2.14) then from (13.2.13) we obtain the following estimate

$$|E_n| \leq \epsilon \sum_{k=0}^{p} |E_n^k|. \qquad (13.2.15)$$

Equation (13.2.13) or (13.2.15) permits us to determine how precisely we must calculate the initial values y_k $(k \leq p)$ in order that E_n does not exceed a predetermined value.

Now we consider the second part of the error E_n'. It must be found from equation (13.2.9) with the initial conditions (13.2.10). We see at once that E_n' is a linear combination of $\alpha_p, \alpha_{p+1}, \ldots, \alpha_{n-1}$:

$$E_n' = \sum_{k=p}^{n-1} E_{n,k} \alpha_k. \qquad (13.2.16)$$

The coefficient $E_{n,k}$ is the influence on E_n' of a rounding error of a unit in the right side of (13.2.3) for $n = k$. The $E_{n,k}$ are Green's functions or functions of influence for the above problem.

In the theory of difference equations[3] an explicit expression for $E_{n,k}$ is obtained in terms of the solutions of the homogeneous equation (13.2.9). We do not give it here because of its complexity.

For our purpose it is useful to note that $E_{n,k}$ is the solution of the equation

$$E_{n+1} = \sum_{i=0}^{p} A_{n,i} E_{n-i} + \delta_{n,k} \qquad (13.2.17)$$

which satisfies the initial conditions

$$E_i = 0, \qquad i = 0, 1, \ldots, p$$

---

[3] See A. A. Markov, *Calculus of Finite Differences*, Part II, Sec. 19, Moscow, 1911 (Russian) or A. O. Gel'fond, *Calculus of Finite Differences*, Part 3, Sec. 3, Moscow, 1936 (Russian).

where $\delta_{n,k}$ is the Kronecker symbol

$$\delta_{n,k} = \begin{cases} 0 & n \neq k \\ 1 & n = k. \end{cases}$$

From the sum (13.2.6) we can determine the number of significant figures which must be used in order that E'_n will not exceed a given value.

Suppose we know that for all steps of the computation the errors α_n do not exceed α:

$$|\alpha_n| \leq \alpha.$$

Then from (13.2.16) we obtain the inequality

$$|E'_n| \leq \alpha \sum_{k=p}^{n-1} |E_{n,k}|. \qquad (13.2.18)$$

The quantities E_n and E'_n depend on the precision of the initial values $y_0, y_1, \ldots, y_p$ and on the number of significant figures carried in the calculations. These quantities can be made as small in absolute value as we desire for each $n \leq N$.

We turn, finally, to the last part of the error E''_n. The difference equation (13.2.11) for E''_n is obtained from (13.2.9) by replacing the constant term α_n by r_n. The initial conditions for both E'_n and E''_n are the same. Therefore an equation similar to (13.2.16) is valid for E''_n with α_n replaced by r_n:

$$E''_n = \sum_{k=p}^{n-1} E_{n,k} r_k. \qquad (13.2.19)$$

The error E''_n depends entirely on the form of the computational formula (13.2.1) or to be more precise on the remainders r_k, the coefficients $A_{n,i}$ and on the number of steps n.

In the next section, where we study the convergence of computational formulas, the sum $\displaystyle\sum_{k=p}^{n-1} E_{n,k} r_k$ will be discussed in more detail.

As an example let us analyze the error of equation (13.1.4) which we used in the last section to evaluate the integral

$$y(x) = \int_0^x e^t \, dt.$$

The expression (13.2.5) for ϵ_n which corresponds to equation (13.1.4) is

$$\epsilon_{n+1} = -4\epsilon_n + 5\epsilon_{n-1} + r_n + \alpha_n.$$

This is a nonhomogeneous difference equation of the second order with constant coefficients and constant term $r_n + \alpha_n$. To solve this equation the initial values ϵ_0 and ϵ_1 of the error must be known.

Let us find the first part of the error E_n which depends on ϵ_0 and ϵ_1. The homogeneous equation for E_n is

$$E_{n+1} = -4E_n + 5E_{n-1}.$$

The solution of this equation for the initial conditions $E_0 = \epsilon_0$ and $E_1 = \epsilon_1$ is

$$E_n = \frac{1}{6}(\epsilon_1 + 5\epsilon_0) + \frac{(-1)^n}{6}(\epsilon_0 - \epsilon_1)5^n.$$

If $\epsilon_0 - \epsilon_1 \neq 0$, E_n grows very rapidly as n increases. For $n = 10$, that is for only 9 steps in the computation, the coefficient of $\epsilon_0 - \epsilon_1$ is $\frac{5^{10}}{6} \approx 1.5 \times 10^6$ which will cause the loss of 6 significant figures in the computations.

Now we investigate E'_n, the error due to rounding. The nonhomogeneous equation (13.2.9) for E'_n is

$$E'_{n+1} = -4E'_n + 5E'_{n-1} + \alpha_n.$$

The solution of this equation for the initial conditions $E'_0 = 0$, $E'_1 = 0$ can be easily found:

$$E' = \frac{1}{6}\sum_{t=0}^{n-1}[1 - (-5)^{n-t-1}]\alpha_{t+1} =$$

$$= \frac{1}{6}\{[1 - (-5)^{n-1}]\alpha_1 + [1 - (-5)^{n-2}]\alpha_2 + \cdots\}.$$

As with E_n, we see that E'_n can grow rapidly as n increases and it can become large even in a small number of steps. A similar remark holds for E''_n.

The rapidity with which the error grows for formula (13.1.4) is illustrated by the computations of the previous section.

Let us again consider the general problem of studying the error ϵ_n. The behavior of ϵ_n as n increases naturally depends on the coefficients $A_{n,i}$.

Let us consider the special case when all the coefficients $A_{n,i}$ are positive. Suppose also that

$$\sum_{i=0}^{p} A_{n,i} = 1. \tag{13.2.20}$$

This means that in a calculation without rounding the formula will be exact when $f(t) = 0$ and $y(x)$ is a constant.

With these assumptions we can find a very simple and effective estimate for ϵ_n. Let us suppose that the initial errors $\epsilon_0, \ldots, \epsilon_p$ do not exceed ϵ in absolute value:

$$|\epsilon_i| \le \epsilon \qquad i = 0, \ldots, p.$$

We can show that for any n the following inequality is valid:

$$|\epsilon_n| \le \epsilon + \sum_{k=p}^{n-1} |\alpha_k + r_k|. \tag{13.2.21}$$

For $n = p + 1$ we easily verify

$$|\epsilon_{p+1}| = \left| \sum_{i=0}^{p} A_{p,i} \epsilon_i + \alpha_p + r_p \right| \le$$

$$\le \sum_{i=0}^{p} A_{p,i} \epsilon + |\alpha_p + r_p| = \epsilon + |\alpha_p + r_p|.$$

Assuming that the inequality is true for all ϵ_i, $i \le n$, we can show that it is also true for ϵ_{n+1}. We have

$$|\epsilon_{n+1}| \le \sum_{i=0}^{p} A_{n,i} |\epsilon_{n-i}| + |\alpha_n + r_n|.$$

Substituting for the $|\epsilon_{n-i}|$ the larger value $\epsilon + \sum_{i=0}^{n-1} |\alpha_i + r_i|$ we obtain

$$|\epsilon_{n+1}| \le \epsilon + \sum_{i=p}^{n-1} |\alpha_i + r_i| + |\alpha_n + r_n|$$

which proves the assertion.

From (13.2.21) we see that in a computational formula with nonnegative coefficients $A_{n,i}$ the errors ϵ_n will "grow slowly" as a function of n. In this respect this type of formula is very well behaved.

13.3. CONVERGENCE AND STABILITY OF THE COMPUTATIONAL PROCESS

First of all we will clarify certain concepts concerning the problem of convergence of a computational process. To simplify the discussion we will assume that the formula is of a certain special form which is most often used in practical problems. We assume that the segment $[x_0, X]$ on which the function $y(x)$ is to be calculated is finite and that values of $y(x)$ are to be found at a set of equally spaced points

$$x_k = x_0 + kh, \qquad k = 0, 1, \ldots, N$$

$$x_0 + Nh \leq X < x_0 + (N + 1)h$$

which we denote by S_h.

Suppose that the coefficients of the computational formula do not depend on n:

$$y(x_{n+1}) = \sum_{i=0}^{p} A_i y(x_i) + \sum_{j=1}^{m} B_{n,j} f(\xi_{n,j}) + r_n. \qquad (13.3.1)$$

The computational method is thus obtained by neglecting the term r_n and rounding the sum to a certain number of significant figures

$$y_{n+1} = \left\{ \sum_{i=0}^{p} A_i y_{n-i} + \sum_{j=1}^{m} B_{n,j} f(\xi_{n,j}) \right\}_n. \qquad (13.3.2)$$

If $y_0, \ldots, y_p$ are known we can find from (13.3.2) the approximate values y_n corresponding to the values $y(x_n)$ on the set S_h.

We define the distance $\rho(y, y_n)$ between $y(x)$ and the function y_n $(n = 0, 1, \ldots, N)$ which is defined on S_h to be the largest absolute value of the error $\epsilon_n = y(x_n) - y_n$:

$$\rho(y, y_n) = \max_n |\epsilon_n| = \max_n |y(x_n) - y_n|.$$

We will say that the computational process converges if, as $h \longrightarrow 0$, we have

$$\rho(y, y_n) \longrightarrow 0. \qquad (13.3.3)$$

The error ϵ_n depends on the errors $\epsilon_0, \epsilon_1, \ldots, \epsilon_p$ of the initial values y_k $(k = 0, 1, \ldots, p)$, the rounding error α_n and the remainder r_n of formula (13.3.1). As in the preceding section we split the error ϵ_n into three parts and discuss how each of these parts influences ϵ_n

$$\epsilon_n = E_n + E_n' + E_n''.$$

In the preceding section we discussed the conditions which E_n, E_n' and E_n'' must satisfy.

Since each of the quantities ϵ_i $(i \leq p)$, α_n and r_n are independent then in order that $\rho(y, y_n) \longrightarrow 0$ as $h \longrightarrow 0$ we must require that the following three conditions be satisfied:

$$\max_n |E_n| \longrightarrow 0, \qquad \max_n |E_n'| \longrightarrow 0, \qquad \max_n |E_n''| \longrightarrow 0. \qquad (13.3.4)$$

The errors E_n and E_n' depend respectively on the ϵ_i $(i \leq p)$ and α_n. Thus it is clear that for any fixed h the precision of the initial values y_i $(i \leq p)$ and the rounding errors can be made as small as we desire so that $\max_n |E_n|$ and $\max_n |E_n'|$ can be made arbitrarily small. Therefore it is only a technical problem to obtain conditions which must be satisfied if the first two conditions (13.3.4) are to be fulfilled. As h decreases we must determine how the accuracy of the initial values y_i $(i \leq n)$ must be increased and how the number of significant figures must be increased so that the error in ϵ_n due to these quantities will tend to zero. Such an investigation gives a criterion for testing the practical suitability of the computational formula and thus will be very valuable. If it turns out that as h decreases the accuracy of these quantities must rapidly increase then such a computational formula must be rejected as being unsuitable in most cases.

With these remarks in mind we must prefer computational formulas for which the precision of the initial values y_i $(i \leq p)$ and the number of significant figures must increase the slowest as $h \longrightarrow 0$. This can also be expressed in another way. Consider for example E_n. Suppose that the initial values y_i $(i \leq p)$ have certain errors ϵ_i. In the computation of the succeeding values y_i $(i > p)$ the error will grow from step to step. The rate of growth clearly depends on the choice of the computational formula. The computational formulas which are of most interest are those for which the rate of growth is minimal. In the theory of the approximate solution of differential equations methods which have the minimal rate of growth of the error are called stable. Thus we will say that the formula is stable with respect to the errors in the initial values if the rate of growth of E_n is minimal. In a similar way we can define stability with respect to the rounding errors α_n, that is the errors in the right side of (13.3.2).

We now discuss the error E_n in more detail. The homogeneous equation for E_n for formula (13.3.2) will be an equation with constant coefficients

$$E_{n+1} = \sum_{i=0}^{p} A_i E_{n-i}. \qquad (13.3.5)$$

As we saw in the last section the solution of this equation, which satisfies the initial conditions $E_i = \epsilon_i$ ($i \le p$), can be written in the form

$$E_n = E_n^0 \epsilon_0 + E_n^1 \epsilon_1 + \cdots + E_n^p \epsilon_p \qquad (13.3.6)$$

where E_n^i is the solution of (13.3.5) for the initial conditions

$$E_k^i = \begin{cases} 0 & k \ne i \\ & \\ 1 & k = i \end{cases} \qquad i, k = 0, 1, \ldots, p.$$

Thus the rate of growth of E_n is related to the rate of growth of E_n^i. If we assume that the initial errors are bounded in absolute value by ϵ

$$|\epsilon_i| \le \epsilon, \qquad i = 0, 1, \ldots, p \qquad (13.3.7)$$

then the following estimate will be valid for E_n

$$|E_n| \le \epsilon \sum_{i=0}^{p} |E_n^i|. \qquad (13.3.8)$$

We will assume that formula (13.3.1) is exact (that is $r_n = 0$) when $f(x) \equiv 0$ and $y(x)$ is a constant. This will be true in most practical cases. Then the coefficients A_i must satisfy

$$\sum_{i=0}^{p} A_i = 1. \qquad (13.3.9)$$

This says that $E_n = 1$ is a solution of the homogeneous equation (13.3.5). This solution is the sum of all the E_n^i:

$$1 = E_n^0 + E_n^1 + \cdots + E_n^p.$$

Thus for each n we have the inequality

$$\sum_{i=0}^{p} |E_n^i| \ge 1.$$

It is possible to give examples for which $\sum_{i=0}^{p} |E_n^i|$ will grow without bound as $n \longrightarrow \infty$ and it can also turn out then that E_n will be unbounded.

The most well behaved formulas with respect to the rate of growth of E_n are clearly those for which the sum $\sum_{i=0}^{p} E_n^i$ is bounded[4] for $n > p$.

---

[4] We will only need to know that this sum is bounded. We will not discuss the problem of finding a bound.

Thus we are led to the following definition:

Equation (13.3.2) is said to be stable with respect to the initial values y_i ($i \leq p$) if there exists a number M such that for any n the following inequality is satisfied

$$|E_n| \leq M \epsilon \qquad (13.3.10)$$

where $|\epsilon_i| \leq \epsilon$, $i = 0, 1, \ldots, p$.

We note that the boundedness of E_n ($n = 0, 1, \ldots$) together with the condition $|\epsilon_i| \leq \epsilon$ is equivalent to the boundedness of all the E_n^i, $i = 0$, $1, \ldots, p$. In fact if all the E_n^i are bounded then from (13.3.6) it follows that E_n is also bounded.

Let us take an arbitrary $k \leq p$ and assume that all the ϵ_k ($k \neq i$, $k \leq p$) are zero. Then

$$E_n = E_n^i \epsilon_i$$

and if E_n is bounded then E_n^i is also bounded.

The most general solution of (13.3.5) is determined by the algebraic equation

$$\lambda^{p+1} = \sum_{i=0}^{p} A_i \lambda^{p-i}.$$

Let $\lambda_1, \lambda_2, \ldots, \lambda_m$ denote the distinct roots of this equation and let k_1, $k_2, \ldots, k_m$ be the multiplicities of these roots. Then the functions

$$\lambda_i^n n^j \qquad (j = 0, 1, \ldots, k_i - 1; \quad i = 1, 2, \ldots, m) \quad (13.3.11)$$

form a complete system of linearly independent solutions.

The solutions E_n^i ($i = 0, 1, \ldots, p$) are obtained from (13.3.11) by a transformation with a nonsingular matrix and therefore the boundedness of all the E_n^i for $i = 0, 1, \ldots$ is equivalent to the boundedness of the solutions (13.3.11). This occurs if and only if there are no λ_i greater than 1 in modulus and if $|\lambda_i| = 1$ then $k_i = 1$. Thus we have established:

Theorem 1. *In order that equation (13.3.2) be stable with respect to the errors in the initial values y_i ($i \leq p$) it is necessary and sufficient that*

1. *The roots of the equation $\lambda^{p+1} = \sum\limits_{i=0}^{p} A_i \lambda^{p-i}$ do not exceed unity in modulus.*

2. *Any root of modulus unity must be simple.*

We now study E_n' which is the error due to the effect of the rounding errors $\alpha_p, \ldots, \alpha_{n-1}$. The error E_n' satisfies equation (13.2.16):

$$E'_n = \sum_{k=p}^{n-1} E_{n,k} \alpha_k.$$

The coefficients $E_{n,k}$, as functions of n, must satisfy (13.2.17) which in the present case is

$$E'_{n+1} = \sum_{i=0}^{p} A_i E'_{n-i} + \delta_{n,k} \qquad (13.3.12)$$

with initial values

$$E_{i,k} = 0, \qquad i = 0, 1, \,\ldots, p. \qquad (13.3.13)$$

We can establish a simple relationship between $E_{n,k}$ and the solution E_n^p which we discussed above. For $n < k$ equation (13.3.12) will be homogeneous and in view of the zero initial conditions $E_{n,k}$ will be zero for each $n \leq k$. In addition $E_{k+1,k} = 1$ which can be seen from (13.3.12) by putting $n = k$. When $n > k$ equation (13.3.12) will also be homogeneous.

Let us consider $E_{n,k}$ for $n \geq k - p + 1$. From the above discussion we can assume that $E_{n,k}$ has the initial values

$$E_{k-p+1,k} = 0, \qquad \ldots, \qquad E_{k,k} = 0, \qquad E_{k+1,k} = 1$$

and that it satisfies the homogeneous equation

$$E'_{n+1} = \sum_{i=0}^{p} A_i E'_{n-i}. \qquad (13.3.14)$$

But we at once see that these same conditions are also satisfied by $E_{n+p-k-1}^p$ and, since the solution is unique for fixed initial conditions, $E_{n,k}$ and $E_{n+p-k-1}^p$ must coincide.

Thus we obtain

$$E'_n = \sum_{k=p}^{n-1} \alpha_k E_{n+p-k-1}^p. \qquad (13.3.15)$$

We will assume that α is an upper bound for the rounding errors α_n for all n, $|\alpha_n| \leq \alpha$. Then

$$|E'_n| \leq \alpha \sum_{k=p}^{n-1} |E_{n+p-k-1}^p|$$

$$\max_n |E'_n| \leq \alpha \sum_{k=p}^{N-1} |E_{N+p-k-1}^p| = \alpha \sum_{k=p}^{N-1} |E_k^p|. \qquad (13.3.16)$$

If we assume that the errors α_n can have any values subject to the condition $|\alpha_n| \leq \alpha$ then the above estimate can not be improved and equality is achieved for $n = N$ when $\alpha_k = \alpha \operatorname{sign} E^p_{N+p-k-1}$. Because $E^p_p = 1$ then for each $N \geq p + 1$ we have $\sum_{k=p}^{N-1} |E^p_k| \geq 1$. As h tends to zero N grows without bound. The value of $\sum_{k=p}^{N-1} |E^p_k|$ will depend on the behavior of the solutions E^p_k as $k \longrightarrow \infty$.

Let us consider the particular solutions of the homogeneous equation (13.3.14)

$$E^p_n, \qquad E^p_{n+1}, \qquad \ldots, \qquad E^p_{n+p}. \qquad (13.3.17)$$

Their initial values for $n = 0, 1, \ldots, p$ form the following matrix

$$\begin{bmatrix} 0 & 0 & \cdots & 0 & 0 & 1 \\ 0 & 0 & \cdots & 0 & 1 & E^p_{p+1} \\ \cdots\cdots\cdots\cdots\cdots\cdots\cdots\cdots\cdots \\ 1 & E^p_{p+1} & \cdots & E^p_{2p-2} & E^p_{2p-1} & E^p_{2p} \end{bmatrix}$$

The determinant of this matrix is different from zero and therefore the solutions of (13.3.17) are linearly independent. Thus these solutions are obtained from the E^i_n $(i = 0, 1, \ldots, p)$ by a nonsingular linear transformation.

Therefore the boundedness of E^i_n $(i = 0, 1, \ldots, p)$ is equivalent to the boundedness of the solutions (13.3.17).

From the assumption (13.3.9) we saw that

$$\sum_{i=0}^{p} |E^i_n| \geq 1, \qquad n = 0, 1, \ldots$$

and in this case the slowest rate of growth of $\sum_{k=p}^{N-1} |E^p_k|$, as $N \longrightarrow \infty$, will occur when all the terms E^p_k in this sum are bounded by a certain number. Then $\sum_{k=p}^{N-1} |E^p_k|$ will be of the order of magnitude $O(N)$. Thus we are led to the following definition:

Equation (13.3.2) *is said to be stable with respect to the rounding errors* α_n *if there exists a number* M_1, *which is independent of h, with the property that for each* $N > p$ *we have*

$$|E_n'| \leq M_1 N\alpha, \qquad (n = p + 1, \ldots, N - 1) \qquad (13.3.18)$$

where $|a_n| \leq \alpha.$

A simple theorem which gives a sufficient condition for stability is:

Theorem 2. *In order that* (13.3.2) *be stable with respect to the rounding error it is sufficient that the following two conditions be fulfilled:*

1. *The equation* $\lambda^{p+1} = \displaystyle\sum_{i=0}^{p} A_i \lambda^{p-i}$ *has no roots of modulus greater than unity.*

2. *Any roots of modulus unity are simple.*

Proof. If the conditions of the theorem are satisfied then the solutions (13.3.11) will be bounded for $n \geq 0$. These solutions are a complete system of solutions and the E_n^p are linear combinations of them. Thus there exists a number M_1 which, for $n \geq 0$, satisfies

$$|E_n^p| \leq M_1.$$

Combining this with (13.3.16) establishes the theorem.

We now study E_n'' which is the part of the error due to the error r_n in (13.3.1). The error ϵ_n will coincide with E_n'' if the computations are carried out using exact initial values $y_k = y(x_k)$ $(k = 0, 1, \ldots, p)$ and if no rounding needs to be performed.

We will say that formula (13.3.1) provides a convergent computational process if

$$\max_{n} |E_n''| \longrightarrow 0 \qquad \text{as} \qquad h \longrightarrow 0. \qquad (13.3.19)$$

Since $E_{n,k} = E_{n+p-k-1}^p$ equation (13.2.19) can be written as

$$E_n'' = \sum_{k=p}^{n-1} r_k E_{n+p-k-1}^p. \qquad (13.3.20)$$

This gives an explicit expression for E_n'' in terms of the errors r_k in the computational formula.

To estimate E_n'' suppose that r is an upper bound for the absolute values of the errors r_n on the entire segment $[x_0, X]$, so that for any n $(0 \leq n \leq N)$

$$|r_n| \leq r. \qquad (13.3.21)$$

Then we have the following estimate for E_n''

$$|E_n''| \leq r \sum_{k=p}^{n-1} |E_{n+p-k-1}^p| = r \sum_{k=p}^{n-1} |E_k^p|. \qquad (13.3.22)$$

Hence

$$\max_n |E_n''| \leq r \sum_{k=p}^{N-1} |E_n^p|. \qquad (13.3.23)$$

The terms r and $\displaystyle\sum_{k=p}^{N-1} |E_n^p|$ on the right of this equation usually depend on the interval size h and if we know how they depend on h we can often predict the behavior of $\max_n |E_n''|$ as $n \longrightarrow \infty$. In particular we can state:

Theorem 3. *If, as $h \longrightarrow 0$,*

$$r \sum_{k=p}^{N-1} |E_k^p| \longrightarrow 0$$

then formula (13.3.2) provides a convergent computational process.

Let us assume that (13.3.2) is stable with respect to the initial values and also with respect to the rounding errors. Thus we assume that the roots of $\lambda^{p+1} = \displaystyle\sum_{i=0}^{p} A_i \lambda^{p-i}$ do not exceed unity in modulus and that any roots with modulus equal to unity are simple.

Then we showed that there exists a number M_1 which for all $n \geq 0$ satisfies $|E_n^p| \leq M_1$. From this and from (13.3.23) we have the following estimate

$$\max_n |E_n''| \leq rM_1(N - p) \leq rM_1 N. \qquad (13.3.24)$$

Thus we have established:

Theorem 4. *If the equation*

$$\lambda^{p+1} = \sum_{i=0}^{p} A_i \lambda^{p-i}$$

has no roots greater than unity in modulus and if the roots of modulus equal to unity are simple then formula (13.3.2) provides a convergent computational process providing that

$$\frac{r}{h} \longrightarrow 0 \qquad \text{as} \qquad h \longrightarrow 0.$$

Let us consider the case which we discussed at the end of the last section in which the coefficients A_k are positive numbers

$$A_k > 0$$

which satisfy the condition

$$\sum_{k=0}^{p} A_k = 1.$$

In this case the error ϵ_n satisfies inequality (13.2.21):

$$|\epsilon_n| \leq \epsilon + \sum_{k=p}^{n-1} |a_k + r_k| \qquad n > p$$

where $\epsilon \geq |\epsilon_i|$, $i = 0, 1, \ldots, p$.

Thus it is easy to obtain an estimate for the summands E_n, E_n' and E_n'' of ϵ_n. We note that if $a_k = 0$ and $r_k = 0$ $(k > p)$ then ϵ_n must coincide with E_n and therefore we have

$$|E_n| \leq \epsilon, \qquad n > p. \tag{13.3.25}$$

Similarly

$$|E_n'| \leq \sum_{k=p}^{n-1} |a_k|, \qquad n > p \tag{13.3.26}$$

$$|E_n''| \leq \sum_{k=p}^{n-1} |r_k|, \qquad n > p. \tag{13.3.27}$$

If $|a_k| \leq a$ and $|r_k| \leq r$ for $p < n \leq N$, then E_n' and E_n'' satisfy the estimates

$$|E_n'| \leq (n - p)a \leq Na \tag{13.3.28}$$

$$|E_n''| \leq (n - p)r \leq Nr. \tag{13.3.29}$$

These inequalities permit us to state:

Theorem 5. *If the coefficients A_k ($k = 0, 1, \ldots, p$) are all positive and satisfy the condition $\sum\limits_{k=0}^{p} A_k = 1$ then equation (13.3.2) is stable with respect to both the errors in the initial values and the rounding errors. If, in addition,*

$$\frac{r}{h} \longrightarrow 0$$

as $h \longrightarrow 0$ then formula (13.3.2) provides a convergent computational process.

CHAPTER 14

Integration of Functions Given in Tabular Form

14.1. ONE METHOD FOR SOLVING THE PROBLEM

Suppose it is necessary to calculate the value of the integral

$$y(x) = y_0 + \int_{x_0}^{x} f(t)\, dt \qquad (14.1.1)$$

for equally spaced points $x_n = x_0 + nh$ on the segment $x_0 \le x \le X$ where $f(x)$ is only known for a set of equally spaced points which includes the x_n. This problem has been widely investigated and many methods for its solution are known. The relationship between this problem and Cauchy's problem for ordinary differential equations has also received much attention. If we are given the equation $y' = f(x, y)$ and we wish to find the solution which satisfies the condition $y(x_0) = y_0$ then this problem can be replaced by the equivalent problem of finding the solution of the integral equation

$$y(x) = y_0 + \int_{x_0}^{x} f(t, y(t))\, dt. \qquad (14.1.2)$$

Thus we can also apply methods for the numerical calculation of an indefinite integral to the solution of first order differential equations[1].

In this chapter we consider one possible method for computing the function (14.1.1). This method leads to a simple computational scheme

---

[1]These problems are different in the following respect. In order to compute the integral (14.1.1) we assume that $f(t)$ is known at all points of the segment $[x_0, X]$ and to find each value of $y(x)$ we can use any values of $f(t)$. In the integral (14.1.2) we will know the values of the function $f(x, y)$ for tabular points preceding x, but the values of $f(x, y)$ for points following x will not be known.

and, as a rule, gives good accuracy if the function is sufficiently smooth on the segment of integration and close to this segment.

Suppose that the computation has been carried up to $x_n = x_0 + nh$. To find the next value $y(x_{n+1})$ of the function (14.1.1) we will use only the immediately preceding value of $y(x)$:

$$y(x_{n+1}) = y(x_n) + \int_{x_n}^{x_{n+1}} f(t)\, dt. \tag{14.1.3}$$

To compute the integral in (14.1.3) we construct an interpolating polynomial for $f(x)$ on the segment $[x_n, x_{n+1}]$. We will use the nodes closest to this segment to construct the interpolating polynomial and will take the same number of nodes on each side of this segment.

We apply Newton's interpolation formula (3.2.6) using the nodes x_n, $x_n + h$, $x_n - h$, $x_n + 2h$, $x_n - 2h$, ... to obtain

$$
\begin{aligned}
f(x) = f(x_n) &+ (x - x_n)f(x_n, x_n + h) + (x - x_n) \times \\
&\times (x - x_n - h)f(x_n, x_n + h, x_n - h) + \\
&+ (x - x_n)(x - x_n - h) \times \\
&\times (x - x_n + h)f(x_n, x_n + h, x_n - h, x_n + 2h) + \cdots .
\end{aligned}
$$

Introducing the new variable u, $x_n = x_0 + uh$, and expressing the divided differences in terms of finite differences gives

$$
\begin{aligned}
f(x_n + uh) = f_n &+ \frac{u}{1!}\,\Delta f_n + \frac{u(u-1)}{2!}\,\Delta^2 f_{n-1} + \\
&+ \frac{(u+1)u(u-1)}{3!}\,\Delta^3 f_{n-1} + \\
&+ \frac{(u+1)u(u-1)(u-2)}{4!}\,\Delta^4 f_{n-2} + \cdots
\end{aligned}
$$

To put this equation in a form which is symmetric with respect to $x_n + \frac{1}{2}h$ we transform the differences of even order using the identities

$$f_n = \frac{1}{2}[f_{n+1} + f_n] - \frac{1}{2}[f_{n+1} - f_n] = \frac{1}{2}[f_{n+1} + f_n] - \frac{1}{2}\Delta f_n$$

$$
\begin{aligned}
\Delta^2 f_{n-1} &= \frac{1}{2}[\Delta^2 f_n + \Delta^2 f_{n-1}] - \frac{1}{2}[\Delta^2 f_n - \Delta^2 f_{n-1}] = \\
&= \frac{1}{2}[\Delta^2 f_n + \Delta^2 f_{n-1}] - \frac{1}{2}\Delta^3 f_{n-1}
\end{aligned}
$$

....

This gives[2]

$$f(x_n + uh) = \frac{f_n + f_{n+1}}{2} + \frac{u - \dfrac{1}{2}}{1!} \, \Delta f + \frac{u(u-1)}{2!} \, \frac{\Delta^2 f_{n-1} + \Delta^2 f_n}{2} +$$

$$+ \frac{\left(u - \dfrac{1}{2}\right) u(u-1)}{3!} \, \Delta^3 f_{n-1} + \cdots +$$

$$+ \frac{(u+k-1)\cdots(u-k)}{(2k)!} \, \frac{\Delta^{2k} f_{n-k} + \Delta^{2k} f_{n-k+1}}{2} + \qquad (14.1.4)$$

$$\frac{\left(u - \dfrac{1}{2}\right)(u+k-1)\cdots(u-k)}{(2k+1)!} \, \Delta^{2k+1} f_{n-k} + r(x).$$

Substituting this representation for $f(t)$ in the integral

$$\int_{x_n}^{x_n + h} f(t)\,dt = h \int_0^1 f(x_n + uh)\,du$$

leads to the following expression for $y(x_{n+1})$:

$$y(x_{n+1}) = y(x_n) + h\left[\frac{f_n + f_{n+1}}{2} - \frac{1}{12} \, \frac{\Delta^2 f_{n-1} + \Delta^2 f_n}{2} + \right.$$

$$+ \frac{11}{720} \, \frac{\Delta^4 f_{n-2} + \Delta^4 f_{n-1}}{2} - \frac{191}{60480} \, \frac{\Delta^6 f_{n-3} + \Delta^6 f_{n-2}}{2} + \qquad (14.1.5)$$

$$\left. + \cdots + C_k \, \frac{\Delta^{2k} f_{n-k} + \Delta^{2k} f_{n-k+1}}{2} \right] + R_{n,k}$$

where

$$C_k = \frac{1}{(2k)!} \int_0^1 (u+k-1)\cdots(u-k)\,du$$

$$R_{n,k} = \int_{x_n}^{x_n + h} r(x)\,dx.$$

A computational formula is thus obtained by selecting some value of k and neglecting the remainder $R_{n,k}$.

Let us consider an example. Suppose we wish to calculate the value of the following integral on the segment $[0, 1]$:

---

[2]In the theory of interpolation this equation is called Bessel's formula.

$$y(x) = \int_0^x J_1(t)\,dt = 1 - J_0(x)$$

where $J_0(t)$ and $J_1(t)$ are Bessel functions of the first kind. We use formula (14.1.5) with $h = 0.2$ and with differences up to and including those of the fourth order

$$y_{n+1} = y_n + 0.2 \left[\frac{f_n + f_{n+1}}{2} - \frac{1}{12} \frac{\Delta^2 f_{n-1} + \Delta^2 f_n}{2} + \frac{11}{720} \frac{\Delta^4 f_{n-2} + \Delta^4 f_{n-1}}{2} \right]$$

$$y_0 = 0, \quad f(x) = J_1(x).$$

The table of differences of $J_1(x)$ which are necessary to use this formula is given below:

x	$J_1(x)$	$\Delta J_1 \cdot 10^7$	$\Delta^2 J_1 \cdot 10^7$	$\Delta^3 J_1 \cdot 10^7$	$\Delta^4 J_1 \cdot 10^7$
$x_{-2} = -0.4$	-0.1960266	965258	$+29750$	-29750	0000
$x_{-1} = -0.2$	-0.0995008	995008	00000	-29750	986
$x_0 = 0.0$	0.0000000	995008	-29750	-28764	1944
$x_1 = 0.2$	0.0995008	965258	-58514	-26820	2830
$x_2 = 0.4$	0.1960266	906744	-85334	-23990	3613
$x_3 = 0.6$	0.2867010	821410	-109324	-20377	4279
$x_4 = 0.8$	0.3688420	712086	-129701	-16098	
$x_5 = 1.0$	0.4400506	582385	-145799		
$x_6 = 1.2$	0.4982891	436586			
$x_7 = 1.4$	0.5419477				

From this table we can calculate the values of the integral $y(x)$. The computation for $y(0.2)$ is:

$$y(0.2) = y(0) + h \left[\frac{f(0) + f(0.2)}{2} - \frac{1}{12} \frac{\Delta^2 f(-0.2) + \Delta^2 f(0)}{2} + \right.$$

$$\left. + \frac{11}{720} \frac{\Delta^4 f(-0.4) + \Delta^4 f(-0.2)}{2} \right] =$$

$$= 0 + 0.2 \left[\frac{0 + 0.0995008}{2} - \frac{1}{12} \frac{0 - 0.0029750}{2} + \right.$$

$$\left. + \frac{11}{720} \frac{0 + 0.0000986}{2} \right] = 0.0099750.$$

The calculated values of $y(x)$ are tabulated below.

x	$\int_0^x J_1(t)\,dt$	x	$\int_0^x J_1(t)\,dt$
0.0	0.0000000	0.6	0.0879951
0.2	0.0099750	0.8	0.1537126
0.4	0.0396017	1.0	0.2348023

All of these values are exact to the seven decimal places which are given, except y (0.4) which has an error of one in the last place.[3]

14.2. THE REMAINDER

The order of the highest finite difference in (14.1.4) is $2k + 1$ and to use this formula we must know the value of $f(x)$ at the $2k + 2$ points $x_n - kh, \ldots, x_n + (k + 1)h$. If we assume that $f(x)$ has a continuous derivative of order $2k + 2$ on $[x_n - kh, x_n + (k + 1)h]$ then the remainder $r(x)$ of the interpolation (14.1.4) can be found from Theorem 4 of Chapter 3:

$$r(x) = \frac{[x - x_n + kh][x - x_n + (k - 1)h] \cdots [x - x_n - (k + 1)h]}{(2k + 2)!} f^{(2k+2)}(\eta) =$$

$$= h^{2k+2} \frac{(u + k)(u + k - 1) \cdots (u - k - 1)}{(2k + 2)!} f^{(2k+2)}(\eta)$$

$$x_n - kh < \eta < x_n + (k + 1)h.$$

Thus we obtain the following expression for $R_{n,k}$ in formula (14.1.5):

$$R_{n,k} = h \int_0^1 r(x_n + uh) du =$$

$$= \frac{h^{2k+3}}{(2k + 2)!} \int_0^1 (u + k)(u + k - 1) \cdots (u - k - 1) f^{(2k+2)}(\eta) du.$$

Since the factor $(u + k) \cdots (u - k - 1)$ does not change sign on $[0, 1]$ the mean value theorem can be applied to the last integral and thus we can make the following assertion:

If $f(x)$ has a continuous derivative of order $2k + 2$ on $[x_n - kh, x_n + (k + 1)h]$ then the remainder $R_{n,k}$ of (14.1.5) has the representation

$$R_{n,k} =$$

$$= h^{2k+3} \frac{f^{(2k+2)}(\xi)}{(2k + 2)!} \int_0^1 (u + k)(u + k - 1) \cdots (u - k - 1) du \quad (14.2.1)$$

where ξ is an interior point of the segment $[x_n - kh, x_n + (k + 1)h]$.

---

[3] See, for example, G. N. Watson, *A Treatise on the Theory of Bessel Functions*, Macmillan, New York, 1944, p. 666.

CHAPTER 15

Calculation of Indefinite Integrals Using a Small Number of Values of the Integrand

15.1. GENERAL ASPECTS OF THE PROBLEM

Here, as in the preceding chapter, we will consider the problem of computing the indefinite integral

$$y(x) = y_0 + \int_{x_0}^{x} f(t)\, dt \tag{15.1.1}$$

for equally spaced values of the argument $x_k = x_0 + kh$ ($k = 0, 1, \ldots$). Here, however, we assume that we may use in the computational formula any nodes for which $f(x)$ is defined.

The largest part of the work in computing the integral (15.1.1) by means of a formula of the form (13.2.1) is usually in calculating the values of the function $f(x)$. There are two ways in which we can reduce this part of the work. We can choose the nodes to achieve a high degree of precision in the formula or we can choose the nodes so that they are used for not just one step in the calculation but for several steps so that for each successive step it is necessary to calculate only a few additional values of $f(x)$.

In the following discussion we will use a combination of these methods to construct formulas. To calculate the value of y_{n+1} we again use only the preceding value of $y(x)$:

$$y_{n+1} = y_n + \int_{x_n}^{x_{n+1}} f(t)\, dt$$

and thus the problem reduces to the computation of the integral in this expression.

If the coefficients of the formula are to be independent of n we must assume that the nodes are situated with period h on the x-axis. We will say that a set of points $\alpha + kh$, for distinct integers k, are similar to the point α.

To calculate the above integral we assume that we will use m nodes $\alpha, \beta, \ldots, \lambda$ on the segment $[x_n, x_{n+1}]$: $x_n \leq \alpha < \beta < \cdots < \lambda < x_{n+1}$. In addition to these basic nodes we will also use the following:

$$a \text{ nodes } \alpha + p_i h \ (i = 1, \ldots, a) \text{ similar to } \alpha$$
$$b \text{ nodes } \beta + q_i h \ (i = 1, \ldots, b) \text{ similar to } \beta$$
$$\cdots\cdots\cdots\cdots\cdots\cdots\cdots\cdots\cdots\cdots\cdots\cdots\cdots\cdots\cdots\cdots$$
$$l \text{ nodes } \lambda + t_i h \ (i = 1, \ldots, l) \text{ similar to } \lambda.$$

The way in which these additional nodes are situated among the points x_k will depend on the numbers $p_i, q_i, \ldots, t_i$ which we assume can be any integers different from zero. We denote the total number of nodes by $N + 1$:

$$m + a + b + \cdots + l = N + 1.$$

Let us consider a formula of the form

$$\int_{x_n}^{x_{n+1}} f(t)\,dt \approx A_0 f(\alpha) + \sum_{i=1}^{a} A_i f(\alpha + p_i h) +$$

$$+ \cdots + L_0 f(\lambda) + \sum_{i=1}^{l} L_i f(\lambda + t_i h). \tag{15.1.2}$$

If we assume that the numbers $p_i, \ldots, t_i$ are given then we must still determine the nodes $\alpha, \ldots, \lambda$ and the coefficients $A_i, \ldots, L_i$ ($i = 0, 1, \ldots$). We wish to choose these quantities so that (15.1.2) has the highest possible algebraic degree of precision.

For each choice of the $\alpha, \ldots, \lambda, p_i, \ldots, t_i$ we can always construct a formula which is exact for all polynomials of degree $\leq N$. We can do this by constructing the Lagrange interpolating polynomial for $f(x)$ using the nodes $\alpha, \alpha + p_i h, \ldots, \lambda, \lambda + t_i h$ and taking as the coefficients in (15.1.2) the integrals of the coefficients of this interpolating polynomial. In this way the coefficients $A_i, \ldots, L_i$ are completely determined. Thus to increase the precision of the formula we have only at our disposal the choice of the nodes $\alpha, \ldots, \lambda$. Below we will show that for any $p_i, \ldots, t_i$ formula (15.1.2) can be made exact for all polynomials of degree $m + N$ by a suitable choice of $\alpha, \ldots, \lambda$ and that this is the highest possible degree of precision.

From the nodes of the formula we construct the following polynomials:

$$\omega(x) = (x - \alpha) \cdots (x - \lambda)$$

$$\omega_\alpha(x) = \prod_{i=1}^{a} (x - \alpha - p_i h), \ \ldots, \ \omega_\lambda(x) = \prod_{i=1}^{l} (x - \lambda - t_i h) \tag{15.1.3}$$

$$\Omega(x) = \omega_\alpha(x) \cdots \omega_\lambda(x).$$

Theorem 1. *No matter how we choose the nodes* $\alpha, \ldots, \lambda$ *and the integers* $p_i, \ldots, t_i$ *the formula* (15.1.2) *can not be exact for all polynomials of degree* $m + N + 1$.

Proof. It is sufficient to consider the polynomial $f(x) = \Omega(x)\omega^2(x)$. The degree of this polynomial is $m + N + 1$. Since all the nodes of the formula are roots of $\Omega(x)\omega^2(x)$ the quadrature sum on the right side of (15.1.2) is zero. The integral $\int_{x_n}^{x_{n+1}} \Omega(x)\omega^2(x)\,dx$, however, is different from zero since the polynomial $\Omega(x)\omega^2(x)$ does not change sign on the segment of integration and it is not identically zero. Therefore (15.1.2) cannot be exact for $f(x) = \Omega(x)\omega^2(x)$.

The algebraic degree of precision of (15.1.2) is always less than $m + N + 1$ and the greatest it can be is $m + N$.

Theorem 2. *In order that formula* (15.1.2) *be exact for all polynomials of degree* $\leq m + N$ *it is necessary and sufficient that the following two conditions be fulfilled:*

1. *The formula must be interpolatory*
2. *For any polynomial* $Q(x)$ *of degree less than* m *we must have*

$$\int_{x_n}^{x_{n+1}} \Omega(x)\omega(x)Q(x)\,dx = 0. \tag{15.1.4}$$

Proof. The necessity of the first condition is evident. To verify the necessity of the second condition let us take an arbitrary polynomial $Q(x)$ of degree less than m and set $f(x) = \Omega(x)\omega(x)Q(x)$. This is a polynomial of degree at most $m + N$ and for it equation (15.1.2) must be exact. But the quadrature sum for $f(x)$ is zero; hence equation (15.1.4) must be satisfied.

Suppose now that both conditions of the theorem are fulfilled and let $f(x)$ be any polynomial of degree $\leq m + N$. Dividing $f(x)$ by $\Omega(x)\omega(x)$ we can represent $f(x)$ in the form $f(x) = \Omega(x)\omega(x)Q(x) + r(x)$ where $Q(x)$ and $r(x)$ are polynomials of degree less than m and $N + 1$ respectively. Since the polynomial $\Omega(x)\omega(x)$ is zero at all the nodes in the formula then at these nodes the polynomials $f(x)$ and $r(x)$ must have the same values. Using the fact that the degree of $r(x)$ is not greater than N and the fact that formula (15.1.2) is interpolatory the following equations must be satisfied:

$$\int_{x_n}^{x_{n+1}} f(x)\,dx = \int_{x_n}^{x_{n+1}} \Omega(x)\omega(x)Q(x)\,dx + \int_{x_n}^{x_{n+1}} r(x)\,dx =$$

$$= A_0 r(\alpha) + \sum_{i=1}^{a} A_i r(\alpha + p_i h) + \cdots =$$

$$= A_0 f(\alpha) + \sum_{i=1}^{a} A_i f(\alpha + p_i h) + \cdots.$$

This establishes the sufficiency of the conditions and completes the proof.

Theorem 2 reduces the question of the existence of quadrature formulas (15.1.2) which have the highest algebraic degree of precision $m + N$ to the question of the existence of nodes $\alpha, \ldots, \lambda$ for which the corresponding polynomial $\Omega(x)\omega(x)$ satisfies the orthogonality condition (15.1.4).

Theorem 3. *For any integers* $p_i, \ldots, t_i$ *we can find nodes* $\alpha, \ldots, \lambda$ *so that the corresponding quadrature formula* (15.1.2) *will have the highest algebraic degree of precision* $m + N$.

Proof. Let us take any system of nodes $\alpha, \beta, \ldots, \lambda$ which satisfy the inequalities

$$x_n \leq \alpha \leq \beta \leq \cdots \leq \lambda \leq x_{n+1} \tag{15.1.5}$$

and construct for these nodes the polynomials $\Omega(x)$ and $\omega(x)$. The polynomial $\Omega(x)$ does not change sign on the segment $[x_n, x_{n+1}]$. We will consider $\Omega(x)$ as a weight function and investigate the system of polynomials $P_k(x)$ which are orthogonal on $[x_n, x_{n+1}]$ with respect to $\Omega(x)$. Let $P_m(x)$ be the mth degree polynomial of this system and let us assume that its leading coefficient is unity:

$$P_m(x) = x^m + p_1 x^{m-1} + p_2 x^{m-2} + \cdots.$$

Any polynomial $Q(x)$ of degree $< m$ satisfies

$$\int_{x_n}^{x_{n+1}} \Omega(x)P_m(x)Q(x)\,dx = 0. \tag{15.1.6}$$

The roots of $P_m(x)$ are all real and simple and they all lie inside the segment $[x_n, x_{n+1}]$. We denote the roots of $P_m(x)$ by $\xi_1, \ldots, \xi_m$ and assume that they are enumerated in increasing order $x_n < \xi_1 < \cdots < \xi_m < x_{n+1}$. If it turns out that $\xi_1 = \alpha, \xi_2 = \beta, \ldots, \xi_m = \lambda$ then $P_m(x)$ coincides with $\omega(x)$ and then $\Omega(x)$ and $\omega(x)$ will satisfy (15.1.4) and the corresponding formula (15.1.2) will have the highest algebraic degree of precision $m + N$.

If the $\xi_1, \ldots, \xi_m$ do not coincide with the $\alpha, \ldots, \lambda$ let us construct a system of linear equations for the coefficients p_k $(k = 1, \ldots, m)$. The orthogonality property (15.1.6) is equivalent to the equations

$$\int_{x_n}^{x_{n+1}} \Omega(x) P_m(x) x^i dx = 0, \qquad i = 0, 1, \ldots, m - 1 \qquad (15.1.7)$$

or if we replace $P_m(x)$ by its expansion in powers of x:

$$c_{m+i} + c_{m+i-1} + c_{m+i-2} p_2 + \cdots + c_i p_m = 0, \qquad i = 0, 1, \ldots, m - 1$$

where

$$c_k = \int_{x_n}^{x_{n+1}} \Omega(x) x^k dx.$$

Since $\Omega(x)$ is a polynomial in $\alpha, \ldots, \lambda$ then the numbers c_k will also be polynomials in $\alpha, \ldots, \lambda$.

The determinant of the system (15.1.7)

$$D = \begin{vmatrix} c_0 & c_1 & \cdots & c_{m-1} \\ c_1 & c_2 & \cdots & c_m \\ \cdots\cdots\cdots\cdots\cdots\cdots \\ c_{m-1} & c_m & \cdots & c_{2m-2} \end{vmatrix}$$

is the determinant of a positive-definite quadratic form

$$\sigma(z_1, \ldots, z_m) = \int_{x_n}^{x_{n+1}} \Omega(x) \left(\sum_{i=1}^{m} x^{i-1} z_i \right)^2 dx$$

and it is known to be different from zero for each set of $\alpha, \ldots, \lambda$ which satisfies (15.1.5). The coefficients p_k $(k = 1, \ldots, m)$ will be rational continuous functions of $\alpha, \ldots, \lambda$.

The roots $\xi_1, \xi_2, \ldots, \xi_m$ of $P_m(x)$ depend continuously on the coefficients p_k and will therefore be continuous functions of $\alpha, \ldots, \lambda$:

$$\xi_1 = \phi_1(\alpha, \ldots, \lambda)$$
$$\cdots\cdots\cdots\cdots\cdots\cdots \qquad (15.1.8)$$
$$\xi_m = \phi_m(\alpha, \ldots, \lambda).$$

These equations can be interpreted in a geometric manner. Consider an m-dimensional Euclidean space of points $(x_1, x_2, \ldots, x_m)$ which we denote by E_m. Equations (15.1.8) can be interpreted as a transformation of the point $(\alpha, \ldots, \lambda)$ of E_m into another point $(\xi_1, \ldots, \xi_m)$ of E_m. Condition (15.1.5), to which $\alpha, \ldots, \lambda$ are subjected, defines an m-dimensional closed simplex[1] in E_m. Since the roots ξ_k satisfy the inequalities $x_n < \xi_1 < \cdots < \xi_m < x_{n+1}$ then equations (15.1.8) define a single-valued, continuous transformation of this simplex onto itself. By the Brouwer fixed-point theorem[2] it is known that there exists an invariant point of this transformation and consequently there exists values $\alpha, \ldots, \lambda$ for which $\xi_1 = \alpha, \ldots, \xi_m = \lambda$ and $P_m(x) = \omega(x)$. Therefore there certainly exists nodes $\alpha, \ldots, \lambda$ which satisfy the inequalities $x_n < \alpha < \cdots < \lambda < x_{n+1}$ for which (15.1.4) is fulfilled. This completes the proof of Theorem 3.

It is not known, in general, whether the points $\alpha, \ldots, \lambda$ will be unique. We now find a representation for the remainder of (15.1.2). Let $[a', b']$ be the segment which contains $[x_n, x_{n+1}]$ and all the nodes of formula (15.1.2).

Theorem 4. *If $f(x)$ has a continuous derivative of order $m + N + 1$ on $[a', b']$ and if formula (15.1.2) has degree of precision $m + N$, then there exists a point ξ in $[a', b']$ with the property that the remainder $R(f)$ of formula (15.1.2) satisfies*

$$R(f) = \frac{f^{(m+N+1)}(\xi)}{(m+N+1)!} \int_{x_n}^{x_{n+1}} \Omega(x)\omega^2(x)\,dx. \tag{15.1.9}$$

Proof. Let us construct an interpolating polynomial for $f(x)$ in the following way. Suppose that at each of the basic nodes $\alpha, \ldots, \lambda$ we are given both the value of $f(x)$ and the value of its derivative $f'(x)$ and at each node of the form $\alpha + p_i h, \ldots, \lambda + t_i h$ we are given only the value of the function $f(x)$. We will have a total of $m + N + 1$ known values. The interpolating polynomial based on these values will be denoted by $H(x)$ and will have degree $\leq m + N$:

$$f(x) = H(x) + r(x).$$

---

[1] An m-dimensional simplex is the generalization of a triangle for two dimensions and a tetrahedron for three dimensions and has $m + 1$ vertices, which do not lie in any $(m - 1)$-dimensional subspace, and is bounded by $m + 1$ $(m - 1)$-dimensional faces.

[2] Brouwer has proved the following theorem: *If we are given any single-valued, continuous transformation of an m-dimensional simplex onto itself then this transformation has at least one invariant point*; L. E. J. Brouwer, "Über Abbildung von Mannigfaltigkeiten," *Math. Annalen*, Vol. 71, 1912, pp. 97–115 or V. V. Nemytskii, "Method of fixed points," *Uspehi Mat. Nauk*, Vol. 1, 1936, p. 153.

By the results of Section 3.3 the remainder $r(x)$ can be represented in the form

$$r(x) = \frac{f^{(m+N+1)}(\eta)}{(m+N+1)!} \, \Omega(x)\omega^2(x)$$

where η is a point inside the segment which contains the nodes of the interpolation and the point x.

It is clear that $R(f) = R(H) + R(r)$. But $H(x)$ has degree $\leq m + N$ so that $R(H) = 0$ and hence $R(f) = R(r)$. The quadrature sum for $r(x)$ is zero since $r(x)$ is zero at all the nodes of the formula. Thus $R(r)$ coincides with the integral of $r(x)$:

$$R(f) = R(r) = \int_{x_n}^{x_{n+1}} r(x)\,dx = \int_{x_n}^{x_{n+1}} \frac{f^{(m+N+1)}(\eta)}{(m+N+1)!} \, \Omega(x)\omega^2(x)\,dx.$$

Since $\Omega(x)\omega^2(x)$ does not change sign on $[x_n, x_{n+1}]$ the assertion of the theorem immediately follows.

15.2. FORMULAS OF SPECIAL FORM

Here we consider formulas for calculating the indefinite integral

$$y(x) = y_0 + \int_{x_0}^{x} f(t)\,dt$$

which use one, two or three values of the integrand $f(x)$ on each step or, in other words, formulas which contain one, two or three basic nodes. We will give numerical values for the nodes and coefficients in these formulas.[3]

All of these formulas can be constructed by a standard method and we describe this method in detail for only one case and in the other cases we only give the final results.

1. We begin with the case of one value of $f(x)$ on each step. These formulas reduce to formulas studied by Gauss and obtained by him in another problem in a different way.

On the segment $[x_n, x_n + h]$ we take the basic node $\alpha_n = x_n + qh$, $0 \leq q < 1$. The nodes are then situated as in Fig. 9.

In order to construct a formula of the form (15.1.2) we use k nodes preceding and following α_n which are similar to α_n. The formula then contains $2k + 1$ nodes. We are free to choose only the parameter q and the

---

[3] The values of the coefficients and nodes of the formulas given in this section were computed by Junior Research Assistant M. A. Filippov of the Leningrad Division of Mat. In-Ta Akad. Nauk SSSR.

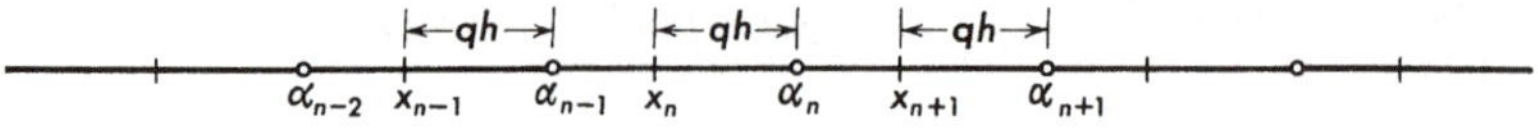

Figure 9.

highest degree of precision of the formula is $2k + 1$. In order to achieve this precision the formula must be interpolatory and it must satisfy the orthogonality condition (15.1.4) which in this case is

$$\int_{x_n}^{x_n+h} \Omega(x)\omega(x)\,dx = 0 \tag{15.2.1}$$

$$\Omega(x)\omega(x) = (x - \alpha_n)(x - \alpha_{n-1})(x - \alpha_{n+1}) \cdots (x - \alpha_{n-k})(x - \alpha_{n+k}) =$$

$$= [x - x_n - qh][(x - x_n - qh)^2 - h^2] \cdots [(x - x_n - qh)^2 - k^2h^2].$$

It is easy to show that (15.2.1) has the solution $q = \dfrac{1}{2}$ and that this solution is unique for $0 \leq q \leq 1$.

We transform the integral (15.2.1) by the transformation $x = x_n + hq + ht$ to obtain

$$\int_{x_n}^{x_n+h} \Omega(x)\omega(x)\,dx = h^{2k+2} \int_{-q}^{1-q} \pi(t)\,dt$$

$$\pi(t) = t(t^2 - 1^2) \cdots (t^2 - k^2).$$

Thus (15.2.1) is equivalent to

$$\phi(q) = \int_{-q}^{1-q} \pi(t)\,dt = 0. \tag{15.2.2}$$

Since $\pi(t)$ is an odd function of t then $\phi\left(\dfrac{1}{2}\right) = \displaystyle\int_{-\frac{1}{2}}^{\frac{1}{2}} \pi(t)\,dt = 0$ and

$q = \dfrac{1}{2}$ is a root of (15.2.2). The derivative of $\phi(q)$ is

$$\phi'(q) = \pi(1 - q) - \pi(-q)$$

and since $\pi(1 - q)$ and $\pi(-q)$ have opposite signs for $0 < q < 1$ then $\phi'(q)$ does not change sign on the interval $0 < q < 1$. Therefore the root $q = \dfrac{1}{2}$ is unique for $0 \leq q \leq 1$ and

$$a_n = x_n + \frac{1}{2}h.$$

To interpolate for $f(x)$ on $[x_n, x_n + h]$ with respect to its values at the nodes a_m $(m = n - k, \ldots, n + k)$ we use Newton's interpolation formula (3.2.6) substituting the nodes in the order

$$a_n, a_n + h, a_n - h, a_n + 2h, a_n - 2h, \ldots.$$

We obtain

$$f(x) = f(a_n) + (x - a_n)f(a_n, a_n + h) +$$
$$+ (x - a_n)(x - a_n - h)f(a_n, a_n + h, a_n - h) + (x - a_n) \times$$
$$\times (x - a_n - h)(x - a_n + h)f(a_n, a_n + h, a_n - h, a_n + 2h) +$$
$$+ \cdots + r(x) =$$

$$= f(a_n) + \frac{x - a_n}{1!h}\Delta f(a_n) +$$

$$+ \frac{(x - a_n)(x - a_n - h)}{2!h^2}\Delta^2 f(a_n - h) + \cdots +$$

$$+ \frac{(x - a_n + kh)\cdots(x - a_n - kh)}{(2k + 1)!h^{2k+1}}\Delta^{2k+1}f(a_n - kh) + r(x).$$

Substituting this expression for $f(x)$ into the equation

$$y_{n+1} = y_n + \int_{x_n}^{x_n+h} f(t)\, dt$$

gives

$$y_{n+1} = y_n + h\left[f(a_n) + \frac{1}{24}\Delta^2 f(a_n - h) - \frac{17}{5760}\Delta^4 f(a_n - 2h) +\right.$$

$$+ \frac{367}{967680}\Delta^6 f(a_n - 3h) - \frac{27859}{464486400}\Delta^8 f(a_n - 4h) +$$

$$+ \frac{1295803}{122624409600}\Delta^{10}f(a_n - 5h) + \cdots +$$

$$\left. + c_k\Delta^{2k}f(a_n - kh)\right] + R_{n,k}.$$

$$c_k = \frac{1}{(2k)!}\int_{-\frac{1}{2}}^{\frac{1}{2}} t^2(t^2 - 1^2)\cdots(t^2 - (k-1)^2)\, dt. \qquad (15.2.3)$$

If $f(x)$ has a continuous derivative of order $2k + 2$ on $[a_n - kh, a_n + kh]$ then, by Theorem 4 of Section 3.2, the remainder of the interpolation can be represented as

$$r(x) = \frac{(x - a_n + kh) \cdots (x - a_n - kh)}{(2k + 2)!} f^{(2k+2)}(\eta)$$

for some interior point η of this segment. To find the remainder $R_{n,k}$ we integrate this expression and make the substitution

$$x = a_n + th = x_n + \frac{1}{2}h + th$$

to obtain

$$R_{n,k} = \frac{f^{(2n+2)}(\xi)}{(2k + 2)!} \int_{-\frac{1}{2}}^{\frac{1}{2}} t^2 (t^2 - 1^2) \cdots (t^2 - k^2)\,dt. \qquad (15.2.4)$$

2. We now consider some of the simplest computational formulas which require two values of $f(x)$ on each step. We use the two basic nodes a_n, β_n on the segment $[x_n, x_n + h]$. The nodes are situated as in Fig. 10.

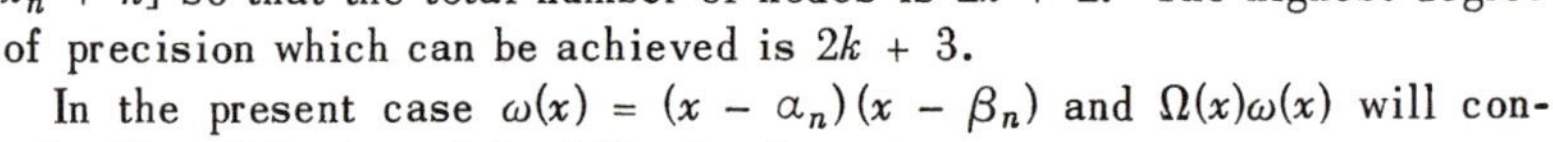

Figure 10.

In addition to a_n and β_n we will also use k nodes on each side of $[x_n, x_n + h]$ so that the total number of nodes is $2k + 2$. The highest degree of precision which can be achieved is $2k + 3$.

In the present case $\omega(x) = (x - a_n)(x - \beta_n)$ and $\Omega(x)\omega(x)$ will contain $2k + 2$ factors of the following form:

$$\Omega(x)\omega(x) = (x - a_n)(x - \beta_n)(x - a_{n+1}) \times$$
$$\times (x - \beta_{n-1})(x - \beta_{n+1})(x - a_{n-1})\cdots.$$

The orthogonality condition (15.1.4) which $\Omega(x)\omega(x)$ must satisfy reduces to the two equations

$$\int_{x_n}^{x_n+h} \Omega(x)\omega(x)\,dx = 0 \qquad \int_{x_n}^{x_n+h} x\Omega(x)\omega(x)\,dx = 0. \qquad (15.2.5)$$

In the first case, $k = 1$, we have four nodes: $a_n, \beta_n, a_{n+1}, \beta_{n-1}$. In order to simplify the problem we make the transformation

$$z = \frac{2}{h}x - \left(1 + \frac{2}{h}x_n\right)$$

which transforms the points $\ldots, x_{n-1}, x_n, x_{n+1}, x_{n+2}, \ldots$ into the points $\ldots, -3, -1, 1, 3, \ldots$ and the midpoint of $[x_n, x_n + h]$ transforms into $z = 0$. The points which $x = \alpha_n$ and $x = \beta_n$ transform into we will denote by p and q. In terms of z we have

$$\omega(x) = (x - \alpha_n)(x - \beta_n) = \frac{h^2}{4}(z - p)(z - q)$$

$$\Omega(x)\omega(x) = (x - \alpha_n)(x - \beta_n)(x - \alpha_{n+1})(x - \beta_{n-1}) =$$

$$= \frac{h^4}{16}(z - q + 2)(z - p)(z - q)(z - p - 2).$$

In terms of z the orthogonality conditions (15.2.5) are

$$\int_{-1}^{1}(z - q + 2)(z - p)(z - q)(z - p - 2)\,dz = 0$$

$$\int_{-1}^{1} z\,(z - q + 2)(z - p)(z - q)(z - p - 2)\,dz = 0$$

or after integrating and collecting terms

$$p^2(1 - 6q + 3q^2) + 2p(3q^2 - 4q - 1) + \left(q^2 + 2q - \frac{17}{5}\right) = 0$$

$$(p + q)\left[\frac{1}{5} + \frac{1}{3}(pq + q - p - 2)\right] = 0. \qquad (15.2.6)$$

From the second of these equations p can have the values

$$p_1 = \frac{q - \dfrac{7}{5}}{1 - q}, \qquad p_2 = -q.$$

Since p and q must satisfy the condition

$$-1 < p < q < 1$$

we see that the solution p_1 must be rejected since it does not satisfy $-1 < p_1$.

From the second solution $p = p_2 = -q$ the first of the equations (15.2.6) gives

$$3q^4 - 12q^3 + 10q^2 + 4q - \frac{17}{5} = 0.$$

Now q must lie in the segment $(0, 1)$ and it is possible to show that this equation has only one root of this form

$$q \approx 0.53332\ 38475.$$

The basic nodes α_n and β_n are then

$$\alpha_n = x_n + \frac{1}{2}(1 - q)h = x_n + (0.23333\ 80763)h$$

$$\beta_n = x_n + \frac{1}{2}(1 + q)h = x_n + (0.76666\ 19237)h.$$

Let us construct the interpolating polynomial for $f(x)$ using its values at the nodes $\beta_{n-1}, \alpha_n, \beta_n, \alpha_{n+1}$:

$$f(x) = \frac{(x - \alpha_n)(x - \beta_n)(x - \alpha_{n+1})}{(\beta_{n-1} - \alpha_n)(\beta_{n-1} - \beta_n)(\beta_{n-1} - \alpha_{n+1})}\, f(\beta_{n-1}) + \cdots + r(x) =$$
$$= P(x) + r(x).$$

Then

$$y_{n+1} = y_n + \int_{x_n}^{x_n+h} f(t)\,dt = y_n + \int_{x_n}^{x_n+h} P(t)\,dt + R_n.$$

Computing the integral of $P(t)$ leads to the following formula:

$$y_{n+1} = y_n + (0.48690\ 23179)h[f(\alpha_n) + f(\beta_n)] +$$
$$+ (0.01309\ 76821)h[f(\beta_{n-1}) + f(\alpha_{n+1})] + R_n. \qquad (15.2.7)$$

The remainder R_n can be found from the representation (15.1.9). Here we must use $m = 2$, $N + 1 = 4$ and

$$\Omega(x)\omega^2(x) = (x - \beta_{n-1})(x - \alpha_n)^2(x - \beta_n)^2(x - \alpha_{n+1}).$$

This leads to

$$R_n = -\frac{14.732017}{4838400}\, h^7 f^{(6)}(\xi) = -0.00000305 h^7 f^{(6)}(\xi) \qquad (15.2.8)$$

$$\beta_{n-1} < \xi < \alpha_{n+1}.$$

We now consider the case $k = 2$. In addition to two basic nodes in the interval $[x_n, x_n + h]$ we also use two nodes in each of the adjoining intervals $[x_n - h, x_n]$ and $[x_n + h, x_n + 2h]$. The nodes are depicted in Fig. 10.

The highest algebraic degree of precision is 7.

The nodes of this quadrature formula are

$$\alpha_n = x_n + (0.23896\ 17210)h, \qquad \beta_n = x_n + (0.76103\ 82790)h.$$

The formula is

$$\begin{aligned}
y_{n+1} = y_n &+ (0.48309\ 24404)h[f(\alpha_n) + f(\beta_n)] + \\
&+ (0.01737\ 14226)h[f(\beta_{n-1}) + f(\alpha_{n+1})] - \\
&- (0.00046\ 38630)h[f(\alpha_{n-1}) + f(\beta_{n+1})] + R_n. \qquad (15.2.9)
\end{aligned}$$

The estimate for the remainder is

$$|R_n| \leq 0.00000008 h^9 M_8$$

$$M_8 = \max_x |f^{(8)}(x)|, \qquad \alpha_{n-1} < x < \beta_{n+1}.$$

In all the cases which we consider below the nodes are situated symmetrically with respect to the middle of the segment $[x_n, x_n + h]$. We will not derive any nonsymmetric formulas of the highest degree of precision.

The case $k = 3$. In addition to two basic nodes in $[x_n, x_n + h]$ we also use two nodes in $[x_n - h, x_n]$ and in $[x_n + h, x_n + 2h]$ and one node in $[x_n - 2h, x_n - h]$ and in $[x_n + 2h, x_n + 3h]$. The nodes are situated as shown in Fig. 11.

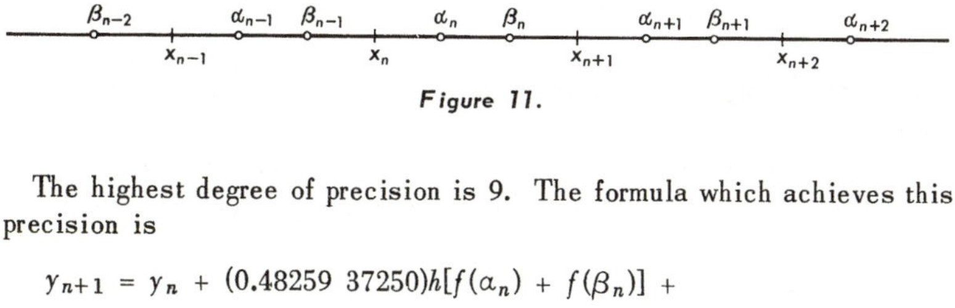

Figure 11.

The highest degree of precision is 9. The formula which achieves this precision is

$$\begin{aligned}
y_{n+1} = y_n &+ (0.48259\ 37250)h[f(\alpha_n) + f(\beta_n)] + \\
&+ (0.01797\ 22221)h[f(\beta_{n-1}) + f(\alpha_{n+1})] - \\
&- (0.00057\ 82647)h[f(\alpha_{n-1}) + f(\beta_{n+1})] + \\
&+ (0.00001\ 23177)h[f(\beta_{n-2}) + f(\alpha_{n+2})] + R_n \qquad (15.2.10)
\end{aligned}$$

where the nodes are

$$\alpha_n = x_n + (0.23963\ 00931)h, \qquad \beta_n = x_n + (0.76036\ 99069)h.$$

The remainder satisfies the estimate

$$|R_n| \leq 0.000000003 h^{11} M_{10}$$

$$M_{10} = \max_x |f^{(10)}(x)|, \qquad \beta_{n-2} < x < \alpha_{n+2}.$$

The case $k = 4$. We use the nodes shown in Fig. 12.

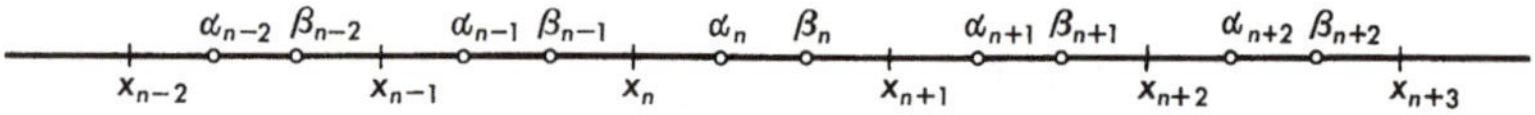

Figure 12.

The highest algebraic degree of precision is 11 and is achieved by the formula

$$
\begin{aligned}
y_{n+1} = y_n &+ (0.47911\ 31668)h[f(\alpha_n) + f(\beta_n)] + \\
&+ (0.02153\ 22932)h[f(\beta_{n-1}) + f(\alpha_{n+1})] - \\
&- (0.00136\ 32927)h[f(\alpha_{n-1}) + f(\beta_{n+1})] + \\
&+ (0.00012\ 36065)h[f(\beta_{n-2}) + f(\alpha_{n+2})] - \\
&- (0.00000\ 57738)h[f(\alpha_{n-2}) + f(\beta_{n+2})] + R_n \qquad (15.2.11)
\end{aligned}
$$

where the nodes are

$$
\alpha_n = x_n + (0.24346\ 00865)h, \qquad \beta_n = x_n + (0.75653\ 99135)h.
$$

The remainder satisfies

$$
|R_n| \leq 0.00000000011 h^{13} M_{12}
$$

$$
M_{12} = \max_x |f^{(12)}(x)|, \qquad \alpha_{n-2} < x < \beta_{n+2}.
$$

3. Finally we give three formulas which use three values of $f(x)$ on each step.

Using three basic nodes and one additional node on each adjacent interval as depicted in Fig. 13 a formula of degree 7 can be constructed.

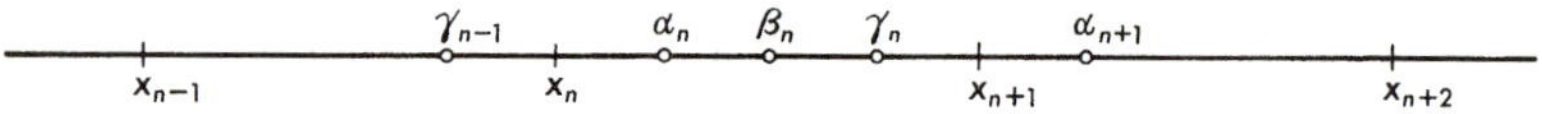

Figure 13.

The formula is

$$
\begin{aligned}
y_{n+1} = y_n &+ (0.40010\ 36566)hf(\beta_n) + \\
&+ (0.29348\ 93491)h[f(\alpha_n) + f(\gamma_n)] + \\
&+ (0.00645\ 88226)h[f(\gamma_{n-1}) + f(\alpha_{n+1})] + R_n \qquad (15.2.12)
\end{aligned}
$$

$$
\begin{aligned}
\alpha_n &= x_n + (0.13518\ 35561)h \\
\beta_n &= x_n + (0.5)h \\
\gamma_n &= x_n + (0.86481\ 64439)h
\end{aligned}
$$

$$|R_n| \leq 0.0000000024h^9 M_8$$

$$M_8 = \max_x |f^{(8)}(x)|, \qquad \gamma_{n-1} < x < \alpha_{n+1}.$$

With the nodes shown in Fig. 14 a formula of degree 9 can be constructed.

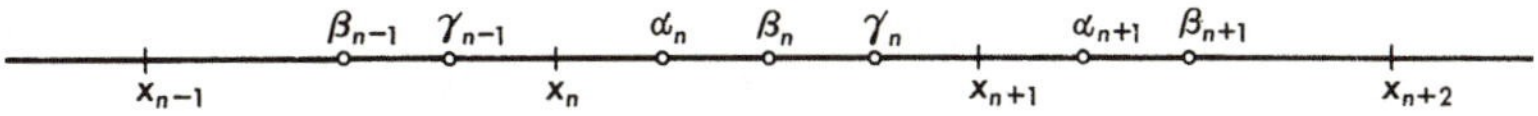

Figure 14.

The formula is

$$
\begin{aligned}
y_{n+1} = y_n &+ (0.38762\ 75418)hf(\beta_n) + \\
&+ (0.29781\ 27562)h[f(\alpha_n) + f(\gamma_n)] + \\
&+ (0.00848\ 08932)h[f(\gamma_{n-1}) + f(\alpha_{n+1})] - \\
&- (0.00010\ 74203)h[f(\beta_{n-1}) + f(\beta_{n+1})] + R_n \qquad (15.2.13)
\end{aligned}
$$

$$\alpha_n = x_n + (0.14145\ 83289)h$$
$$\beta_n = x_n + (0.5)h$$
$$\gamma_n = x_n + (0.85854\ 16711)h$$

$$|R_n| \leq 0.000000000002h^{11} M_{10}$$

$$M_{10} = \max_x |f^{(10)}(x)|, \qquad \beta_{n-1} < x < \beta_{n+1}.$$

The last formula we give uses nodes situated as in Fig. 15 and has degree 11.

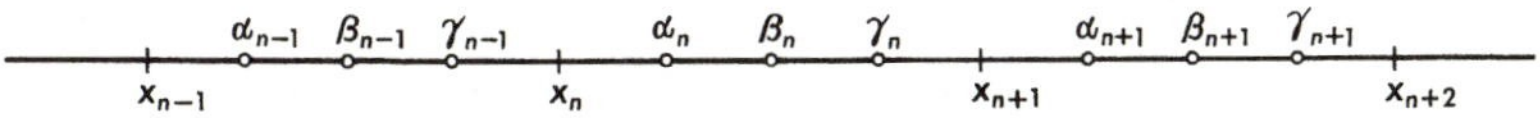

Figure 15.

The formula is

$$
\begin{aligned}
y_{n+1} = y_n &+ (0.38134\ 28493)hf(\beta_n) + \\
&+ (0.29986\ 68413)h[f(\alpha_n) + f(\gamma_n)] + \\
&+ (0.00967\ 80471)h[f(\gamma_{n-1}) + f(\alpha_{n+1})] - \\
&- (0.00022\ 28947)h[f(\beta_{n-1}) + f(\beta_{n+1})] + \\
&+ (0.00000\ 65816)h[f(\alpha_{n-1}) + f(\gamma_{n+1})] + R_n \qquad (15.2.14)
\end{aligned}
$$

$$\alpha_n = x_n + (0.14469\ 85558)h$$

$$\beta_n = x_n + (0.5)h$$

$$\gamma_n = x_n + (0.85530\ 14442)h$$

$$|R_n| \leq 0.0000000000003 h^{13} M_{12}$$

$$M_{12} = \max_x |f^{(12)}(x)|, \qquad \alpha_{n-1} < x < \gamma_{n+1}.$$

Example. Let us calculate the elliptic integral of the first kind

$$y(x) = \int_0^x \frac{dt}{\sqrt{(1 - t^2)(1 - k^2 t^2)}}$$

for $k^2 = 0.5$.

For this calculation we use (15.2.7) with step size $h = 0.1$. This formula contains 4 nodes and its degree of precision is 5. For each additional step in the calculation we must compute two new values of the integrand.

As a comparison the above integral was also calculated by formula (14.1.5):

$$y_{n+1} = y_n + h\left[\frac{f_n + f_{n+1}}{2} - \frac{1}{12}\frac{\Delta^2 f_{n-1} + \Delta^2 f_n}{2} + \cdots\right].$$

Here the step size was taken to be $h = 0.05$ so that for each step of length 0.1 two new values of $f(x)$ would also be required. Two forms of this formula were used:

1. with four nodes $x_{n-1}, x_n, x_{n+1}, x_{n+2}$;
2. with six nodes $x_{n-2}, x_{n-1}, \ldots, x_{n+3}$.

In the first case formula (14.1.5) contains the same number of nodes as (15.2.7) and in the second case the formula has the same algebraic degree of precision as (15.2.7). The exact values of the integrals were taken from the table of Legendre.[4]

x	Exact value of $y(x)$	Formula (15.2.7)	Error $\times 10^{10}$
0.0	0.00000 00000	0.00000 00000	0
0.1	0.10025 11946	0.10025 11947	−1
0.2	0.20203 89248	0.20203 89251	−3
0.3	0.30705 49305	0.30705 49312	−7
0.4	0.41734 51597	0.41734 51612	−15
0.5	0.53562 27328	0.53562 27370	−42
0.6	0.66584 78254	0.66584 78390	−136
0.7	0.81448 92840	0.81448 93476	−636
0.8	0.99390 71263	0.99390 73932	−1669

---

[4] A. M. Legendre, *Legendres Tafeln der Elliptischen Normalintegrale erster und zweiter Gattung*, Stuttgart, 1931.

x	(14.1.5) with 4 nodes	Error $\times 10^{10}$	(14.1.5) with 6 nodes	Error $\times 10^{10}$
0.00	0.00000 00000		0.00000 00000	
0.05	0.05003 12182		0.05003 12881	
0.10	0.10025 10523	1423	0.10025 11965	−19
0.15	0.15085 26647		0.15085 28927	
0.20	0.20203 86018	3230	0.20203 89297	−49
0.25	0.25402 62515		0.25402 67037	
0.30	0.30705 43281	6024	0.30705 49416	−111
0.35	0.36139 09430		0.36139 17730	
0.40	0.41734 40530	11067	0.41734 51861	−264
0.45	0.47527 55059		0.47527 70768	
0.50	0.53562 05749	21579	0.53562 28057	−729
0.55	0.59891 61532		0.59891 94255	
0.60	0.66584 30644	47610	0.66584 80773	−2519
0.65	0.73729 25074		0.73735 06471	
0.70	0.81447 62323	130517	0.81449 05609	−12769
0.75	0.89912 20667		0.89920 04674	
0.80	0.99385 27113	544150	0.99392 09917	−138654

CHAPTER 16

Methods Which Use
Several Previous Values
of the Integral

16.1. INTRODUCTION

In the last two chapters we have discussed separate problems on the approximate evaluation of indefinite integrals and in both cases we used only one preceding value of the integral in order to approximate the next value. Computational methods of this type are always stable with respect to the errors of the initial values and the rounding errors providing that they are exact for $f(x) \equiv 0$ and $y(x) \equiv 1$. In such cases the formula must have the form

$$y_{n+1} = y_{n-k} + \sum_{j=1}^{m} B_{n,j} f(\xi_{n,j}) + r_n$$

and thus is a formula with positive coefficients A_i and therefore is stable with respect to the errors in the initial values and the rounding errors as we showed in Section 13.3.

It is a more difficult problem to derive formulas which use more than one preceding value of $y(x)$ since formulas of the highest algebraic degree of precision of this type which are stable cannot always be found.

In this chapter we discuss one method for constructing stable formulas of the highest degree of precision.

Suppose we wish to calculate the integral

$$y(x) = y_0 + \int_{x_0}^{x} f(t) \, dt$$

for an arbitrary set of points

$$x_k \qquad (k = 0, 1, \ldots, N; \qquad x_k < x_{k+1})$$

on the segment $[x_0, X]$. We assume that the calculation has been carried up to the point x_n and that we have computed $y(x_n)$. In order to compute $y(x_{n+1})$ we can use any of the previously computed values y_k $(k \leq n)$ of $y(x)$ and any values of $y'(x) = f(x)$. Here we assume that the derivative $y'(x)$ is known everywhere on $[x_0, X]$ and that we may use any nodes whatsoever on this segment at which to evaluate $f(x)$.

This is a problem of interpolation to find the value of $y(x)$ at one of the fixed nodes x_{n+1} in terms of values of the same function and of its derivative $y'(x) = f(x)$. We will see below that it will be useful to divide the nodes in the formula into three classes.

Let $y(z)$ be any function which is defined and differentiable on a certain segment $[a, b]$. On this segment we take $r + s + u$ distinct points

$$\begin{array}{cccc}
\xi_1, & \xi_2, & \cdots, & \xi_r \\
\xi_{r+1}, & \xi_{r+2}, & \cdots, & \xi_{r+s} \\
\xi_{r+s+1}, & \xi_{r+s+2}, & \cdots, & \xi_{r+s+u}
\end{array} \qquad (16.1.1)$$

At the first r of these nodes we assume that we know the value of $y(z)$:

$$y(\xi_1), \ldots, y(\xi_r).$$

At the next s nodes we assume that we know the value of both the function and its derivative:

$$y(\xi_j), \qquad y'(\xi_j) \qquad j = r+1, \ldots, r+s.$$

At the last u nodes we assume that we know only the value of the derivative

$$y'(\xi_j) \quad j = r+s+1, \ldots, r+s+u.$$

We will call the nodes written in the first line of (16.1.1) the simple nodes, those in the second line the double nodes and those in the last line the auxiliary nodes. In addition we let x denote a point of $[a, b]$ which does not coincide with any of the simple or double nodes, but which may coincide with an auxiliary node.

We select certain $r + s$ numbers a_j $(j = 1, 2, \ldots, r + s)$ and $s + u$ numbers β_j $(j = r + 1, \ldots, r + s + u)$ which will be defined later and consider the expression

$$y(x) = \sum_{j=1}^{r+s} a_j \, y(\xi_j) + \sum_{j=r+1}^{r+s+u} \beta_j y'(\xi_j) + R. \qquad (16.1.2)$$

Neglecting the remainder R gives an approximate expression for $y(x)$:

$$y(x) \approx \sum_{j=1}^{r+s} a_j y(\xi_j) + \sum_{j=r+1}^{r+s+u} \beta_j y'(\xi_j). \qquad (16.1.3)$$

The degree of precision of this equation is defined in the usual way: we say that (16.1.3) has algebraic degree of precision m if it is exact for all monomials $y(z) = z^k$ $(k = 0, 1, \ldots, m)$ and is not exact for $y(z) = z^{m+1}$. We will determine what the highest degree of precision of (16.1.3) may be and under what conditions it is achieved.

Theorem 1. *For any a_j, β_j and any disposition of points ξ_j and x the degree of precision m of* (16.1.3) *is always less than $r + 2s + 2u$:*

$$m < r + 2s + 2u.$$

Proof. It suffices to show that there always exists a polynomial of degree not exceeding $r + 2s + 2u$ for which equation (16.1.3) is not exact.

We assume, at first, that none of the auxiliary nodes coincide with x and consider the polynomial

$$y(z) = (z - \xi_1) \cdots (z - \xi_r)(z - \xi_{r+1})^2 \cdots (z - \xi_{r+s+u})^2 = A(z) \qquad (16.1.4)$$

of degree $r + 2s + 2u$.

It is obvious that $A(\xi_j) = 0$ for $j = 1, \ldots, r + s + u$ and $A'(\xi_j) = 0$ for $j \geq r + 1$. Thus the right side of (16.1.3) is zero for this function. The left side $y(x) = A(x) \neq 0$ because $x \neq \xi_j$ and (16.1.3) can not be exact.

Now assume that one of the auxiliary nodes, for example ξ_{r+s+u}, coincides with x. We introduce the polynomial $B(z)$ of degree $r + 2s + 2u - 2$:

$$B(z) = \frac{A(z)}{(z - \xi_{r+s+u})^2}.$$

If $B'(x) \neq 0$ we put

$$y(z) = B(z) \left[z - x - \frac{B(x)}{B'(x)} \right].$$

This is a polynomial of degree $r + 2s + 2u - 1$ for which the right side of (16.1.3) is zero. The left side is $y(x) = -B^2(x)/B'(x) \neq 0$ and (16.1.3) is not satisfied. In this case the degree of precision of (16.1.3) is less than $r + 2s + 2u - 1$.

If $B'(x) = 0$ then (16.1.3) is not satisfied for $y(z) = B(z)$. In this case the degree of precision is less than $r + 2s + 2u - 2$. This proves Theorem 1.

This theorem shows that the greatest possible degree of precision of (16.1.3) is $r + 2s + 2u - 1$. Later we show that this degree of precision can be achieved by an appropriate choice of the coefficients a_j and β_j and auxiliary nodes $\xi_j (j > r + s)$.

From the proof of Theorem 1 we see that when one of the nodes $\xi_j (j > r + s)$ coincides with x the degree of precision of (16.1.3) is less than $r + 2s + 2u - 1$ and can not achieve its highest value. Therefore we will always assume that all the ξ_j are different from the point x.

16.2. CONDITIONS UNDER WHICH THE HIGHEST DEGREE OF PRECISION IS ACHIEVED

If we require that (16.1.3) be exact for the monomials x^k, $k = 0, 1, \dots,$ $r + 2s + 2u - 1$, we obtain the following system of $r + 2s + 2u$ equations

$$\sum_{j=1}^{r+s} a_j \xi_j^k + \sum_{j=r+1}^{r+s+u} \beta_j k \xi_j^{k-1} = x^k \qquad (k = 0, 1, \dots, r + 2s + 2u - 1). \quad (16.2.1)$$

This system can be studied by comparing it with the related interpolation problem.

At the simple nodes let us be given the values of the function

$$y(\xi_j) \qquad j = 1, \dots, r. \qquad (16.2.2)$$

At all the double and auxillary nodes let us be given both the value of the function and also its derivative

$$y(\xi_j), \qquad y'(\xi_j) \qquad j = r + 1, \dots, r + s + u. \qquad (16.2.3)$$

Using these values consider the problem of interpolating for the value $y(x)$. Taking certain numbers $a_j' \; (j = 1, \dots, r + s + u)$ and $\beta_j' \; (j = r + 1, \dots, r + s + u)$ we construct the approximate equation

$$y(x) \approx \sum_{j=1}^{r+s+u} a_j' y(\xi_j) + \sum_{j=r+1}^{r+s+u} \beta_j' y'(\xi_j). \qquad (16.2.4)$$

Here it is possible to determine the $r + 2s + 2u$ coefficients a_j' and β_j' if we require that (16.2.4) be exact for the monomials $y(z) = z^k$, $k = 0, 1, \dots,$ $r + 2s + 2u - 1$. This gives a system of $r + 2s + 2u$ linear equations for the a_j', β_j':

$$\sum_{j=1}^{r+s+u} a_j' \xi_j^k + \sum_{j=r+1}^{r+s+u} \beta_j' k \xi_j^{k-1} = x^k \qquad (k = 0, 1, \dots, r + 2s + 2u - 1). \quad (16.2.5)$$

The determinant of this system is

$$\Delta = \begin{vmatrix} 1 & 1 & \cdots & 1 & 0 & \cdots & 0 \\ \xi_1 & \xi_2 & \cdots & \xi_{r+s+u} & 1 & \cdots & 1 \\ \xi_1^2 & \xi_2^2 & \cdots & \xi_{r+s+u}^2 & 2\xi_{r+1} & \cdots & 2\xi_{r+s+u} \\ \multicolumn{7}{c}{\cdots\cdots\cdots\cdots\cdots\cdots\cdots\cdots\cdots\cdots\cdots\cdots} \end{vmatrix}$$

where we have written the coefficients of the a'_j in the first $r + s + u$ columns and the coefficients of the β'_j in the last $s + u$ columns. This determinant is different from zero[1] if $\xi_i \neq \xi_j$ $(i, j = 1, \ldots, r + s + u; i \neq j)$. In the above case this will be true.

Thus the system (16.2.5) has a unique solution provided that the points ξ_j are all distinct. The relationship between the systems (16.2.1) and (16.2.5) is given by the following two assertions which are easily verified.

1. *Suppose the system (16.2.1) has a solution. Then this solution is unique and the unknowns α'_j and β'_j in (16.2.5) satisfy the following relationships:*

$$\begin{aligned} a'_j &= a_j & j = 1, \ldots, r + s \\ a'_{r+s+1} &= 0, \ldots, & a'_{r+s+u} = 0 \qquad (16.2.6) \\ \beta'_j &= \beta_j & j = r+1, \ldots, r+s. \end{aligned}$$

Indeed, suppose that the numbers a_j $(j = 1, \ldots, r + s)$ and β_j $(j = r + 1, \ldots, r + s + u)$ are a solution of the system (16.2.1). Together with these numbers we also take u numbers $a_{r+s+1} = 0, \ldots, a_{r+s+u} = 0$. The resulting system of $r + 2s + 2u$ numbers will clearly satisfy (16.2.5) and since (16.2.5) has a unique solution the relationships (16.2.6) must be valid.

2. *If the numbers α'_j and β'_j are a solution of the system (16.2.5) and if*

$$a'_{r+s+1} = 0, \ldots, \qquad a'_{r+s+u} = 0 \qquad (16.2.7)$$

then the system (16.2.1) has a solution and the following relationships

----

[1] We can easily see this if we take the Vandermondian determinant of order $r + 2s + 2u$ in the parameters $\xi_1, \ldots, \xi_{r+2s+2u}$

$$W(\xi_1, \ldots, \xi_{r+2s+2u}) = \prod_{i>j} (\xi_i - \xi_j) \qquad (*)$$

and calculate the mixed derivative with respect to $\xi_{r+s+u+1}, \ldots, \xi_{r+2s+2u}$ and then set $\xi_{r+s+u+1} = \xi_{r+1}, \ldots, \xi_{r+2s+2u} = \xi_{r+s+u}$. After performing these operations we obtain Δ. But this is the same as striking out the factors $(\xi_{r+s+u+j} - \xi_{r+j})$, $j = 1, \ldots, s + u$, from the product (*). What remains is a new product which is clearly different from zero.

are valid:

$$a_j = a_j' \qquad j = 1, \ldots, r+s$$
$$\beta_j = \beta_j' \qquad j = r+1, \ldots, r+s+u \tag{16.2.8}$$

Here we need only note that if the conditions (16.2.7) are fulfilled then the equations (16.2.5) coincide with (16.2.1).

The solution of the system (16.2.5) is easily constructed from the theory of interpolation. Let $y(z)$ be a polynomial of degree $\le r + 2s + 2u - 1$. When the a_j' and β_j' satisfy equations (16.2.5) then (16.2.4) must be exact for any value of x. This is an interpolation formula for the value $y(x)$ in terms of the values of this polynomial at the points ξ_j $(j \le r+s+u)$ and the values of its derivative at the points ξ_j $(r+1 \le j \le r+s+u)$. This is an interpolation with r simple nodes ξ_j $(j \le r)$ and $s+u$ double nodes ξ_j $(r < j \le r+s+u)$. The interpolating polynomial can be represented by Hermite's formula (3.3.8) which in the present case is

$$
\begin{aligned}
y(x) = &\sum_{j=1}^{r} \frac{A(x)}{(x-\xi_j)A'(\xi_j)} \, y(\xi_j) + \\
&+ \sum_{j=r+1}^{r+s+u} \frac{A_j(x)}{A_j(\xi_j)} \left[1 - (x-\xi_j)\frac{A_j'(\xi_j)}{A_j(\xi_j)} \right] y(\xi_j) + \\
&+ \sum_{j=r+1}^{r+s+u} \frac{A_j(x)}{A_j(\xi_j)} \, (x-\xi_j) \, y'(\xi_j) \\
&\qquad A_j(x) = A(x)/(x-\xi_j)^2 .
\end{aligned}
\tag{16.2.9}
$$

The right sides of (16.2.4) and (16.2.9) must coincide and since the values $y(\xi_j)$ and $y'(\xi_j)$ are arbitrary the coefficients a_j' and β_j' must be equal to the corresponding coefficients of (16.2.9).

The conditions (16.2.7) which must hold for (16.2.1) to be solvable are

$$\frac{A_j(x)}{A_j(\xi_j)} \left[1 - \frac{A_j'(\xi_j)}{A_j(\xi_j)} (x-\xi_j) \right] = 0 \qquad j = r+s+1, \ldots, r+s+u.$$

Since $A_j(x)/A_j(\xi_j) \ne 0$ the expression in brackets must be zero and dividing this expression by $x - \xi_j$ gives

$$\sum_{k=r+s+1}^{r+s+u}{}^{*} \frac{2}{\xi_j - \xi_k} + \sum_{k=1}^{r} \frac{2}{\xi_j - \xi_k} + \sum_{k=r+1}^{r+s} \frac{2}{\xi_j - \xi_k} + \frac{1}{\xi_j - x} = 0. \tag{16.2.10}$$

Here in the first sum the sumbol * indicates that the term for $k = j$ is omitted.

The results of this section can be summarized in the following theorem.

Theorem 2. *The following two conditions are necessary and sufficient for (16.1.3) to have the highest degree of precision $r + 2s + 2u - 1$:*

1. *The points ξ_j and x must satisfy the system of u equations (16.2.10).*

2. *The coefficients a_j and β_j must have the values*

$$a_j = \frac{A(x)}{(x - \xi_j) A'(\xi_j)} \qquad j = 1, \dots, r$$

$$a_j = \frac{A_j(x)}{A_j(\xi_j)} \left[1 - (x - \xi_j) \frac{A_j'(\xi_j)}{A_j(\xi_j)} \right] \qquad j = r+1, \dots, r+x \qquad (16.2.11)$$

$$\beta_j = \frac{A_j(x)}{A_j(\xi_j)} (x - \xi_j) \qquad j = r+1, \dots, r+s+u.$$

16.3. THE NUMBER OF INTERPOLATING POLYNOMIALS OF THE HIGHEST DEGREE OF PRECISION

Equations (16.2.10) can be studied in an intuitive way by using an electrostatic analogy similar to the analogy of Section 11.4. We take two points z_1 and z_2 in the complex plane and place at these points particles with charges e_1 and e_2. We assume that these particles exert on each other a force which is inversely proportional to the distance between them and directly proportional to the size of their charges.

Assuming that the coefficient of proportionality is unity then the force which z_1 exerts on z_2 is

$$\frac{e_1 e_2}{z_2 - z_1}.$$

Suppose that in the plane we take $r + s + 1$ points $x, \xi_1, \dots, \xi_{r+s}$. At each of the points $x, \xi_1, \dots, \xi_r$ we put particles with unit charge and at each of the points $\xi_{r+1}, \dots, \xi_{r+s}$ particles with charge two and we fix these particles at these points. Together with these particles we take u free particles of charge 2 at the points $\xi_{r+s+1}, \dots, \xi_{r+s+u}$.

When the free particles are at equilibrium the sum of the forces on each free particle must be zero

$$\sum_{k=r+s+1}^{r+s+u}{}^{*} \frac{4}{\xi_j - \xi_k} + \sum_{k=1}^{r} \frac{2}{\xi_j - \xi_k} + \sum_{k=r+1}^{r+s} \frac{4}{\xi_j - \xi_k} +$$

$$+ \frac{2}{\xi_j - x} = 0, \qquad j = r+s+1, \dots, r+s+u$$

These equations differ only by the multiple of 2 from equations

(16.2.10) which are the conditions for which (16.1.3) has the highest degree of precision.

From this analogy the following assertions concerning (16.2.10) are evident.[2]

1. If $x, \xi_1, \ldots, \xi_{r+s}$ are any complex numbers and if $\xi_{r+s+1}, \ldots,$ ξ_{r+s+u} satisfy (16.2.10) then the points $\xi_{r+s+1}, \ldots, \xi_{r+s+u}$ lie in the smallest convex polygon containing $x, \xi_1, \ldots, \xi_{r+s}$. In particular when $x, \xi_1, \ldots, \xi_{r+s}$ are real numbers the points $\xi_{r+s+1}, \ldots, \xi_{r+s+u}$ lie inside the smallest segment containing x and ξ_j $(j = 1, \ldots, r+s)$.

2. Let $x, \xi_1, \ldots, \xi_{r+s}$ be real and distinct. These points divide the real axis into $r+s$ adjacent intervals. Suppose that we have indicated beforehand how many of the auxiliary nodes should belong to each of these intervals. The number of such ways in which these auxiliary nodes may be arranged is $\dfrac{(r+s+u-1)!}{u!\,(r+s-1)!}$. For each of these arrangements of the auxiliary nodes there exists a solution of the system (16.2.10).

3. If we consider solutions which differ only by permutations of the nodes $\xi_{r+s+1}, \ldots, \xi_{r+s+u}$ as a single solution then for each arrangement of these nodes among the points $x, \xi_1, \ldots, \xi_{r+s}$ there will be one and only one solution for (16.2.10).

These results can be expressed in the following theorem.

Theorem 3. *For any set of real and distinct points* $x, \xi_1, \ldots, \xi_{r+s}$ *the auxiliary nodes* $\xi_{r+s+1}, \ldots, \xi_{r+s+u}$ *can be selected in* $\dfrac{(r+s+u-1)!}{u!\,(r+s-1)!}$ *ways which will make* (16.1.3) *have the highest algebraic degree of precision* $m = r + 2s + 2u - 1$. *For each partitioning of the auxiliary nodes among the* $r+s$ *intervals formed by* $x, \xi_1, \ldots, \xi_{r+s}$ *there exists one and only one solution of this type.*

16.4. THE REMAINDER OF THE INTERPOLATION AND MINIMIZATION OF ITS ESTIMATE

Consider the remainder of the interpolation formula (16.1.3)

$$R(y) = y(x) - \sum_{j=1}^{r+s} \alpha_j y(\xi_j) - \sum_{j=r+1}^{r+s+u} \beta_j y'(\xi_j) \qquad (16.4.1)$$

---

[2] These can be proved by an arithmetic argument similar to that used by T. Stieltjes in an analogous case concerning the existence of polynomial solutions to differential equations (Collected Works, Groningen, 1914, v. 1, p. 434–439). The relationship between the interpolation problem and differential equations is studied in Section 16.6.

for which we assume (16.2.10) is satisfied so that it has the highest algebraic degree of precision $r + 2s + 2u - 1$. Then the interpolation (16.1.3) coincides with (16.2.4) and they have the same precision. Equation (16.2.4) has r simple nodes and $s + u$ double nodes. Assuming that $y(z)$ has a continuous derivative of order $r + 2s + 2u$ on the segment $[a, b]$ then, by Theorem 6 of Chapter 3, we can find a point $\xi \in [a, b]$ for which

$$R(y) = \frac{A(x)}{(r + 2s + 2u)!} y^{(r+2s+2u)}(\xi) \qquad (16.4.2)$$

$$A(x) = (x - \xi_1) \cdots (x - \xi_r)(x - \xi_{r+1})^2 \cdots (x - \xi_{r+s+u})^2. \qquad (16.4.3)$$

This remainder then coincides with the remainder (16.4.1).

In the class of functions defined by the inequality

$$|f^{(r+2s+2u)}(z)| \leq M \qquad z \in [a, b]$$

the remainder has the precise estimate

$$|R(y)| \leq \frac{|A(x)|}{(r + 2s + 2u)!} M. \qquad (16.4.4)$$

The highest degree of precision $r + 2s + 2u - 1$ can be achieved by $\dfrac{(r + s + u - 1)!}{u!(r + s - 1)!}$ different choices of the auxiliary nodes and it is natural to ask which of these formulas will minimize the right side of (16.4.4).

The only term in (16.4.4) which depends on the nodes is $|A(x)|$ and we will determine which choice of nodes makes this a minimum. The problem formulated in Section 16.1 was that of calculating $y(x)$ at the node x_{n+1}. The simple and double nodes ξ_j $(j = 1, \ldots, r + s)$ must be points of the fixed set $x_0, x_1, x_2, \ldots, x_n$.

We must therefore select the simple nodes and indicate how the auxiliary nodes are distributed among them. We assume that the auxiliary nodes are enumerated in increasing order.

From the discussion of the previous section we can solve the problem of minimizing $|A(x)|$ by an intuitive argument. We prefer this line of reasoning since it is much shorter than a constructive arithmetic proof.

Assume that the simple and double nodes have been chosen and that we wish to determine the distribution of the auxiliary nodes. Consider the factors

$$(x_{n+1} - \xi_{r+s+1}), \ldots, (x_{n+1} - \xi_{r+s+u}) \qquad (16.4.5)$$

of $A(x_{n+1})$. These are distances between x_{n+1} and the nodes ξ_{r+s+1}, $\ldots, \xi_{r+s+u}$. These distances will be a minimum when all the auxiliary nodes lie in the segment adjacent to x_{n+1}. This is true for any choice of the simple and double nodes.

Now consider the factors

$$(x_{n+1} - \xi_1), \; \ldots, \; (x_{n+1} - \xi_r), \quad (x_{n+1} - \xi_{r+1})^2, \; \ldots, \; (x_{n+1} - \xi_{r+s})^2.$$

Since the terms containing the double nodes are raised to the second power it is clear that the double nodes must be the points of the set x_k which are closest to x_{n+1}: the double nodes ξ_j $(j = r + 1, \ldots, r + s)$ must coincide with the points $x_n, x_{n-1}, \ldots, x_{n-s+1}$. The simple nodes then must be the points $x_{n-s}, \ldots, x_{n-s-r+1}$.

Thus we can state:

The estimate of the remainder (16.4.4) is a minimum when the nodes are chosen in the following way:

1. *The auxiliary nodes are situated in the segment* $[x_n, x_{n+1}]$;
2. *The double nodes are taken at the points* $x_n, x_{n-1}, \ldots, x_{n-s+1}$;
3. *The simple nodes are taken at the points* $x_{n-s}, \ldots, x_{n-s-r+1}$.

16.5. CONDITIONS FOR WHICH THE COEFFICIENTS a_j ARE POSITIVE

Formulas with positive coefficients a_j play an important role in the theory of indefinite integration because, as we showed in Section 13.3, such formulas are stable with respect to the errors in the initial values and the rounding errors. In this section we will see which formulas of the form (16.1.3) of the highest degree of precision have positive a_j. We take $x = x_{n+1}$ and assume that the ξ_j $(j = 1, \ldots, r + s + u)$ are chosen as described at the end of the previous section.

Let ξ_j $(1 \le j \le r)$ be one of the simple nodes. The coefficient a_j which corresponds to this node is, by (16.2.11):

$$a_j = \frac{A(x_{n+1})}{(x_{n+1} - \xi_j)A'(\xi_j)}.$$

The term $A(x_{n+1})$ is positive since all its factors $(x_{n+1} - \xi_j)$ are positive and thus a_j has the same sign as $A'(\xi_j)$:

$$A'(\xi_j) = (\xi_j - \xi_1) \cdots (\xi_j - \xi_{j-1})(\xi_j - \xi_{j+1}) \cdots (\xi_j - \xi_r) \times$$
$$\times (\xi_j - \xi_{r+1})^2 \cdots (\xi_j - \xi_{r+s+u})^2.$$

Thus for two adjacent simple nodes the values of $A'(\xi_j)$ will have opposite signs.

Therefore the a_j for the simple nodes can all be positive only in the two cases $r = 0$, in which there are no simple nodes, and $r = 1$, in which there is one simple node.

We now consider a double node ξ_j $(r < j \le r + s)$. By (16.2.11) the corresponding a_j is

$$a_j = \frac{A_j(x_{n+1})(\xi_j - x_{n+1})}{A_j(\xi_j)} \left[\frac{1}{\xi_j - x_{n+1}} + \frac{A_j'(\xi_j)}{A_j(\xi_j)} \right]$$

$$A_j(z) = A(z)/(z - \xi_j)^2.$$

Since ξ_j lies to the left of x_{n+1} and the simple nodes all lie to the left of all the double nodes then the term outside the brackets is negative. Then a_j will be positive if the following inequality is satisfied:

$$\frac{1}{\xi_j - x_{n+1}} + \frac{A_j'(\xi_j)}{A_j(\xi_j)} = \frac{1}{\xi_j - x_{n+1}} + \sum_{k=1}^{r} \frac{1}{\xi_j - \xi_k} +$$

$$+ \sum_{k=r+1}^{r+s+u} {}^* \frac{2}{\xi_j - \xi_k} < 0. \tag{16.5.1}$$

This equation has a simple physical interpretation if we use the electrostatic analogy of Section 16.3. At ξ_j is a particle with charge 2 and the left side of (16.5.1) is the resultant of all the repulsive forces which act on this particle from all the other particles of this system.

Inequality (16.5.1) states that this resultant must be directed towards the left.

The point x_{n+1} and the auxiliary nodes always lie to the right of ξ_j and the particles situated at these points exert a leftward directed force on ξ_j. Thus it is clear that we can make the following assertion concerning the existence of formulas with positive coefficients for all the double nodes:

For any r and s there exists a number u_0 with the property that if $u \ge u_0$ then all the coefficients a_j $(j = r + 1, \ldots, r + s)$ will be positive.

Suppose we calculate $y(x)$ for a set of equally spaced points $x_k = x_0 + kh$ $(k = 0, 1, \ldots)$. Let us take one double node $(s = 1)$ at $\xi_2 = x_n$ and one simple node $(r = 1)$ at $\xi_1 = x_{n-1}$. We have shown that the coefficient a_1 for the simple node is positive. The condition that a_2 be positive is

$$\frac{1}{x_n - x_{n+1}} + \frac{1}{x_n - x_{n-1}} + \sum_{j=3}^{u+2} \frac{2}{x_n - \xi_j} = \sum_{j=3}^{u+2} \frac{2}{x_n - \xi_j} < 0.$$

Since all the $\xi_j > x_n$ this inequality is satisfied for all $u \ge 1$.

Now let us take two double nodes $(s = 2)$ at x_n and x_{n-1} and no simple nodes $(r = 0)$. The conditions that a_1 and a_2 be positive are:

$$\frac{1}{x_{n-1} - x_{n+1}} + \frac{2}{x_{n-1} - x_n} + \sum_{j=3}^{u+2} \frac{2}{x_{n-1} - \xi_j} =$$

$$= -\frac{1}{2h} - \frac{2}{h} + \sum_{j=3}^{u+2} \frac{2}{x_{n-1} - \xi_j} < 0$$

$$\frac{1}{x_n - x_{n+1}} + \frac{2}{x_n - x_{n-1}} + \sum_{j=3}^{u+2} \frac{2}{x_n - \xi_j} =$$

$$= -\frac{1}{h} + \frac{2}{h} + \sum_{j=3}^{u+2} \frac{2}{x_n - \xi_j} < 0.$$

These are also satisfied for all $u \geq 1$.

16.6. CONNECTION WITH THE EXISTENCE OF A POLYNOMIAL SOLUTION TO A CERTAIN DIFFERENTIAL EQUATION

In this section we show that equation (16.2.10) is equivalent to the existence of a polynomial solution to a certain differential equation.

Equation (16.2.10) was obtained from the equation which preceded it which can be written in the form

$$A_j(\xi_j) + (\xi_j - x)A_j'(\xi_j) = \left\{ \frac{d}{dz} [(z - x)A_j(z)] \right\}_{z=\xi_j} = 0$$

or since $A_j(z) = A(z)/(z - \xi_j)^2$

$$\frac{d}{dz} \left[\frac{(z - x)A(z)}{(z - \xi_j)^2} \right]_{z=\xi_j} = 0 \qquad (j = r + s + 1, \ldots, r + s + u). \quad (16.6.1)$$

It will be convenient to write this in another form. We introduce the polynomials $\sigma(z)$ and $\Pi_u(z)$ corresponding to the double and auxiliary nodes:

$$\sigma(z) = (z - \xi_{r+1}) \cdots (z - \xi_{r+s})$$

$$\Pi_u(z) = (z - \xi_{r+s+1}) \cdots (z - \xi_{r+s+u}).$$

We also form

$$p(z) = (z - x)(z - \xi_1) \cdots (z - \xi_r)$$

$$(z - x)A(z) = p(z)\sigma^2(z)\Pi_u^2(z)$$

so that (16.6.1) can be written as

$$\frac{d}{dz}\left[\frac{p(z)\sigma^2(z)\Pi_u^2(z)}{(z - \xi_j)^2}\right]_{z=\xi_j} = 0$$

or

$$\Pi_u'(\xi_j)\left\{\frac{d}{dz}\left[p(z)\sigma^2(z)\frac{d\Pi_u(z)}{dz}\right]\right\}_{z=\xi_j} = 0.$$

Since $z = \xi_j$ $(j > r + s)$ is a simple root of $\Pi_u(z)$ then $\Pi_u'(\xi_j) \neq 0$ and we must have

$$\left\{\frac{d}{dz}\left[p(z)\sigma^2(z)\frac{d\Pi_u}{dz}\right]\right\}_{z=\xi_j} = 0 \qquad (j > r + s). \qquad (16.6.2)$$

This says that each root of $\Pi_u(z)$ is also a root of $(p\sigma^2\Pi_u')'$. Because the roots of $\Pi_u(z)$ are simple the polynomial $(p\sigma^2\Pi_u')'$ must be divisible by $\Pi_u(z)$.

The polynomial $(p\sigma^2\Pi_u')'$ must also be divisible by $\sigma(z)$ and since the roots of $\sigma(z)$ are distinct from the roots of $\Pi_u(z)$ then $(p\sigma^2\Pi_u')'$ must be divisible by $\sigma(z)\Pi_u(z)$. Since the degree of $(p\sigma^2\Pi_u')'$ is $r + 2s + u - 1$ then there is a polynomial $\rho(z)$ of degree $r + s - 1$ for which

$$[p(z)\sigma^2(z)\Pi_u'(z)]' = \rho(z)\sigma(z)\Pi_u(z). \qquad (16.6.3)$$

This equation can be considered as a second order differential equation with respect to $\Pi_u(z)$ and we can make the following assertion:

If equation (16.2.10) is satisfied then there exists a polynomial $\rho(z)$ of degree $r + s - 1$ which will make $\Pi_u(z)$ the solution of the differential equation (16.6.3).

The proof of the converse assertion requires certain preliminary remarks on the analytic properties of the solution of (16.6.3).

If we perform the differentiation in (16.6.3) and divide both sides by $p(z)\sigma^2(z)$ we obtain

$$\Pi_u''(z) + \left(\sum_{k=1}^{r}\frac{1}{z - \xi_k} + \sum_{k=r+1}^{r+s}\frac{2}{z - \xi_k} + \frac{1}{z - x}\right)\Pi_u'(z) +$$

$$+ \left(\sum_{k=1}^{r+s}\frac{a_k}{z - \xi_k} + \frac{a_u}{z - x}\right)\Pi_u(z) = 0 \quad (16.6.4)$$

$$\sum_{k=0}^{r+s} a_k = 0.$$

The singular points of this equation are $x, \xi_1, \ldots, \xi_{r+s}, \infty$. These are regular singular points[3] and we consider any one of the points ξ_k or x.

The analytic construction of the canonical solution of (16.6.4) in a neighborhood of this point depends on the roots of an algebraic equation which is either $\alpha(\alpha - 1) + \alpha = \alpha^2 = 0$ or $\alpha(\alpha - 1) + 2\alpha = \alpha(\alpha + 1) = 0$. These have for solutions the double root $\alpha = 0$ or the roots $\alpha = 0$, $\alpha = -1$. In both cases one of the canonical solutions will be holomorphic at the singular point and different from zero there; the other solution is unbounded in a neighborhood of this point.

We will now assume that $\rho(z)$ has the property that (16.6.3) has a polynomial of degree u as a solution. For all the singular points x and ξ_k this will be a holomorphic canonical solution and therefore is known to be different from zero at each of these points. The roots of $\Pi_u(z)$ are distinct from x and ξ_k and thus they are simple because (16.6.4) can have no multiple roots other than at the singular points. Let the roots of $\Pi_u(z)$ be $\xi_{r+s+1}, \ldots, \xi_{r+s+u}$.

If in (16.6.4) we set z equal to one of the roots ξ_j $(j > r + s)$ then the term in $\Pi_u(z)$ vanishes. Then dividing by $\Pi'_u(\xi_j)$ gives

$$\frac{\Pi''_u(\xi_j)}{\Pi'_u(\xi_j)} + \sum_{k=1}^{r} \frac{1}{\xi_j - \xi_k} + \sum_{k=r+1}^{r+s} \frac{2}{\xi_j - \xi_k} + \frac{1}{\xi_j - x} = 0$$

$$(j = r + s + 1, \ldots, r + s + u).$$

This equation coincides with (16.2.10). Hence we can make the assertion:

If $p(z)$ is a polynomial of degree $r + s - 1$ for which (16.6.3) has a polynomial of degree u as a solution then the roots $\xi_{r+s+1}, \ldots, \xi_{r+s+u}$ of this solution satisfy the system of equations (16.2.10).

The nodes of the interpolation (16.1.3) which has the highest algebraic degree of precision can be determined by finding the solution of (16.6.3) which is a polynomial of degree u and then finding its roots.

16.7. SOME PARTICULAR FORMULAS

In this section we tabulate certain formulas of the highest degree of precision for low values of r, s and u.[4] The nodes are chosen to minimize the remainder in the manner described at the end of Section 16.4. The coefficients of the formulas were calculated using (16.2.11). The

---

[3] See, for example, V. I. Smirnov, *Course of Higher Mathematics*, Gostekhizdat, Moscow, 1949, Vol. 3, part 2, sec. 98 (Russian).

[4] The coefficients and nodes in these formulas were calculated by research assistant K. E. Chernin of the Leningrad section of the Mathematical Institute of the Academy of Sciences of the U.S.S.R. These values are exact to within a unit in the last place.

nodes were found by means of the differential equation[5] (16.6.3). The set of points x_n are assumed to be equally spaced with an interval h.

1. $r = 1$, $s = 0$. Here we use the value of $y(x)$ at the point x_n and u values of the derivative at auxiliary nodes between x_n and x_{n+1}. The auxiliary nodes are chosen so that the formula has the highest algebraic degree of precision. This is equivalent to representing $y(x_{n+1})$ in terms of $y(x_n)$ by

$$y(x_{n+1}) = y(x_n) + \int_{x_n}^{x_{n+1}} f(t)\,dt \qquad (16.7.1)$$

and calculating this integral by a Gauss quadrature formula

$$y_{n+1} = y_n + h[B_1 f(x_1 + t_1 h) + \cdots + B_u f(x_u + t_u h)]$$

where the B_k and t_k are the Gauss coefficients and nodes for the segment $[0, 1]$.

2. $r = 0$, $s = 1$. We use the value of $y(x)$ and $f(x)$ at the point x_n and, in addition, u values of $f(x)$ at auxiliary nodes between x_n and x_{n+1}. The highest degree of precision is $2u + 1$. The formula corresponds to the Markov (or Radau) formula with one fixed node at x_n and u nodes between x_n and x_{n+1}. Values of the coefficients and nodes for $u = 1, 2, \ldots, 6$ are given in Section 9.2 for the segment $[-1, 1]$.

3. $r = 1$, $s = 1$. We use the value of $y(x)$ at x_{n-1}, the values of $y(x)$ and $f(x)$ at x_n and the value of $f(x)$ at u auxiliary nodes between x_n and x_{n+1}:

$$y_{n+1} = A_{-1} y_{n-1} + A_0 y_0 + h\left[B_0 f(x_n) + \sum_{j=1}^{u} B_j f(x_n + t_j h) \right] + R.$$

The degree of precision is $2u + 2$ and the remainder has the estimate

$$R = \theta \frac{2h^{2u+3}}{(2u + 3)!} \left[\frac{u!(u + 1)!}{(2u + 1)!} \right]^2 f^{(2u+2)}(\xi)$$

$$0 < \theta < 1, \qquad x_{n-1} < \xi < x_{n+1}.$$

---

[5] See: V. I. Krylov, "Interpolation of the highest order of accuracy in the problem of indefinite integration," *Trudy Mat. Inst. Steklov*, Vol. 38, 1951, pp. 97–145 (Russian).

The nodes and coefficients for $u = 1, 2, 3, 4$ are tabulated below.

$$u = 1$$

$$A_{-1} = 0.02943725$$
$$A_0 = 0.97056275 \qquad B_0 = 0.3431458$$
$$t_1 = 0.7071068 \qquad\qquad\qquad B_1 = 0.6862915$$

$$u = 2$$

$$A_{-1} = 0.001113587$$
$$A_0 = 0.998886413 \qquad B_0 = 0.1334818$$
$$t_1 = 0.3879073 \qquad\qquad\qquad B_1 = 0.5221058$$
$$t_2 = 0.8593118 \qquad\qquad\qquad B_2 = 0.3455260$$

$$u = 3$$

$$A_{-1} = 0.00004136036$$
$$A_0 = 0.99995863964 \qquad B_0 = 0.07095688$$
$$t_1 = 0.2312666 \qquad\qquad\qquad B_1 = 0.3458379$$
$$t_2 = 0.6124982 \qquad\qquad\qquad B_2 = 0.3776724$$
$$t_3 = 0.9177954 \qquad\qquad\qquad B_3 = 0.2055739$$

$$u = 4$$

$$A_{-1} = 0.000001479556$$
$$A_0 = 0.999998520444 \qquad B_0 = 0.04407358$$
$$t_1 = 0.1507625 \qquad\qquad\qquad B_1 = 0.2361168$$
$$t_2 = 0.4352756 \qquad\qquad\qquad B_2 = 0.3128314$$
$$t_3 = 0.7366581 \qquad\qquad\qquad B_3 = 0.2713300$$
$$t_4 = 0.9462337 \qquad\qquad\qquad B_4 = 0.1356498$$

4. $r = 0$, $s = 2$. We use the formula

$$y_{n+1} = A_{-1}y_{n-1} + A_0 y_n +$$

$$+ h\left[B_{-1}f(x_{n-1}) + B_0 f(x_0) + \sum_{j=1}^{u} B_j f(x_n + t_j h) \right] + R.$$

for which the highest degree of precision is $2u + 3$. The remainder has the estimate

$$R = \theta \frac{4h^{2u+4}}{(2u+4)!} \left[\frac{u!(u+1)!}{(2u+1)!} \right]^2 f^{(2u+3)}(\xi)$$

$$0 < \theta < 1, \qquad x_{n-1} < \xi < x_{n+1}.$$

The nodes and coefficients for $u = 1, 2, 3, 4$ are tabulated below.

$$u = 1$$

$$A_{-1} = 0.16250915 \qquad B_{-1} = 0.044532584$$
$$A_0 = 0.83749085 \qquad B_0 = 0.49218941$$
$$t_1 = 0.74031242 \qquad\qquad\quad B_1 = 0.62578716$$

$$u = 2$$

$$A_{-1} = 0.007766326 \qquad B_{-1} = 0.001560689$$
$$A_0 = 0.99223367 \qquad B_0 = 0.1640716$$
$$t_1 = 0.4207573 \qquad\qquad\qquad\qquad B_1 = 0.5242954$$
$$t_2 = 0.8717520 \qquad\qquad\qquad\qquad B_2 = 0.3178386$$

$$u = 3$$

$$A_{-1} = 0.0003626295 \qquad B_{-1} = 0.00005699653$$
$$A_0 = 0.9996373705 \qquad B_0 = 0.08143433$$
$$t_1 = 0.2515111 \qquad\qquad\qquad\qquad B_1 = 0.3609307$$
$$t_2 = 0.6333509 \qquad\qquad\qquad\qquad B_2 = 0.3658920$$
$$t_3 = 0.9235139 \qquad\qquad\qquad\qquad B_3 = 0.1920487$$

$$u = 4$$

$$A_{-1} = 0.00001576632 \qquad B_{-1} = 0.000002030488$$
$$A_0 = 0.99998423368 \qquad B_0 = 0.04885024$$
$$t_1 = 0.1627293 \qquad\qquad\qquad\qquad B_1 = 0.2491361$$
$$t_2 = 0.4540978 \qquad\qquad\qquad\qquad B_2 = 0.3124621$$
$$t_3 = 0.7493776 \qquad\qquad\qquad\qquad B_3 = 0.2613448$$
$$t_4 = 0.9492874 \qquad\qquad\qquad\qquad B_4 = 0.1282206$$

APPENDIX A

GAUSSIAN QUADRATURE FORMULAS FOR CONSTANT WEIGHT FUNCTION

Here we give values of the $A_k^{(n)}$ and $x_k^{(n)}$ which make the approximation

$$\int_{-1}^{1} f(x)\,dx \approx \sum_{k=1}^{n} A_k^{(n)} f(x_k^{(n)})$$

exact whenever $f(x)$ is a polynomial of degree $\leq 2n - 1$. These formulas are discussed in Section 7.2. The $A_k^{(n)}$ and $x_k^{(n)}$ are symmetric with respect to $x = 0$:

$$A_k^{(n)} = A_{n-k+1}^{(n)}, \qquad x_k^{(n)} = -x_{n-k+1}^{(n)}$$

and the tables give only the values corresponding to $0 \leq x_k^{(n)} \leq 1$.

The tabulated values are taken from

H. J. Gawlik, "Zeros of Legendre Polynomials of orders 2-64 and weight coefficients of Gauss quadrature formulae," Armament Research and Development Establishment, Memorandum (B) 77/58. Fort Halstead, Kent, 1958.

where the $A_k^{(n)}$ and $x_k^{(n)}$ are given to 20 decimal places. Values for $n = 2, 4, 8, 16, 24, 32, 40, 48, 64, 80, 96$ of the same accuracy are given in the two tables

P. Davis and P. Rabinowitz, "Abscissas and weights for Gaussian quadratures of high order," *J. Res. Nat. Bur. Standards*, Vol. 56, 1956, pp. 35-37.

P. Davis and P. Rabinowitz, "Additional abscissas and weights for Gaussian quadrature of high order: values for $n = 64, 80$ and 96," *J. Res. Nat. Bur. Standards*, Vol. 60, 1958, pp. 613-14.

These three tables are the most extensive values of the $A_k^{(n)}$ and $x_k^{(n)}$ which are known and are believed to be accurate to within a few units in the last significant figures.

$x_k^{(n)}$	$A_k^{(n)}$
n = 2	
0.57735 02691 89625 76451	1.00000 00000 00000 00000
n = 3	
0.77459 66692 41483 37704	0.55555 55555 55555 55556
0.00000 00000 00000 00000	0.88888 88888 88888 88889
n = 4	
0.86113 63115 94052 57522	0.34785 48451 37453 85737
0.33998 10435 84856 26480	0.65214 51548 62546 14263

APPENDIX A (*Continued*)

$x_k^{(n)}$	$A_k^{(n)}$

n = 5

0.90617 98459 38663 99280	0.23692 68850 56189 08751
0.53846 93101 05683 09104	0.47862 86704 99366 46804
0.00000 00000 00000 00000	0.56888 88888 88888 88889

n = 6

0.93246 95142 03152 02781	0.17132 44923 79170 34504
0.66120 93864 66264 51366	0.36076 15730 48138 60757
0.23861 91860 83196 90863	0.46791 39345 72691 04739

n = 7

0.94910 79123 42758 52453	0.12948 49661 68869 69327
0.74153 11855 99394 43986	0.27970 53914 89276 66790
0.40584 51513 77397 16691	0.38183 00505 05118 94495
0.00000 00000 00000 00000	0.41795 91836 73469 38776

n = 8

0.96028 98564 97536 23168	0.10122 85362 90376 25915
0.79666 64774 13626 73959	0.22238 10344 53374 47054
0.52553 24099 16328 98582	0.31370 66458 77887 28734
0.18343 46424 95649 80494	0.36268 37833 78361 98297

n = 9

0.96816 02395 07626 08984	0.08127 43883 61574 41197
0.83603 11073 26635 79430	0.18064 81606 94857 40406
0.61337 14327 00590 39731	0.26061 06964 02935 46232
0.32425 34234 03808 92904	0.31234 70770 40002 84007
0.00000 00000 00000 00000	0.33023 93550 01259 76316

n = 10

0.97390 65285 17171 72008	0.06667 13443 08688 13759
0.86506 33666 88984 51073	0.14945 13491 50580 59315
0.67940 95682 99024 40623	0.21908 63625 15982 04400
0.43339 53941 29247 19080	0.26926 67193 09996 35509
0.14887 43389 81631 21089	0.29552 42247 14752 87017

n = 11

0.97822 86581 46056 99280	0.05566 85671 16173 66648
0.88706 25997 68095 29908	0.12558 03694 64904 62464
0.73015 20055 74049 32409	0.18629 02109 27734 25143
0.51909 61292 06811 81593	0.23319 37645 91990 47992
0.26954 31559 52344 97233	0.26280 45445 10246 66218
0.00000 00000 00000 00000	0.27292 50867 77900 63071

n = 12

0.98156 06342 46719 25069	0.04717 53363 86511 82719
0.90411 72563 70474 85668	0.10693 93259 95318 43096
0.76990 26741 94304 68704	0.16007 83285 43346 22633
0.58731 79542 86617 44730	0.20316 74267 23065 92175
0.36783 14989 98180 19375	0.23349 25365 38354 80876
0.12523 34085 11468 91547	0.24914 70458 13402 78500

APPENDIX A (*Continued*)

$$x_k^{(n)} \qquad\qquad\qquad A_k^{(n)}$$

n = 13

$x_k^{(n)}$	$A_k^{(n)}$
0.98418 30547 18588 14947	0.04048 40047 65315 87952
0.91759 83992 22977 96521	0.09212 14998 37728 44792
0.80157 80907 33309 91279	0.13887 35102 19787 23846
0.64234 93394 40340 22064	0.17814 59807 61945 73828
0.44849 27510 36446 85288	0.20781 60475 36888 50231
0.23045 83159 55134 79407	0.22628 31802 62897 23841
0.00000 00000 00000 00000	0.23255 15532 30873 91019

n = 14

$x_k^{(n)}$	$A_k^{(n)}$
0.98628 38086 96812 33884	0.03511 94603 31751 86303
0.92843 48836 63573 51734	0.08015 80871 59760 20981
0.82720 13150 69764 99319	0.12151 85706 87903 18469
0.68729 29048 11685 47015	0.15720 31671 58193 53457
0.51524 86363 58154 09197	0.18553 83974 77937 81374
0.31911 23689 27889 76044	0.20519 84637 21295 60397
0.10805 49487 07343 66207	0.21526 38534 63157 79020

n = 15

$x_k^{(n)}$	$A_k^{(n)}$
0.98799 25180 20485 42849	0.03075 32419 96117 26835
0.93727 33924 00705 90431	0.07036 60474 88108 12471
0.84820 65834 10427 21620	0.10715 92204 67171 93501
0.72441 77313 60170 04742	0.13957 06779 26154 31445
0.57097 21726 08538 84754	0.16626 92058 16993 93355
0.39415 13470 77563 36990	0.18616 10000 15562 21103
0.20119 40939 97434 52230	0.19843 14853 27111 57646
0.00000 00000 00000 00000	0.20257 82419 25561 27288

n = 16

$x_k^{(n)}$	$A_k^{(n)}$
0.98940 09349 91649 93260	0.02715 24594 11754 09485
0.94457 50230 73232 57608	0.06225 35239 38647 89286
0.86563 12023 87831 74388	0.09515 85116 82492 78481
0.75540 44083 55003 03390	0.12462 89712 55533 87205
0.61787 62444 02643 74845	0.14959 59888 16576 73208
0.45801 67776 57227 38634	0.16915 65193 95002 53819
0.28160 35507 79258 91323	0.18260 34150 44923 58887
0.09501 25098 37637 44019	0.18945 06104 55068 49629

n = 20

$x_k^{(n)}$	$A_k^{(n)}$
0.99312 85991 85094 92479	0.01761 40071 39152 11831
0.96397 19272 77913 79127	0.04060 14298 00386 94133
0.91223 44282 51325 90587	0.06267 20483 34109 06357
0.83911 69718 22218 82339	0.08327 67415 76704 74873
0.74633 19064 60150 79261	0.10193 01198 17240 43504
0.63605 36807 26515 02545	0.11819 45319 61518 41731
0.51086 70019 50827 09800	0.13168 86384 49176 62690
0.37370 60887 15419 56067	0.14209 61093 18382 05133
0.22778 58511 41645 07808	0.14917 29864 72603 74679
0.07652 65211 33497 33375	0.15275 33871 30725 85070

APPENDIX A (*Continued*)

$x_k^{(n)}$	$A_k^{(n)}$

$n = 24$

0.99518 72199 97021 36018	0.01234 12297 99987 19955
0.97472 85559 71309 49820	0.02853 13886 28933 66318
0.93827 45520 02732 75852	0.04427 74388 17419 80617
0.88641 55270 04401 03421	0.05929 85849 15436 78075
0.82000 19859 73902 92195	0.07334 64814 11080 30573
0.74012 41915 78554 36424	0.08619 01615 31953 27592
0.64809 36519 36975 56925	0.09761 86521 04113 88827
0.54542 14713 88839 53566	0.10744 42701 15965 63478
0.43379 35076 26045 13849	0.11550 56680 53725 60135
0.31504 26796 96163 37439	0.12167 04729 27803 39120
0.19111 88674 73616 30916	0.12583 74563 46828 29612
0.06405 68928 62605 62609	0.12793 81953 46752 15697

$n = 28$

0.99644 24975 73954 44995	0.00912 42825 93094 51774
0.98130 31653 70872 75369	0.02113 21125 92771 25975
0.95425 92806 28938 19725	0.03290 14277 82304 37998
0.91563 30263 92132 07387	0.04427 29347 59004 22784
0.86589 25225 74395 04894	0.05510 73456 75716 74543
0.80564 13709 17179 17145	0.06527 29239 66999 59579
0.73561 08780 13631 77203	0.07464 62142 34568 77902
0.65665 10940 38864 96122	0.08311 34172 28901 21839
0.56972 04718 11401 71931	0.09057 17443 93032 84094
0.47587 42249 55118 26103	0.09693 06579 97929 91585
0.37625 15160 89078 71022	0.10211 29675 78060 76981
0.27206 16276 35178 07768	0.10605 57659 22846 41791
0.16456 92821 33380 77128	0.10871 11922 58294 13525
0.05507 92898 84034 27043	0.11004 70130 16475 19628

$n = 32$

0.99726 38618 49481 56354	0.00701 86100 09470 09660
0.98561 15115 45268 33540	0.01627 43947 30905 67061
0.96476 22555 87506 43077	0.02539 20653 09262 05945
0.93490 60759 37739 68917	0.03427 38629 13021 43310
0.89632 11557 66052 12397	0.04283 58980 22226 68066
0.84936 76137 32569 97013	0.05099 80592 62376 17620
0.79448 37959 67942 40696	0.05868 40934 78535 54714
0.73218 21187 40289 68039	0.06582 22227 76361 84684
0.66304 42669 30215 20098	0.07234 57941 08848 50623
0.58771 57572 40762 32904	0.07819 38957 87070 30647
0.50689 99089 32229 39002	0.08331 19242 26946 75522
0.42135 12761 30635 34536	0.08765 20930 04403 81114
0.33186 86022 82127 64978	0.09117 38786 95763 88471
0.23928 73622 52137 07454	0.09384 43990 80804 56564
0.14447 19615 82796 49349	0.09563 87200 79274 85942
0.04830 76656 87738 31623	0.09654 00885 14727 80057

APPENDIX A (*Continued*)

$x_k^{(n)}$ $A_k^{(n)}$

$n = 36$

$x_k^{(n)}$	$A_k^{(n)}$
0.99783 04624 84085 83620	0.00556 57196 64245 04536
0.98858 64789 02212 23807	0.01291 59472 84065 57441
0.97202 76910 49697 94934	0.02018 15152 97735 47153
0.94827 29843 99507 54520	0.02729 86214 98568 77909
0.91749 77745 15659 06608	0.03421 38107 70307 22992
0.87992 98008 90397 13198	0.04087 57509 23644 89547
0.83584 71669 92475 30642	0.04723 50834 90265 97842
0.78557 62301 32206 51283	0.05324 47139 77759 91909
0.72948 91715 93556 58209	0.05886 01442 45324 81731
0.66800 12365 85521 06210	0.06403 97973 55015 48956
0.60156 76581 35980 53508	0.06874 53238 35736 44261
0.53068 02859 26245 16164	0.07294 18850 05653 06135
0.45586 39444 33420 26721	0.07659 84106 45870 67453
0.37767 25471 19689 21632	0.07968 78289 12071 60191
0.29668 49953 44028 27050	0.08218 72667 04339 70952
0.21350 08923 16865 57894	0.08407 82189 79661 93493
0.12873 61038 09384 78865	0.08534 66857 39338 62749
0.04301 81984 73708 60723	0.08598 32756 70394 74749

$n = 40$

$x_k^{(n)}$	$A_k^{(n)}$
0.99823 77097 10559 20035	0.00452 12770 98533 19126
0.99072 62386 99457 00645	0.01049 82845 31152 81362
0.97725 99499 83774 26266	0.01642 10583 81907 88871
0.95791 68192 13791 65580	0.02224 58491 94166 95726
0.93281 28082 78676 53336	0.02793 70069 80023 40110
0.90209 88069 68874 29673	0.03346 01952 82547 84739
0.86595 95032 12259 50382	0.03878 21679 74472 01764
0.82461 22308 33311 66320	0.04387 09081 85673 27199
0.77830 56514 26519 38769	0.04869 58076 35072 23206
0.72731 82551 89927 10328	0.05322 78469 83936 82436
0.67195 66846 14179 54838	0.05743 97690 99391 55137
0.61255 38896 67980 23795	0.06130 62424 92928 93917
0.54946 71250 95128 20208	0.06480 40134 56601 03807
0.48307 58016 86178 71291	0.06791 20458 15233 90383
0.41377 92043 71605 00152	0.07061 16473 91286 77970
0.34199 40908 25758 47301	0.07288 65823 95804 05906
0.26815 21850 07253 68114	0.07472 31690 57968 26420
0.19269 75807 01371 09972	0.07611 03619 00626 24237
0.11608 40706 75255 20848	0.07703 98181 64247 96559
0.03877 24175 06050 82193	0.07750 59479 78424 81126

APPENDIX A (*Continued*)

$$x_k^{(n)} \qquad\qquad A_k^{(n)}$$

$$n = 48$$

$x_k^{(n)}$	$A_k^{(n)}$
0.99877 10072 52426 11860	0.00315 33460 52305 83862
0.99353 01722 66350 75755	0.00732 75539 01276 26210
0.98412 45837 22826 85774	0.01147 72345 79234 53948
0.97059 15925 46247 25046	0.01557 93157 22943 84873
0.95298 77031 60430 86072	0.01961 61604 57355 52781
0.93138 66907 06554 33311	0.02357 07608 39324 37914
0.90587 91367 15569 67282	0.02742 65097 08356 94820
0.87657 20202 74247 88591	0.03116 72278 32798 08890
0.84358 82616 24393 53071	0.03477 72225 64770 43889
0.80706 62040 29442 62708	0.03824 13510 65830 70632
0.76715 90325 15740 33925	0.04154 50829 43464 74921
0.72403 41309 23814 65467	0.04467 45608 56694 28042
0.67787 23796 32663 90521	0.04761 66584 92490 47482
0.62886 73967 76513 62400	0.05035 90355 53854 47496
0.57722 47260 83972 70382	0.05289 01894 85193 66710
0.52316 09747 22233 03368	0.05519 95036 99984 16287
0.46690 29047 50958 40454	0.05727 72921 00403 21570
0.40868 64819 90716 72992	0.05911 48396 98395 63575
0.34875 58862 92160 73816	0.06070 44391 65893 88005
0.28736 24873 55455 57674	0.06203 94231 59892 66390
0.22476 37903 94689 06122	0.06311 41922 86254 02566
0.16122 23560 68891 71806	0.06392 42385 84648 18662
0.09700 46992 09462 69893	0.06446 61644 35950 08221
0.03238 01709 62869 36203	0.06473 76968 12683 92250

APPENDIX B

GAUSSIAN-HERMITE QUADRATURE FORMULAS

Here we give values of the $A_k^{(n)}$ and $x_k^{(n)}$ which make the approximation

$$\int_{-\infty}^{\infty} e^{-x^2} f(x)\, dx \approx \sum_{k=1}^{n} A_k^{(n)} f(x_k^{(n)})$$

exact whenever $f(x)$ is a polynomial of degree $\leq 2n - 1$. These formulas are discussed in Section 7.4. The $A_k^{(n)}$ and $x_k^{(n)}$ are symmetric with respect to $x = 0$ and the tables give only the values corresponding to $0 \leq x_k^{(n)}$.

We give here values for $n = 1\,(1)\,20$ given by

H. E. Salzer, R. Zucker and R. Capuano, "Table of the zeros and weight factors of the first twenty Hermite polynomials," *J. Res. Nat. Bur. Standards*, Vol. 48, 1952, pp. 111–116.

These are the most extensive values of these quadrature formulas which are known. A number in parenthesis before a value of a coefficient is the power of 10 by which the tabulated value must be multiplied; for example, $(-1)0.8131\ldots$ means that the coefficient is $0.08131\ldots$.

$x_k^{(n)}$	$A_k^{(n)}$
n = 1	
0.00000 00000 00000	1.77245 38509 055
n = 2	
0.70710 67811 86548	0.88622 69254 528
n = 3	
0.00000 00000 00000	1.18163 59006 037
1.22474 48713 91589	0.29540 89751 509
n = 4	
0.52464 76232 75290	0.80491 40900 055
1.65068 01238 85785	(−1)0.81312 83544 725
n = 5	
0.00000 00000 00000	0.94530 87204 829
0.95857 24646 13819	0.39361 93231 522
2.02018 28704 56086	(−1)0.19953 24205 905
n = 6	
0.43607 74119 27617	0.72462 95952 244
1.33584 90740 13697	0.15706 73203 229
2.35060 49736 74492	(−2)0.45300 09905 509
n = 7	
0.00000 00000 00000	0.81026 46175 568
0.81628 78828 58965	0.42560 72526 101
1.67355 16287 67471	(−1)0.54515 58281 913
2.65196 13568 35233	(−3)0.97178 12450 995

343

APPENDIX B (*Continued*)

$x_k^{(n)}$	$A_k^{(n)}$

$n = 8$

$x_k^{(n)}$	$A_k^{(n)}$
0.38118 69902 07322	0.66114 70125 582
1.15719 37124 46780	0.20780 23258 149
1.98165 67566 95843	(−1)0.17077 98300 741
2.93063 74202 57244	(−3)0.19960 40722 114

$n = 9$

$x_k^{(n)}$	$A_k^{(n)}$
0.00000 00000 00000	0.72023 52156 061
0.72355 10187 52838	0.43265 15590 026
1.46855 32892 16668	(−1)0.88474 52739 438
2.26658 05845 31843	(−2)0.49436 24275 537
3.19099 32017 81528	(−4)0.39606 97726 326

$n = 10$

$x_k^{(n)}$	$A_k^{(n)}$
0.34290 13272 23705	0.61086 26337 353
1.03661 08297 89514	0.24013 86110 823
1.75668 36492 99882	(−1)0.33874 39445 548
2.53273 16742 32790	(−2)0.13436 45746 781
3.43615 91188 37738	(−5)0.76404 32855 233

$n = 11$

$x_k^{(n)}$	$A_k^{(n)}$
0.00000 00000 00000	0.65475 92869 146
0.65680 95668 82100	0.42935 97523 561
1.32655 70844 94933	0.11722 78751 677
2.02594 80158 25755	(−1)0.11911 39544 491
2.78329 00997 81652	(−3)0.34681 94663 233
3.66847 08465 59583	(−5)0.14395 60393 714

$n = 12$

$x_k^{(n)}$	$A_k^{(n)}$
0.31424 03762 54359	0.57013 52362 625
0.94778 83912 40164	0.26049 23102 642
1.59768 26351 52605	(−1)0.51607 98561 588
2.27950 70805 01060	(−2)0.39053 90584 629
3.02063 70251 20890	(−4)0.85736 87043 588
3.88972 48978 69782	(−6)0.26585 51684 356

$n = 13$

$x_k^{(n)}$	$A_k^{(n)}$
0.00000 00000 00000	0.60439 31879 211
0.60576 38791 71060	0.42161 62968 985
1.22005 50365 90748	0.14032 33206 870
1.85310 76516 01512	(−1)0.20862 77529 617
2.51973 56856 78238	(−2)0.12074 59992 719
3.24660 89783 72410	(−4)0.20430 36040 271
4.10133 75961 78640	(−7)0.48257 31850 073

$n = 14$

$x_k^{(n)}$	$A_k^{(n)}$
0.29174 55106 7256	0.53640 59097 121
0.87871 37873 2940	0.27310 56090 642
1.47668 27311 4114	(−1)0.68505 53422 347
2.09518 32585 0772	(−2)0.78500 54726 458
2.74847 07249 8540	(−3)0.35509 26135 519
3.46265 69336 0227	(−5)0.47164 84355 019
4.30444 85704 7363	(−8)0.86285 91168 125

APPENDIX B (*Continued*)

$x_k^{(n)}$	$A_k^{(n)}$
	$n = 15$
0.00000 00000 0000	0.56410 03087 264
0.56506 95832 5558	0.41202 86874 989
1.13611 55852 1092	0.15848 89157 959
1.71999 25751 8649	(−1)0.30780 03387 255
2.32573 24861 7386	(−2)0.27780 68842 913
2.96716 69279 0560	(−3)0.10000 44412 325
3.66995 03734 0445	(−5)0.10591 15547 711
4.49999 07073 0939	(−8)0.15224 75804 254
	$n = 16$
0.27348 10461 3815	0.50792 94790 166
0.82295 14491 4466	0.28064 74585 285
1.38025 85391 9888	(−1)0.83810 04139 899
1.95178 79909 1625	(−1)0.12880 31153 551
2.54620 21578 4748	(−3)0.93228 40086 242
3.17699 91619 7996	(−4)0.27118 60092 538
3.86944 79048 6012	(−6)0.23209 80844 865
4.68873 89393 0582	(−9)0.26548 07474 011
	$n = 17$
0.00000 00000 000	0.53091 79376 249
0.53163 30013 427	0.40182 64694 704
1.06764 87257 435	0.17264 82976 701
1.61292 43142 212	(−1)0.40920 03414 976
2.17350 28266 666	(−2)0.50673 49957 628
2.75776 29157 039	(−3)0.29864 32866 978
3.37893 20911 415	(−5)0.71122 89140 021
4.06194 66758 755	(−7)0.49770 78981 631
4.87134 51936 744	(−10)0.45805 78930 799
	$n = 18$
0.25826 77505 191	0.48349 56947 255
0.77668 29192 674	0.28480 72856 700
1.30092 08583 896	(−1)0.97301 74764 132
1.83553 16042 616	(−1)0.18640 04238 754
2.38629 90891 667	(−2)0.18885 22630 268
2.96137 75055 316	(−4)0.91811 26867 929
3.57376 90684 863	(−5)0.18106 54481 093
4.24811 78735 681	(−7)0.10467 20579 579
5.04836 40088 745	(−11)0.78281 99772 116
	$n = 19$
0.00000 00000 000	0.50297 48882 762
0.50352 01634 239	0.39160 89886 130
1.01036 83871 343	0.18363 27013 070
1.52417 06193 935	(−1)0.50810 38690 905
2.04923 17098 506	(−2)0.79888 66777 723
2.59113 37897 945	(−3)0.67087 75214 072
3.15784 88183 476	(−4)0.27209 19776 316
3.76218 73519 640	(−6)0.44882 43147 223
4.42853 28066 038	(−8)0.21630 51009 864
5.22027 16905 375	(−11)0.13262 97094 499

APPENDIX B (*Continued*)

$x_k^{(n)}$	$A_k^{(n)}$
$n = 20$	
0.24534 07083 009	0.46224 36696 006
0.73747 37285 454	0.28667 55053 628
1.23407 62153 953	0.10901 72060 200
1.73853 77121 166	(−1)0.24810 52088 746
2.25497 40020 893	(−2)0.32437 73342 238
2.78880 60584 281	(−3)0.22833 86360 163
3.34785 45673 832	(−5)0.78025 56478 532
3.94476 40401 156	(−6)0.10860 69370 769
4.60368 24495 507	(−9)0.43993 40992 273
5.38748 08900 112	(−12)0.22293 93645 534

APPENDIX C

GAUSSIAN-LAGUERRE QUADRATURE FORMULAS

Here we give values of the $A_k^{(n)}$ and $x_k^{(n)}$ which make the approximation

$$\int_0^\infty x^a e^{-x} f(x)\,dx \approx \sum_{k=1}^n A_k^{(n)} f(x_k^{(n)})$$

exact whenever $f(x)$ is a polynomial of degree $\leq 2n-1$. These formulas are discussed in Section 7.5.

We give the values for $a = 0$, $n = 4\,(4)\,32$ tabulated by

P. Rabinowitz and G. Weiss, "Tables of abscissas and weights for nu-

merical evaluation of integrals of the form $\int_0^\infty e^{-x} x^a f(x)\,dx$,"

Math. Tables Aids Comput., Vol. 13, 1959, pp. 285–93.

and also the values for $a = 0$, $n = 1\,(1)\,15$ tabulated by

H. E. Salzer and R. Zucker, "Table of the zeros and weight factors of the first fifteen Laguerre polynomials," *Bull. Amer. Math. Soc.*, Vol. 55, 1949, pp. 1004–12.

except for the three cases $n = 4, 8, 12$ where we give the more accurate values given by Rabinowitz and Weiss. Rabinowitz and Weiss also give values of the $A_k^{(n)}$ and $x_k^{(n)}$ for $a = 1\,(1)\,5$, $n = 4\,(4)\,16$ which we have not included here.

$x_k^{(n)}$	$A_k^{(n)}$
$n = 1$	
1.00000 00000 00	1.00000 00000 00
$n = 2$	
0.58578 64376 27	0.85355 33905 93
3.41421 35623 73	0.14644 66094 07
$n = 3$	
0.41577 45567 83	0.71109 30099 29
2.29428 03602 79	0.27851 77335 69
6.28994 50829 37	(−1) 0.10389 25650 16
$n = 4$	
0.32254 76896 19392 312	0.60315 41043 41633 602
1.74576 11011 58346 58	0.35741 86924 37799 687
4.53662 02969 21127 98	(−1) 0.38887 90851 50053 843
9.39507 09123 01133 13	(−3) 0.53929 47055 61327 450
$n = 5$	
0.26356 03197 18	0.52175 56105 83
1.41340 30591 07	0.39866 68110 83
3.59642 57710 41	(−1) 0.75942 44968 17
7.08581 00058 59	(−2) 0.36117 58679 92
12.64080 08442 76	(−4) 0.23369 97238 58

APPENDIX C (*Continued*)

$x_k^{(n)}$	$A_k^{(n)}$
	$n = 6$
0.22284 66041 79	0.45896 46739 50
1.18893 21016 73	0.41700 08307 72
2.99273 63260 59	0.11337 33820 74
5.77514 35691 05	(-1) 0.10399 19745 31
9.83746 74183 83	(-3) 0.26101 72028 15
15.98287 39806 02	(-6) 0.89854 79064 30
	$n = 7$
0.19304 36765 60	0.40931 89517 01
1.02666 48953 39	0.42183 12778 62
2.56787 67449 51	0.14712 63486 58
4.90035 30845 26	(-1) 0.20633 51446 87
8.18215 34445 63	(-2) 0.10740 10143 28
12.73418 02917 98	(-4) 0.15865 46434 86
19.39572 78622 63	(-7) 0.31703 15479 00
	$n = 8$
0.17027 96323 05101 000	0.36918 85893 41637 530
0.90370 17767 99379 912	0.41878 67808 14342 956
2.25108 66298 66130 69	0.17579 49866 37171 806
4.26670 01702 87658 79	(-1) 0.33343 49226 12156 515
7.04590 54023 93465 70	(-2) 0.27945 36235 22567 252
10.75851 60101 80995 2	(-4) 0.90765 08773 35821 310
15.74067 86412 78004 6	(-6) 0.84857 46716 27253 154
22.86313 17368 89264 1	(-8) 0.10480 01174 87151 038
	$n = 9$
0.15232 22277 32	0.33612 64217 98
0.80722 00227 42	0.41121 39804 24
2.00513 51556 19	0.19928 75253 71
3.78347 39733 31	(-1) 0.47460 56276 57
6.20495 67778 77	(-2) 0.55996 26610 79
9.37298 52516 88	(-3) 0.30524 97670 93
13.46623 69110 92	(-5) 0.65921 23026 08
18.83359 77889 92	(-7) 0.41107 69330 35
26.37407 18909 27	(-10) 0.32908 74030 35
	$n = 10$
0.13779 34705 40	0.30844 11157 65
0.72945 45495 03	0.40111 99291 55
1.80834 29017 40	0.21806 82876 12
3.40143 36978 55	(-1) 0.62087 45609 87
5.55249 61400 64	(-2) 0.95015 16975 18
8.33015 27467 64	(-3) 0.75300 83885 88
11.84378 58379 00	(-4) 0.28259 23349 60
16.27925 78313 78	(-6) 0.42493 13984 96
21.99658 58119 81	(-8) 0.18395 64823 98
29.92069 70122 74	(-12) 0.99118 27219 61

APPENDIX C (*Continued*)

$x_k^{(n)}$	$A_k^{(n)}$
	$n = 11$
0.12579 64421 88	0.28493 32128 94
0.66541 82558 39	0.38972 08895 28
1.64715 05458 72	0.23278 18318 49
3.09113 81430 35	(−1) 0.76564 45354 62
5.02928 44015 80	(−1) 0.14393 28276 74
7.50988 78638 07	(−2) 0.15188 80846 48
10.60595 09995 47	(−4) 0.85131 22435 47
14.43161 37580 64	(−5) 0.22924 03879 57
19.17885 74032 15	(−7) 0.24863 53702 77
25.21770 93396 78	(−10) 0.77126 26933 69
33.49719 28471 76	(−13) 0.28837 75868 32
	$n = 12$
0.11572 21173 58020 675	0.26473 13710 55443 190
0.61175 74845 15130 665	0.37775 92758 73137 982
1.51261 02697 76418 79	0.24408 20113 19877 564
2.83375 13377 43507 23	(−1) 0.90449 22221 16809 307
4.59922 76394 18348 48	(−1) 0.20102 38115 46340 965
6.84452 54531 15177 35	(−2) 0.26639 73541 86531 588
9.62131 68424 56867 04	(−3) 0.20323 15926 62999 392
13.00605 49933 06347 7	(−5) 0.83650 55856 81979 875
17.11685 51874 62255 7	(−6) 0.16684 93876 54091 026
22.15109 03793 97005 7	(−8) 0.13423 91030 51500 415
28.48796 72509 84000 3	(−11) 0.30616 01635 03502 078
37.09912 10444 66920 3	(−15) 0.81480 77467 42624 168
	$n = 13$
0.10714 23884 72	0.24718 87084 30
0.56613 18990 40	0.36568 88229 01
1.39856 43364 51	0.25256 24200 58
2.61659 71084 06	0.10347 07580 24
4.23884 59290 17	(−1) 0.26432 75441 56
6.29225 62711 40	(−2) 0.42203 96040 27
8.81500 19411 87	(−3) 0.41188 17704 73
11.86140 35888 11	(−4) 0.23515 47398 15
15.51076 20377 04	(−6) 0.73173 11620 25
19.88463 56638 80	(−7) 0.11088 41625 70
25.18526 38646 78	(−10) 0.67708 26692 21
31.80038 63019 47	(−12) 0.11599 79959 91
40.72300 86692 66	(−16) 0.22450 93203 89
	$n = 14$
0.09974 75070 33	0.23181 55771 45
0.52685 76488 52	0.35378 46915 98
1.30062 91212 51	0.25873 46102 45
2.43080 10787 31	0.11548 28935 57
3.93210 28222 93	(−1) 0.33192 09215 93
5.82553 62183 02	(−2) 0.61928 69437 01
8.14024 01415 65	(−3) 0.73989 03778 67
10.91649 95073 66	(−4) 0.54907 19466 84

(*contd.*)

APPENDIX C (*Continued*)

$x_k^{(n)}$	$A_k^{(n)}$

$n = 14$

$x_k^{(n)}$	$A_k^{(n)}$
14.21080 50111 61	(−5) 0.24095 85764 09
18.10489 22202 18	(−7) 0.58015 43981 68
22.72338 16282 69	(−9) 0.68193 14692 49
28.27298 17232 48	(−11) 0.32212 07751 89
35.14944 36605 92	(−14) 0.42213 52440 52
44.36608 17111 17	(−18) 0.60523 75022 29

$n = 15$

$x_k^{(n)}$	$A_k^{(n)}$
0.09330 78120 17	0.21823 48859 40
0.49269 17403 02	0.34221 01779 23
1.21559 54120 71	0.26302 75779 42
2.26994 95262 04	0.12642 58181 06
3.66762 27217 51	(−1) 0.40206 86492 10
5.42533 66274 14	(−2) 0.85638 77803 61
7.56591 62266 13	(−2) 0.12124 36147 21
10.12022 85680 19	(−3) 0.11167 43923 44
13.13028 24821 76	(−5) 0.64599 26762 02
16.65440 77083 30	(−6) 0.22263 16907 10
20.77647 88994 49	(−8) 0.42274 30384 98
25.62389 42267 29	(−10) 0.39218 97267 04
31.40751 91697 54	(−12) 0.14565 15264 07
38.53068 33064 86	(−15) 0.14830 27051 11
48.02608 55726 86	(−19) 0.16005 94906 21

$n = 16$

$x_k^{(n)}$	$A_k^{(n)}$
0.08764 94104 78927 8403	0.20615 17149 57800 994
0.46269 63289 15080 832	0.33105 78549 50884 166
1.14105 77748 31226 86	0.26579 57776 44214 153
2.12928 36450 98380 62	0.13629 69342 96377 540
3.43708 66338 93206 65	(−1) 0.47328 92869 41252 190
5.07801 86145 49767 91	(−1) 0.11299 90008 03394 532
7.07033 85350 48234 13	(−2) 0.18490 70943 52631 086
9.43831 43363 91938 78	(−3) 0.20427 19153 08278 460
12.21422 33688 66158 7	(−4) 0.14844 58687 39812 988
15.44152 73687 81617 1	(−6) 0.68283 19330 87119 956
19.18015 68567 53134 9	(−7) 0.18810 24841 07967 321
23.51590 56939 91908 5	(−9) 0.28623 50242 97388 162
28.57872 97428 82140 4	(−11) 0.21270 79033 22410 297
34.58339 87022 86625 8	(−14) 0.62979 67002 51786 779
41.94045 26476 88332 6	(−17) 0.50504 73700 03551 282
51.70116 03395 43318 4	(−21) 0.41614 62370 37285 519

$n = 20$

$x_k^{(n)}$	$A_k^{(n)}$
0.07053 98896 91988 7534	0.16874 68018 51113 862
0.37212 68180 01611 444	0.29125 43620 06068 282
0.91658 21024 83273 565	0.26668 61028 67001 289
1.70730 65310 28343 88	0.16600 24532 69506 840
2.74919 92553 09432 13	(−1) 0.74826 06466 87923 705
4.04892 53138 50886 92	(−1) 0.24964 41730 92832 211
5.61517 49708 61616 51	(−2) 0.62025 50844 57223 685
7.45901 74536 71063 31	(−2) 0.11449 62386 47690 824
9.59439 28695 81096 77	(−3) 0.15574 17730 27811 975

(*contd.*)

APPENDIX C (*Continued*)

$x_k^{(n)}$	$A_k^{(n)}$

$n = 20$

$x_k^{(n)}$	$A_k^{(n)}$
12.03880 25469 64316 3	(−4) 0.15401 44086 52249 157
14.81429 34426 30740 0	(−5) 0.10864 86366 51798 235
17.94889 55205 19376 0	(−7) 0.53301 20909 55671 475
21.47878 82402 85011 0	(−8) 0.17579 81179 05058 200
25.45170 27931 86905 5	(−10) 0.37255 02402 51232 087
29.93255 46317 00612 0	(−12) 0.47675 29251 57819 052
35.01343 42404 79000 0	(−14) 0.33728 44243 36243 841
40.83305 70567 28571 1	(−16) 0.11550 14339 50039 883
47.61999 40473 46502 1	(−19) 0.15395 22140 58234 355
55.81079 57500 63898 9	(−23) 0.52864 42725 56915 783
66.52441 65256 15753 8	(−27) 0.16564 56612 49902 330

$n = 24$

$x_k^{(n)}$	$A_k^{(n)}$
0.05901 98521 81507 9770	0.14281 19733 34781 851
0.31123 91461 98483 727	0.25877 41075 17423 903
0.76609 69055 45936 646	0.25880 67072 72869 802
1.42559 75908 03613 09	0.18332 26889 77778 025
2.29256 20586 32190 29	(−1) 0.98166 27262 99188 922
3.37077 42642 08997 72	(−1) 0.40732 47815 14086 460
4.66508 37034 67170 79	(−1) 0.13226 01940 51201 567
6.18153 51187 36765 41	(−2) 0.33693 49058 47830 355
7.92753 92471 72152 18	(−3) 0.67216 25640 93547 890
9.91209 80150 77706 02	(−3) 0.10446 12146 59275 180
12.14610 27117 29765 6	(−4) 0.12544 72197 79933 332
14.64273 22895 96674 3	(−5) 0.11513 15812 73727 992
17.41799 26465 08978 7	(−7) 0.79608 12959 13363 026
20.49146 00826 16424 7	(−8) 0.40728 58987 54999 966
23.88732 98481 69733 2	(−9) 0.15070 08226 29258 492
27.63593 71743 32717 4	(−11) 0.39177 36515 05845 138
31.77604 13523 74723 3	(−13) 0.68941 81052 95808 569
36.35840 58016 51621 7	(−15) 0.78198 00382 45944 847
41.45172 04848 70767 0	(−17) 0.53501 88813 01003 760
47.15310 64451 56323 0	(−19) 0.20105 17464 55550 347
53.60857 45446 95069 8	(−22) 0.36057 65864 55295 904
61.05853 14472 18761 6	(−25) 0.24518 18845 87840 269
69.96224 00351 05030 4	(−29) 0.40883 01593 68065 782
81.49827 92339 48885 4	(−33) 0.55753 45788 32835 675

$n = 28$

$x_k^{(n)}$	$A_k^{(n)}$
0.05073 46248 49873 8876	0.12377 88439 54286 428
0.26748 72686 40741 084	0.23227 92769 00901 161
0.65813 66283 54791 519	0.24751 18960 36477 212
1.22397 18083 84907 72	0.19230 71131 32382 827
1.96676 76124 73777 70	0.11640 53617 21130 006
2.88888 33260 30321 89	(−1) 0.56345 90536 44773 065
3.99331 16592 50114 14	(−1) 0.22066 36432 62588 079
5.28373 60628 43442 56	(−2) 0.70258 87635 58386 773
6.76460 34042 43505 15	(−2) 0.18206 07892 69585 487
8.44121 63282 71324 49	(−3) 0.38334 43038 57123 177
10.31985 04629 93260 1	(−4) 0.65350 87080 69439 831
12.40790 34144 60671 7	(−5) 0.89713 62053 41076 834

(*contd.*)

APPENDIX C (*Continued*)

$x_k^{(n)}$	$A_k^{(n)}$

$n = 28$

$x_k^{(n)}$	$A_k^{(n)}$
14.71408 51641 35748 8	(−6) 0.98470 12256 24928 887
17.24866 34156 08056 3	(−7) 0.85640 75852 67304 245
20.02378 33299 51712 7	(−8) 0.58368 38763 13834 429
23.05389 01350 30296 0	(−9) 0.30756 38877 84230 228
26.35629 73744 01317 6	(−10) 0.12325 90952 72442 282
29.95196 68335 96182 1	(−12) 0.36821 73674 10831 200
33.86660 55165 84459 2	(−14) 0.79987 90575 96890 965
38.13225 44101 94646 8	(−15) 0.12249 22500 32408 341
42.78967 23707 72576 3	(−17) 0.12711 24295 03067 374
47.89207 16336 22743 7	(−20) 0.84885 93367 68654 320
53.51129 79596 64294 2	(−22) 0.34024 55379 42551 185
59.74879 60846 41240 8	(−25) 0.74201 56588 86748 513
66.75697 72839 06469 6	(−28) 0.76004 13205 80173 769
74.78677 81523 39161 8	(−31) 0.28739 10317 94039 581
84.31783 71072 27043 1	(−35) 0.25418 22903 88931 800
96.58242 06275 27319 1	(−40) 0.16613 75878 02903 396

$n = 32$

$x_k^{(n)}$	$A_k^{(n)}$
0.04448 93658 33267 0184	0.10921 83419 52384 971
0.23452 61095 19618 537	0.21044 31079 38813 234
0.57688 46293 01886 426	0.23521 32296 69848 005
1.07244 87538 17817 63	0.19590 33359 72881 043
1.72240 87764 44645 44	0.12998 37862 86071 761
2.52833 67064 25794 88	(−1) 0.70578 62386 57174 415
3.49221 32730 21994 49	(−1) 0.31760 91250 91750 703
4.61645 67697 49767 39	(−1) 0.11918 21483 48385 571
5.90395 85041 74243 95	(−2) 0.37388 16294 61152 479
7.35812 67331 86241 11	(−3) 0.98080 33066 14955 132
8.98294 09242 12596 10	(−3) 0.21486 49188 01364 188
10.78301 86325 39972 1	(−4) 0.39203 41967 98794 720
12.76369 79867 42725 1	(−5) 0.59345 41612 86863 288
14.93113 97555 22557 3	(−6) 0.74164 04578 66755 222
17.29245 43367 15314 8	(−7) 0.76045 67879 12078 148
19.85586 09403 36054 7	(−8) 0.63506 02226 62580 674
22.63088 90131 96774 5	(−9) 0.42813 82971 04092 888
25.62863 60224 59247 8	(−10) 0.23058 99491 89133 608
28.86210 18163 23474 7	(−12) 0.97993 79288 72709 406
32.34662 91539 64737 0	(−13) 0.32378 01657 72926 646
36.10049 48057 51973 8	(−15) 0.81718 23443 42071 943
40.14571 97715 39441 5	(−16) 0.15421 33833 39382 337
44.50920 79957 54938 0	(−18) 0.21197 92290 16361 861
49.22439 49873 08639 2	(−20) 0.20544 29673 78804 543
54.33372 13333 96907 3	(−22) 0.13469 82586 63739 516
59.89250 91621 34018 2	(−25) 0.56612 94130 39735 937
65.97537 72879 35052 8	(−27) 0.14185 60545 46303 691
72.68762 80906 62708 6	(−30) 0.19133 75494 45422 431
80.18744 69779 13523 1	(−33) 0.11922 48760 09822 236
88.73534 04178 92398 7	(−37) 0.26715 11219 24013 699
98.82954 28682 83972 6	(−41) 0.13386 16942 10625 628
111.75139 80979 37695	(−47) 0.45105 36193 89897 424

INDEX

A CATALOG OF SELECTED
DOVER BOOKS
IN SCIENCE AND MATHEMATICS

Astronomy

BURNHAM'S CELESTIAL HANDBOOK, Robert Burnham, Jr. Thorough guide to the stars beyond our solar system. Exhaustive treatment. Alphabetical by constellation: Andromeda to Cetus in Vol. 1; Chamaeleon to Orion in Vol. 2; and Pavo to Vulpecula in Vol. 3. Hundreds of illustrations. Index in Vol. 3. 2,000pp. 6⅛ x 9¼.
Vol. I: 0-486-23567-X
Vol. II: 0-486-23568-8
Vol. III: 0-486-23673-0

EXPLORING THE MOON THROUGH BINOCULARS AND SMALL TELESCOPES, Ernest H. Cherrington, Jr. Informative, profusely illustrated guide to locating and identifying craters, rills, seas, mountains, other lunar features. Newly revised and updated with special section of new photos. Over 100 photos and diagrams. 240pp. 8¼ x 11. 0-486-24491-1

THE EXTRATERRESTRIAL LIFE DEBATE, 1750–1900, Michael J. Crowe. First detailed, scholarly study in English of the many ideas that developed from 1750 to 1900 regarding the existence of intelligent extraterrestrial life. Examines ideas of Kant, Herschel, Voltaire, Percival Lowell, many other scientists and thinkers. 16 illustrations. 704pp. 5⅜ x 8½. 0-486-40675-X

THEORIES OF THE WORLD FROM ANTIQUITY TO THE COPERNICAN REVOLUTION, Michael J. Crowe. Newly revised edition of an accessible, enlightening book recreates the change from an earth-centered to a sun-centered conception of the solar system. 242pp. 5⅜ x 8½. 0-486-41444-2

A HISTORY OF ASTRONOMY, A. Pannekoek. Well-balanced, carefully reasoned study covers such topics as Ptolemaic theory, work of Copernicus, Kepler, Newton, Eddington's work on stars, much more. Illustrated. References. 521pp. 5⅜ x 8½.
0-486-65994-1

A COMPLETE MANUAL OF AMATEUR ASTRONOMY: TOOLS AND TECHNIQUES FOR ASTRONOMICAL OBSERVATIONS, P. Clay Sherrod with Thomas L. Koed. Concise, highly readable book discusses: selecting, setting up and maintaining a telescope; amateur studies of the sun; lunar topography and occultations; observations of Mars, Jupiter, Saturn, the minor planets and the stars; an introduction to photoelectric photometry; more. 1981 ed. 124 figures. 25 halftones. 37 tables. 335pp. 6½ x 9¼. 0-486-40675-X

AMATEUR ASTRONOMER'S HANDBOOK, J. B. Sidgwick. Timeless, comprehensive coverage of telescopes, mirrors, lenses, mountings, telescope drives, micrometers, spectroscopes, more. 189 illustrations. 576pp. 5⅝ x 8¼. (Available in U.S. only.)
0-486-24034-7

STARS AND RELATIVITY, Ya. B. Zel'dovich and I. D. Novikov. Vol. 1 of *Relativistic Astrophysics* by famed Russian scientists. General relativity, properties of matter under astrophysical conditions, stars, and stellar systems. Deep physical insights, clear presentation. 1971 edition. References. 544pp. 5⅝ x 8¼. 0-486-69424-0

Chemistry

THE SCEPTICAL CHYMIST: THE CLASSIC 1661 TEXT, Robert Boyle. Boyle defines the term "element," asserting that all natural phenomena can be explained by the motion and organization of primary particles. 1911 ed. viii+232pp. 5⅜ x 8½.
0-486-42825-7

RADIOACTIVE SUBSTANCES, Marie Curie. Here is the celebrated scientist's doctoral thesis, the prelude to her receipt of the 1903 Nobel Prize. Curie discusses establishing atomic character of radioactivity found in compounds of uranium and thorium; extraction from pitchblende of polonium and radium; isolation of pure radium chloride; determination of atomic weight of radium; plus electric, photographic, luminous, heat, color effects of radioactivity. ii+94pp. 5⅜ x 8½.   0-486-42550-9

CHEMICAL MAGIC, Leonard A. Ford. Second Edition, Revised by E. Winston Grundmeier. Over 100 unusual stunts demonstrating cold fire, dust explosions, much more. Text explains scientific principles and stresses safety precautions. 128pp. 5⅜ x 8½.   0-486-67628-5

THE DEVELOPMENT OF MODERN CHEMISTRY, Aaron J. Ihde. Authoritative history of chemistry from ancient Greek theory to 20th-century innovation. Covers major chemists and their discoveries. 209 illustrations. 14 tables. Bibliographies. Indices. Appendices. 851pp. 5⅜ x 8½.   0-486-64235-6

CATALYSIS IN CHEMISTRY AND ENZYMOLOGY, William P. Jencks. Exceptionally clear coverage of mechanisms for catalysis, forces in aqueous solution, carbonyl- and acyl-group reactions, practical kinetics, more. 864pp. 5⅜ x 8½.
0-486-65460-5

ELEMENTS OF CHEMISTRY, Antoine Lavoisier. Monumental classic by founder of modern chemistry in remarkable reprint of rare 1790 Kerr translation. A must for every student of chemistry or the history of science. 539pp. 5⅜ x 8½. 0-486-64624-6

THE HISTORICAL BACKGROUND OF CHEMISTRY, Henry M. Leicester. Evolution of ideas, not individual biography. Concentrates on formulation of a coherent set of chemical laws. 260pp. 5⅜ x 8½.   0-486-61053-5

A SHORT HISTORY OF CHEMISTRY, J. R. Partington. Classic exposition explores origins of chemistry, alchemy, early medical chemistry, nature of atmosphere, theory of valency, laws and structure of atomic theory, much more. 428pp. 5⅜ x 8½. (Available in U.S. only.)   0-486-65977-1

GENERAL CHEMISTRY, Linus Pauling. Revised 3rd edition of classic first-year text by Nobel laureate. Atomic and molecular structure, quantum mechanics, statistical mechanics, thermodynamics correlated with descriptive chemistry. Problems. 992pp. 5⅜ x 8½.   0-486-65622-5

FROM ALCHEMY TO CHEMISTRY, John Read. Broad, humanistic treatment focuses on great figures of chemistry and ideas that revolutionized the science. 50 illustrations. 240pp. 5⅜ x 8½.   0-486-28690-8

Engineering

DE RE METALLICA, Georgius Agricola. The famous Hoover translation of greatest treatise on technological chemistry, engineering, geology, mining of early modern times (1556). All 289 original woodcuts. 638pp. 6¾ x 11.　　0-486-60006-8

FUNDAMENTALS OF ASTRODYNAMICS, Roger Bate et al. Modern approach developed by U.S. Air Force Academy. Designed as a first course. Problems, exercises. Numerous illustrations. 455pp. 5⅜ x 8½.　　0-486-60061-0

DYNAMICS OF FLUIDS IN POROUS MEDIA, Jacob Bear. For advanced students of ground water hydrology, soil mechanics and physics, drainage and irrigation engineering and more. 335 illustrations. Exercises, with answers. 784pp. 6⅛ x 9¼.
0-486-65675-6

THEORY OF VISCOELASTICITY (Second Edition), Richard M. Christensen. Complete consistent description of the linear theory of the viscoelastic behavior of materials. Problem-solving techniques discussed. 1982 edition. 29 figures. xiv+364pp. 6⅛ x 9¼.　　0-486-42880-X

MECHANICS, J. P. Den Hartog. A classic introductory text or refresher. Hundreds of applications and design problems illuminate fundamentals of trusses, loaded beams and cables, etc. 334 answered problems. 462pp. 5⅜ x 8½.　　0-486-60754-2

MECHANICAL VIBRATIONS, J. P. Den Hartog. Classic textbook offers lucid explanations and illustrative models, applying theories of vibrations to a variety of practical industrial engineering problems. Numerous figures. 233 problems, solutions. Appendix. Index. Preface. 436pp. 5⅜ x 8½.　　0-486-64785-4

STRENGTH OF MATERIALS, J. P. Den Hartog. Full, clear treatment of basic material (tension, torsion, bending, etc.) plus advanced material on engineering methods, applications. 350 answered problems. 323pp. 5⅜ x 8½.　　0-486-60755-0

A HISTORY OF MECHANICS, René Dugas. Monumental study of mechanical principles from antiquity to quantum mechanics. Contributions of ancient Greeks, Galileo, Leonardo, Kepler, Lagrange, many others. 671pp. 5⅜ x 8½. 0-486-65632-2

STABILITY THEORY AND ITS APPLICATIONS TO STRUCTURAL MECHANICS, Clive L. Dym. Self-contained text focuses on Koiter postbuckling analyses, with mathematical notions of stability of motion. Basing minimum energy principles for static stability upon dynamic concepts of stability of motion, it develops asymptotic buckling and postbuckling analyses from potential energy considerations, with applications to columns, plates, and arches. 1974 ed. 208pp. 5⅜ x 8½.
0-486-42541-X

METAL FATIGUE, N. E. Frost, K. J. Marsh, and L. P. Pook. Definitive, clearly written, and well-illustrated volume addresses all aspects of the subject, from the historical development of understanding metal fatigue to vital concepts of the cyclic stress that causes a crack to grow. Includes 7 appendixes. 544pp. 5⅜ x 8½. 0-486-40927-9

ROCKETS, Robert Goddard. Two of the most significant publications in the history of rocketry and jet propulsion: "A Method of Reaching Extreme Altitudes" (1919) and "Liquid Propellant Rocket Development" (1936). 128pp. 5⅜ x 8½.   0-486-42537-1

STATISTICAL MECHANICS: PRINCIPLES AND APPLICATIONS, Terrell L. Hill. Standard text covers fundamentals of statistical mechanics, applications to fluctuation theory, imperfect gases, distribution functions, more. 448pp. 5⅜ x 8½.
0-486-65390-0

ENGINEERING AND TECHNOLOGY 1650–1750: ILLUSTRATIONS AND TEXTS FROM ORIGINAL SOURCES, Martin Jensen. Highly readable text with more than 200 contemporary drawings and detailed engravings of engineering projects dealing with surveying, leveling, materials, hand tools, lifting equipment, transport and erection, piling, bailing, water supply, hydraulic engineering, and more. Among the specific projects outlined-transporting a 50-ton stone to the Louvre, erecting an obelisk, building timber locks, and dredging canals. 207pp. 8⅜ x 11¼.
0-486-42232-1

THE VARIATIONAL PRINCIPLES OF MECHANICS, Cornelius Lanczos. Graduate level coverage of calculus of variations, equations of motion, relativistic mechanics, more. First inexpensive paperbound edition of classic treatise. Index. Bibliography. 418pp. 5⅜ x 8½.   0-486-65067-7

PROTECTION OF ELECTRONIC CIRCUITS FROM OVERVOLTAGES, Ronald B. Standler. Five-part treatment presents practical rules and strategies for circuits designed to protect electronic systems from damage by transient overvoltages. 1989 ed. xxiv+434pp. 6⅛ x 9¼.   0-486-42552-5

ROTARY WING AERODYNAMICS, W. Z. Stepniewski. Clear, concise text covers aerodynamic phenomena of the rotor and offers guidelines for helicopter performance evaluation. Originally prepared for NASA. 537 figures. 640pp. 6⅛ x 9¼.
0-486-64647-5

INTRODUCTION TO SPACE DYNAMICS, William Tyrrell Thomson. Comprehensive, classic introduction to space-flight engineering for advanced undergraduate and graduate students. Includes vector algebra, kinematics, transformation of coordinates. Bibliography. Index. 352pp. 5⅜ x 8½.   0-486-65113-4

HISTORY OF STRENGTH OF MATERIALS, Stephen P. Timoshenko. Excellent historical survey of the strength of materials with many references to the theories of elasticity and structure. 245 figures. 452pp. 5⅜ x 8½.   0-486-61187-6

ANALYTICAL FRACTURE MECHANICS, David J. Unger. Self-contained text supplements standard fracture mechanics texts by focusing on analytical methods for determining crack-tip stress and strain fields. 336pp. 6⅛ x 9¼.   0-486-41737-9

STATISTICAL MECHANICS OF ELASTICITY, J. H. Weiner. Advanced, self-contained treatment illustrates general principles and elastic behavior of solids. Part 1, based on classical mechanics, studies thermoelastic behavior of crystalline and polymeric solids. Part 2, based on quantum mechanics, focuses on interatomic force laws, behavior of solids, and thermally activated processes. For students of physics and chemistry and for polymer physicists. 1983 ed. 96 figures. 496pp. 5⅜ x 8½.
0-486-42260-7

Mathematics

FUNCTIONAL ANALYSIS (Second Corrected Edition), George Bachman and Lawrence Narici. Excellent treatment of subject geared toward students with background in linear algebra, advanced calculus, physics and engineering. Text covers introduction to inner-product spaces, normed, metric spaces, and topological spaces; complete orthonormal sets, the Hahn-Banach Theorem and its consequences, and many other related subjects. 1966 ed. 544pp. 6⅛ x 9¼. 0-486-40251-7

ASYMPTOTIC EXPANSIONS OF INTEGRALS, Norman Bleistein & Richard A. Handelsman. Best introduction to important field with applications in a variety of scientific disciplines. New preface. Problems. Diagrams. Tables. Bibliography. Index. 448pp. 5⅜ x 8½. 0-486-65082-0

VECTOR AND TENSOR ANALYSIS WITH APPLICATIONS, A. I. Borisenko and I. E. Tarapov. Concise introduction. Worked-out problems, solutions, exercises. 257pp. 5⅝ x 8¼. 0-486-63833-2

AN INTRODUCTION TO ORDINARY DIFFERENTIAL EQUATIONS, Earl A. Coddington. A thorough and systematic first course in elementary differential equations for undergraduates in mathematics and science, with many exercises and problems (with answers). Index. 304pp. 5⅜ x 8½. 0-486-65942-9

FOURIER SERIES AND ORTHOGONAL FUNCTIONS, Harry F. Davis. An incisive text combining theory and practical example to introduce Fourier series, orthogonal functions and applications of the Fourier method to boundary-value problems. 570 exercises. Answers and notes. 416pp. 5⅜ x 8½. 0-486-65973-9

COMPUTABILITY AND UNSOLVABILITY, Martin Davis. Classic graduate-level introduction to theory of computability, usually referred to as theory of recurrent functions. New preface and appendix. 288pp. 5⅜ x 8½. 0-486-61471-9

ASYMPTOTIC METHODS IN ANALYSIS, N. G. de Bruijn. An inexpensive, comprehensive guide to asymptotic methods–the pioneering work that teaches by explaining worked examples in detail. Index. 224pp. 5⅜ x 8½ 0-486-64221-6

APPLIED COMPLEX VARIABLES, John W. Dettman. Step-by-step coverage of fundamentals of analytic function theory–plus lucid exposition of five important applications: Potential Theory; Ordinary Differential Equations; Fourier Transforms; Laplace Transforms; Asymptotic Expansions. 66 figures. Exercises at chapter ends. 512pp. 5⅜ x 8½. 0-486-64670-X

INTRODUCTION TO LINEAR ALGEBRA AND DIFFERENTIAL EQUATIONS, John W. Dettman. Excellent text covers complex numbers, determinants, orthonormal bases, Laplace transforms, much more. Exercises with solutions. Undergraduate level. 416pp. 5⅜ x 8½. 0-486-65191-6

RIEMANN'S ZETA FUNCTION, H. M. Edwards. Superb, high-level study of landmark 1859 publication entitled "On the Number of Primes Less Than a Given Magnitude" traces developments in mathematical theory that it inspired. xiv+315pp. 5⅜ x 8½. 0-486-41740-9

CALCULUS OF VARIATIONS WITH APPLICATIONS, George M. Ewing. Applications-oriented introduction to variational theory develops insight and promotes understanding of specialized books, research papers. Suitable for advanced undergraduate/graduate students as primary, supplementary text. 352pp. 5⅜ x 8½.
0-486-64856-7

COMPLEX VARIABLES, Francis J. Flanigan. Unusual approach, delaying complex algebra till harmonic functions have been analyzed from real variable viewpoint. Includes problems with answers. 364pp. 5⅜ x 8½.
0-486-61388-7

AN INTRODUCTION TO THE CALCULUS OF VARIATIONS, Charles Fox. Graduate-level text covers variations of an integral, isoperimetrical problems, least action, special relativity, approximations, more. References. 279pp. 5⅜ x 8½.
0-486-65499-0

COUNTEREXAMPLES IN ANALYSIS, Bernard R. Gelbaum and John M. H. Olmsted. These counterexamples deal mostly with the part of analysis known as "real variables." The first half covers the real number system, and the second half encompasses higher dimensions. 1962 edition. xxiv+198pp. 5⅜ x 8½. 0-486-42875-3

CATASTROPHE THEORY FOR SCIENTISTS AND ENGINEERS, Robert Gilmore. Advanced-level treatment describes mathematics of theory grounded in the work of Poincaré, R. Thom, other mathematicians. Also important applications to problems in mathematics, physics, chemistry and engineering. 1981 edition. References. 28 tables. 397 black-and-white illustrations. xvii + 666pp. 6⅛ x 9¼.
0-486-67539-4

INTRODUCTION TO DIFFERENCE EQUATIONS, Samuel Goldberg. Exceptionally clear exposition of important discipline with applications to sociology, psychology, economics. Many illustrative examples; over 250 problems. 260pp. 5⅜ x 8½.
0-486-65084-7

NUMERICAL METHODS FOR SCIENTISTS AND ENGINEERS, Richard Hamming. Classic text stresses frequency approach in coverage of algorithms, polynomial approximation, Fourier approximation, exponential approximation, other topics. Revised and enlarged 2nd edition. 721pp. 5⅜ x 8½.
0-486-65241-6

INTRODUCTION TO NUMERICAL ANALYSIS (2nd Edition), F. B. Hildebrand. Classic, fundamental treatment covers computation, approximation, interpolation, numerical differentiation and integration, other topics. 150 new problems. 669pp. 5⅜ x 8½.
0-486-65363-3

THREE PEARLS OF NUMBER THEORY, A. Y. Khinchin. Three compelling puzzles require proof of a basic law governing the world of numbers. Challenges concern van der Waerden's theorem, the Landau-Schnirelmann hypothesis and Mann's theorem, and a solution to Waring's problem. Solutions included. 64pp. 5⅜ x 8½.
0-486-40026-3

THE PHILOSOPHY OF MATHEMATICS: AN INTRODUCTORY ESSAY, Stephan Körner. Surveys the views of Plato, Aristotle, Leibniz & Kant concerning propositions and theories of applied and pure mathematics. Introduction. Two appendices. Index. 198pp. 5⅜ x 8½.
0-486-25048-2

INTRODUCTORY REAL ANALYSIS, A.N. Kolmogorov, S. V. Fomin. Translated by Richard A. Silverman. Self-contained, evenly paced introduction to real and functional analysis. Some 350 problems. 403pp. 5⅜ x 8½. 0-486-61226-0

APPLIED ANALYSIS, Cornelius Lanczos. Classic work on analysis and design of finite processes for approximating solution of analytical problems. Algebraic equations, matrices, harmonic analysis, quadrature methods, much more. 559pp. 5⅜ x 8½. 0-486-65656-X

AN INTRODUCTION TO ALGEBRAIC STRUCTURES, Joseph Landin. Superb self-contained text covers "abstract algebra": sets and numbers, theory of groups, theory of rings, much more. Numerous well-chosen examples, exercises. 247pp. 5⅜ x 8½. 0-486-65940-2

QUALITATIVE THEORY OF DIFFERENTIAL EQUATIONS, V. V. Nemytskii and V.V. Stepanov. Classic graduate-level text by two prominent Soviet mathematicians covers classical differential equations as well as topological dynamics and ergodic theory. Bibliographies. 523pp. 5⅜ x 8½. 0-486-65954-2

THEORY OF MATRICES, Sam Perlis. Outstanding text covering rank, nonsingularity and inverses in connection with the development of canonical matrices under the relation of equivalence, and without the intervention of determinants. Includes exercises. 237pp. 5⅜ x 8½. 0-486-66810-X

INTRODUCTION TO ANALYSIS, Maxwell Rosenlicht. Unusually clear, accessible coverage of set theory, real number system, metric spaces, continuous functions, Riemann integration, multiple integrals, more. Wide range of problems. Undergraduate level. Bibliography. 254pp. 5⅜ x 8½. 0-486-65038-3

MODERN NONLINEAR EQUATIONS, Thomas L. Saaty. Emphasizes practical solution of problems; covers seven types of equations. ". . . a welcome contribution to the existing literature...."–*Math Reviews*. 490pp. 5⅜ x 8½. 0-486-64232-1

MATRICES AND LINEAR ALGEBRA, Hans Schneider and George Phillip Barker. Basic textbook covers theory of matrices and its applications to systems of linear equations and related topics such as determinants, eigenvalues and differential equations. Numerous exercises. 432pp. 5⅜ x 8½. 0-486-66014-1

LINEAR ALGEBRA, Georgi E. Shilov. Determinants, linear spaces, matrix algebras, similar topics. For advanced undergraduates, graduates. Silverman translation. 387pp. 5⅜ x 8½. 0-486-63518-X

ELEMENTS OF REAL ANALYSIS, David A. Sprecher. Classic text covers fundamental concepts, real number system, point sets, functions of a real variable, Fourier series, much more. Over 500 exercises. 352pp. 5⅜ x 8½. 0-486-65385-4

SET THEORY AND LOGIC, Robert R. Stoll. Lucid introduction to unified theory of mathematical concepts. Set theory and logic seen as tools for conceptual understanding of real number system. 496pp. 5⅜ x 8¼. 0-486-63829-4

CATALOG OF DOVER BOOKS

TENSOR CALCULUS, J.L. Synge and A. Schild. Widely used introductory text covers spaces and tensors, basic operations in Riemannian space, non-Riemannian spaces, etc. 324pp. 5⅜ x 8¼.   0-486-63612-7

ORDINARY DIFFERENTIAL EQUATIONS, Morris Tenenbaum and Harry Pollard. Exhaustive survey of ordinary differential equations for undergraduates in mathematics, engineering, science. Thorough analysis of theorems. Diagrams. Bibliography. Index. 818pp. 5⅜ x 8½.   0-486-64940-7

INTEGRAL EQUATIONS, F. G. Tricomi. Authoritative, well-written treatment of extremely useful mathematical tool with wide applications. Volterra Equations, Fredholm Equations, much more. Advanced undergraduate to graduate level. Exercises. Bibliography. 238pp. 5⅜ x 8½.   0-486-64828-1

FOURIER SERIES, Georgi P. Tolstov. Translated by Richard A. Silverman. A valuable addition to the literature on the subject, moving clearly from subject to subject and theorem to theorem. 107 problems, answers. 336pp. 5⅜ x 8½.   0-486-63317-9

INTRODUCTION TO MATHEMATICAL THINKING, Friedrich Waismann. Examinations of arithmetic, geometry, and theory of integers; rational and natural numbers; complete induction; limit and point of accumulation; remarkable curves; complex and hypercomplex numbers, more. 1959 ed. 27 figures. xii+260pp. 5⅜ x 8½.   0-486-63317-9

POPULAR LECTURES ON MATHEMATICAL LOGIC, Hao Wang. Noted logician's lucid treatment of historical developments, set theory, model theory, recursion theory and constructivism, proof theory, more. 3 appendixes. Bibliography. 1981 edition. ix + 283pp. 5⅜ x 8½.   0-486-67632-3

CALCULUS OF VARIATIONS, Robert Weinstock. Basic introduction covering isoperimetric problems, theory of elasticity, quantum mechanics, electrostatics, etc. Exercises throughout. 326pp. 5⅜ x 8½.   0-486-63069-2

THE CONTINUUM: A CRITICAL EXAMINATION OF THE FOUNDATION OF ANALYSIS, Hermann Weyl. Classic of 20th-century foundational research deals with the conceptual problem posed by the continuum. 156pp. 5⅜ x 8½.
     0-486-67982-9

CHALLENGING MATHEMATICAL PROBLEMS WITH ELEMENTARY SOLUTIONS, A. M. Yaglom and I. M. Yaglom. Over 170 challenging problems on probability theory, combinatorial analysis, points and lines, topology, convex polygons, many other topics. Solutions. Total of 445pp. 5⅜ x 8½. Two-vol. set.
     Vol. I: 0-486-65536-9   Vol. II: 0-486-65537-7

INTRODUCTION TO PARTIAL DIFFERENTIAL EQUATIONS WITH APPLICATIONS, E. C. Zachmanoglou and Dale W. Thoe. Essentials of partial differential equations applied to common problems in engineering and the physical sciences. Problems and answers. 416pp. 5⅜ x 8½.   0-486-65251-3

THE THEORY OF GROUPS, Hans J. Zassenhaus. Well-written graduate-level text acquaints reader with group-theoretic methods and demonstrates their usefulness in mathematics. Axioms, the calculus of complexes, homomorphic mapping, p-group theory, more. 276pp. 5⅜ x 8½.   0-486-40922-8

Math–Decision Theory, Statistics, Probability

ELEMENTARY DECISION THEORY, Herman Chernoff and Lincoln E. Moses. Clear introduction to statistics and statistical theory covers data processing, probability and random variables, testing hypotheses, much more. Exercises. 364pp. 5⅜ x 8½. 0-486-65218-1

STATISTICS MANUAL, Edwin L. Crow et al. Comprehensive, practical collection of classical and modern methods prepared by U.S. Naval Ordnance Test Station. Stress on use. Basics of statistics assumed. 288pp. 5⅜ x 8½. 0-486-60599-X

SOME THEORY OF SAMPLING, William Edwards Deming. Analysis of the problems, theory and design of sampling techniques for social scientists, industrial managers and others who find statistics important at work. 61 tables. 90 figures. xvii +602pp. 5⅜ x 8½. 0-486-64684-X

LINEAR PROGRAMMING AND ECONOMIC ANALYSIS, Robert Dorfman, Paul A. Samuelson and Robert M. Solow. First comprehensive treatment of linear programming in standard economic analysis. Game theory, modern welfare economics, Leontief input-output, more. 525pp. 5⅜ x 8½. 0-486-65491-5

PROBABILITY: AN INTRODUCTION, Samuel Goldberg. Excellent basic text covers set theory, probability theory for finite sample spaces, binomial theorem, much more. 360 problems. Bibliographies. 322pp. 5⅜ x 8½. 0-486-65252-1

GAMES AND DECISIONS: INTRODUCTION AND CRITICAL SURVEY, R. Duncan Luce and Howard Raiffa. Superb nontechnical introduction to game theory, primarily applied to social sciences. Utility theory, zero-sum games, n-person games, decision-making, much more. Bibliography. 509pp. 5⅜ x 8½. 0-486-65943-7

INTRODUCTION TO THE THEORY OF GAMES, J. C. C. McKinsey. This comprehensive overview of the mathematical theory of games illustrates applications to situations involving conflicts of interest, including economic, social, political, and military contexts. Appropriate for advanced undergraduate and graduate courses; advanced calculus a prerequisite. 1952 ed. x+372pp. 5⅜ x 8½. 0-486-42811-7

FIFTY CHALLENGING PROBLEMS IN PROBABILITY WITH SOLUTIONS, Frederick Mosteller. Remarkable puzzlers, graded in difficulty, illustrate elementary and advanced aspects of probability. Detailed solutions. 88pp. 5⅜ x 8½. 65355-2

PROBABILITY THEORY: A CONCISE COURSE, Y. A. Rozanov. Highly readable, self-contained introduction covers combination of events, dependent events, Bernoulli trials, etc. 148pp. 5⅝ x 8¼. 0-486-63544-9

STATISTICAL METHOD FROM THE VIEWPOINT OF QUALITY CONTROL, Walter A. Shewhart. Important text explains regulation of variables, uses of statistical control to achieve quality control in industry, agriculture, other areas. 192pp. 5⅜ x 8½. 0-486-65232-7

Math–Geometry and Topology

ELEMENTARY CONCEPTS OF TOPOLOGY, Paul Alexandroff. Elegant, intuitive approach to topology from set-theoretic topology to Betti groups; how concepts of topology are useful in math and physics. 25 figures. 57pp. 5⅜ x 8½. 0-486-60747-X

COMBINATORIAL TOPOLOGY, P. S. Alexandrov. Clearly written, well-organized, three-part text begins by dealing with certain classic problems without using the formal techniques of homology theory and advances to the central concept, the Betti groups. Numerous detailed examples. 654pp. 5⅜ x 8½. 0-486-40179-0

EXPERIMENTS IN TOPOLOGY, Stephen Barr. Classic, lively explanation of one of the byways of mathematics. Klein bottles, Moebius strips, projective planes, map coloring, problem of the Koenigsberg bridges, much more, described with clarity and wit. 43 figures. 210pp. 5⅜ x 8½. 0-486-25933-1

THE GEOMETRY OF RENÉ DESCARTES, René Descartes. The great work founded analytical geometry. Original French text, Descartes's own diagrams, together with definitive Smith-Latham translation. 244pp. 5⅜ x 8½. 0-486-60068-8

EUCLIDEAN GEOMETRY AND TRANSFORMATIONS, Clayton W. Dodge. This introduction to Euclidean geometry emphasizes transformations, particularly isometries and similarities. Suitable for undergraduate courses, it includes numerous examples, many with detailed answers. 1972 ed. viii+296pp. 6⅛ x 9¼. 0-486-43476-1

PRACTICAL CONIC SECTIONS: THE GEOMETRIC PROPERTIES OF ELLIPSES, PARABOLAS AND HYPERBOLAS, J. W. Downs. This text shows how to create ellipses, parabolas, and hyperbolas. It also presents historical background on their ancient origins and describes the reflective properties and roles of curves in design applications. 1993 ed. 98 figures. xii+100pp. 6½ x 9¼. 0-486-42876-1

THE THIRTEEN BOOKS OF EUCLID'S ELEMENTS, translated with introduction and commentary by Sir Thomas L. Heath. Definitive edition. Textual and linguistic notes, mathematical analysis. 2,500 years of critical commentary. Unabridged. 1,414pp. 5⅜ x 8½. Three-vol. set.
Vol. I: 0-486-60088-2 Vol. II: 0-486-60089-0 Vol. III: 0-486-60090-4

SPACE AND GEOMETRY: IN THE LIGHT OF PHYSIOLOGICAL, PSYCHOLOGICAL AND PHYSICAL INQUIRY, Ernst Mach. Three essays by an eminent philosopher and scientist explore the nature, origin, and development of our concepts of space, with a distinctness and precision suitable for undergraduate students and other readers. 1906 ed. vi+148pp. 5⅜ x 8½. 0-486-43909-7

GEOMETRY OF COMPLEX NUMBERS, Hans Schwerdtfeger. Illuminating, widely praised book on analytic geometry of circles, the Moebius transformation, and two-dimensional non-Euclidean geometries. 200pp. 5⅝ x 8¼. 0-486-63830-8

DIFFERENTIAL GEOMETRY, Heinrich W. Guggenheimer. Local differential geometry as an application of advanced calculus and linear algebra. Curvature, transformation groups, surfaces, more. Exercises. 62 figures. 378pp. 5⅜ x 8½. 0-486-63433-7

History of Math

THE WORKS OF ARCHIMEDES, Archimedes (T. L. Heath, ed.). Topics include the famous problems of the ratio of the areas of a cylinder and an inscribed sphere; the measurement of a circle; the properties of conoids, spheroids, and spirals; and the quadrature of the parabola. Informative introduction. clxxxvi+326pp. 5⅜ x 8½.
0-486-42084-1

A SHORT ACCOUNT OF THE HISTORY OF MATHEMATICS, W. W. Rouse Ball. One of clearest, most authoritative surveys from the Egyptians and Phoenicians through 19th-century figures such as Grassman, Galois, Riemann. Fourth edition. 522pp. 5⅜ x 8½.
0-486-20630-0

THE HISTORY OF THE CALCULUS AND ITS CONCEPTUAL DEVELOP-MENT, Carl B. Boyer. Origins in antiquity, medieval contributions, work of Newton, Leibniz, rigorous formulation. Treatment is verbal. 346pp. 5⅜ x 8½.   0-486-60509-4

THE HISTORICAL ROOTS OF ELEMENTARY MATHEMATICS, Lucas N. H. Bunt, Phillip S. Jones, and Jack D. Bedient. Fundamental underpinnings of modern arithmetic, algebra, geometry and number systems derived from ancient civiliza-tions. 320pp. 5⅜ x 8½.
0-486-25563-8

A HISTORY OF MATHEMATICAL NOTATIONS, Florian Cajori. This classic study notes the first appearance of a mathematical symbol and its origin, the com-petition it encountered, its spread among writers in different countries, its rise to pop-ularity, its eventual decline or ultimate survival. Original 1929 two-volume edition presented here in one volume. xxviii+820pp. 5⅜ x 8½.
0-486-67766-4

GAMES, GODS & GAMBLING: A HISTORY OF PROBABILITY AND STATISTICAL IDEAS, F. N. David. Episodes from the lives of Galileo, Fermat, Pascal, and others illustrate this fascinating account of the roots of mathematics. Features thought-provoking references to classics, archaeology, biography, poetry. 1962 edition. 304pp. 5⅜ x 8½. (Available in U.S. only.)
0-486-40023-9

OF MEN AND NUMBERS: THE STORY OF THE GREAT MATHEMATICIANS, Jane Muir. Fascinating accounts of the lives and accom-plishments of history's greatest mathematical minds–Pythagoras, Descartes, Euler, Pascal, Cantor, many more. Anecdotal, illuminating. 30 diagrams. Bibliography. 256pp. 5⅜ x 8½.
0-486-28973-7

HISTORY OF MATHEMATICS, David E. Smith. Nontechnical survey from ancient Greece and Orient to late 19th century; evolution of arithmetic, geometry, trigonometry, calculating devices, algebra, the calculus. 362 illustrations. 1,355pp. 5⅜ x 8½. Two-vol. set.   Vol. I: 0-486-20429-4   Vol. II: 0-486-20430-8

A CONCISE HISTORY OF MATHEMATICS, Dirk J. Struik. The best brief his-tory of mathematics. Stresses origins and covers every major figure from ancient Near East to 19th century. 41 illustrations. 195pp. 5⅜ x 8½.   0-486-60255-9

Physics

OPTICAL RESONANCE AND TWO-LEVEL ATOMS, L. Allen and J. H. Eberly. Clear, comprehensive introduction to basic principles behind all quantum optical resonance phenomena. 53 illustrations. Preface. Index. 256pp. 5⅜ x 8½. 0-486-65533-4

QUANTUM THEORY, David Bohm. This advanced undergraduate-level text presents the quantum theory in terms of qualitative and imaginative concepts, followed by specific applications worked out in mathematical detail. Preface. Index. 655pp. 5⅜ x 8½. 0-486-65969-0

ATOMIC PHYSICS (8th EDITION), Max Born. Nobel laureate's lucid treatment of kinetic theory of gases, elementary particles, nuclear atom, wave-corpuscles, atomic structure and spectral lines, much more. Over 40 appendices, bibliography. 495pp. 5⅜ x 8½. 0-486-65984-4

A SOPHISTICATE'S PRIMER OF RELATIVITY, P. W. Bridgman. Geared toward readers already acquainted with special relativity, this book transcends the view of theory as a working tool to answer natural questions: What is a frame of reference? What is a "law of nature"? What is the role of the "observer"? Extensive treatment, written in terms accessible to those without a scientific background. 1983 ed. xlviii+172pp. 5⅜ x 8½. 0-486-42549-5

AN INTRODUCTION TO HAMILTONIAN OPTICS, H. A. Buchdahl. Detailed account of the Hamiltonian treatment of aberration theory in geometrical optics. Many classes of optical systems defined in terms of the symmetries they possess. Problems with detailed solutions. 1970 edition. xv + 360pp. 5⅜ x 8½. 0-486-67597-1

PRIMER OF QUANTUM MECHANICS, Marvin Chester. Introductory text examines the classical quantum bead on a track: its state and representations; operator eigenvalues; harmonic oscillator and bound bead in a symmetric force field; and bead in a spherical shell. Other topics include spin, matrices, and the structure of quantum mechanics; the simplest atom; indistinguishable particles; and stationary-state perturbation theory. 1992 ed. xiv+314pp. 6⅛ x 9¼. 0-486-42878-8

LECTURES ON QUANTUM MECHANICS, Paul A. M. Dirac. Four concise, brilliant lectures on mathematical methods in quantum mechanics from Nobel Prize-winning quantum pioneer build on idea of visualizing quantum theory through the use of classical mechanics. 96pp. 5⅜ x 8½. 0-486-41713-1

THIRTY YEARS THAT SHOOK PHYSICS: THE STORY OF QUANTUM THEORY, George Gamow. Lucid, accessible introduction to influential theory of energy and matter. Careful explanations of Dirac's anti-particles, Bohr's model of the atom, much more. 12 plates. Numerous drawings. 240pp. 5⅜ x 8½. 0-486-24895-X

ELECTRONIC STRUCTURE AND THE PROPERTIES OF SOLIDS: THE PHYSICS OF THE CHEMICAL BOND, Walter A. Harrison. Innovative text offers basic understanding of the electronic structure of covalent and ionic solids, simple metals, transition metals and their compounds. Problems. 1980 edition. 582pp. 6⅛ x 9¼. 0-486-66021-4

HYDRODYNAMIC AND HYDROMAGNETIC STABILITY, S. Chandrasekhar. Lucid examination of the Rayleigh-Benard problem; clear coverage of the theory of instabilities causing convection. 704pp. 5⅝ x 8¼. 0-486-64071-X

INVESTIGATIONS ON THE THEORY OF THE BROWNIAN MOVEMENT, Albert Einstein. Five papers (1905–8) investigating dynamics of Brownian motion and evolving elementary theory. Notes by R. Fürth. 122pp. 5⅜ x 8½. 0-486-60304-0

THE PHYSICS OF WAVES, William C. Elmore and Mark A. Heald. Unique overview of classical wave theory. Acoustics, optics, electromagnetic radiation, more. Ideal as classroom text or for self-study. Problems. 477pp. 5⅜ x 8½. 0-486-64926-1

GRAVITY, George Gamow. Distinguished physicist and teacher takes reader-friendly look at three scientists whose work unlocked many of the mysteries behind the laws of physics: Galileo, Newton, and Einstein. Most of the book focuses on Newton's ideas, with a concluding chapter on post-Einsteinian speculations concerning the relationship between gravity and other physical phenomena. 160pp. 5⅜ x 8½. 0-486-42563-0

PHYSICAL PRINCIPLES OF THE QUANTUM THEORY, Werner Heisenberg. Nobel Laureate discusses quantum theory, uncertainty, wave mechanics, work of Dirac, Schroedinger, Compton, Wilson, Einstein, etc. 184pp. 5⅜ x 8½. 0-486-60113-7

ATOMIC SPECTRA AND ATOMIC STRUCTURE, Gerhard Herzberg. One of best introductions; especially for specialist in other fields. Treatment is physical rather than mathematical. 80 illustrations. 257pp. 5⅜ x 8½. 0-486-60115-3

AN INTRODUCTION TO STATISTICAL THERMODYNAMICS, Terrell L. Hill. Excellent basic text offers wide-ranging coverage of quantum statistical mechanics, systems of interacting molecules, quantum statistics, more. 523pp. 5⅜ x 8½. 0-486-65242-4

THEORETICAL PHYSICS, Georg Joos, with Ira M. Freeman. Classic overview covers essential math, mechanics, electromagnetic theory, thermodynamics, quantum mechanics, nuclear physics, other topics. First paperback edition. xxiii + 885pp. 5⅜ x 8½. 0-486-65227-0

PROBLEMS AND SOLUTIONS IN QUANTUM CHEMISTRY AND PHYSICS, Charles S. Johnson, Jr. and Lee G. Pedersen. Unusually varied problems, detailed solutions in coverage of quantum mechanics, wave mechanics, angular momentum, molecular spectroscopy, more. 280 problems plus 139 supplementary exercises. 430pp. 6½ x 9¼. 0-486-65236-X

THEORETICAL SOLID STATE PHYSICS, Vol. 1: Perfect Lattices in Equilibrium; Vol. II: Non-Equilibrium and Disorder, William Jones and Norman H. March. Monumental reference work covers fundamental theory of equilibrium properties of perfect crystalline solids, non-equilibrium properties, defects and disordered systems. Appendices. Problems. Preface. Diagrams. Index. Bibliography. Total of 1,301pp. 5⅜ x 8½. Two volumes. Vol. I: 0-486-65015-4 Vol. II: 0-486-65016-2

WHAT IS RELATIVITY? L. D. Landau and G. B. Rumer. Written by a Nobel Prize physicist and his distinguished colleague, this compelling book explains the special theory of relativity to readers with no scientific background, using such familiar objects as trains, rulers, and clocks. 1960 ed. vi+72pp. 5⅜ x 8½. 0-486-42806-0

CATALOG OF DOVER BOOKS

A TREATISE ON ELECTRICITY AND MAGNETISM, James Clerk Maxwell. Important foundation work of modern physics. Brings to final form Maxwell's theory of electromagnetism and rigorously derives his general equations of field theory. 1,084pp. 5⅜ x 8½. Two-vol. set. Vol. I: 0-486-60636-8 Vol. II: 0-486-60637-6

QUANTUM MECHANICS: PRINCIPLES AND FORMALISM, Roy McWeeny. Graduate student-oriented volume develops subject as fundamental discipline, opening with review of origins of Schrödinger's equations and vector spaces. Focusing on main principles of quantum mechanics and their immediate consequences, it concludes with final generalizations covering alternative "languages" or representations. 1972 ed. 15 figures. xi+155pp. 5⅜ x 8½. 0-486-42829-X

INTRODUCTION TO QUANTUM MECHANICS With Applications to Chemistry, Linus Pauling & E. Bright Wilson, Jr. Classic undergraduate text by Nobel Prize winner applies quantum mechanics to chemical and physical problems. Numerous tables and figures enhance the text. Chapter bibliographies. Appendices. Index. 468pp. 5⅜ x 8½. 0-486-64871-0

METHODS OF THERMODYNAMICS, Howard Reiss. Outstanding text focuses on physical technique of thermodynamics, typical problem areas of understanding, and significance and use of thermodynamic potential. 1965 edition. 238pp. 5⅜ x 8½. 0-486-69445-3

THE ELECTROMAGNETIC FIELD, Albert Shadowitz. Comprehensive undergraduate text covers basics of electric and magnetic fields, builds up to electromagnetic theory. Also related topics, including relativity. Over 900 problems. 768pp. 5⅜ x 8¼. 0-486-65660-8

GREAT EXPERIMENTS IN PHYSICS: FIRSTHAND ACCOUNTS FROM GALILEO TO EINSTEIN, Morris H. Shamos (ed.). 25 crucial discoveries: Newton's laws of motion, Chadwick's study of the neutron, Hertz on electromagnetic waves, more. Original accounts clearly annotated. 370pp. 5⅜ x 8½. 0-486-25346-5

EINSTEIN'S LEGACY, Julian Schwinger. A Nobel Laureate relates fascinating story of Einstein and development of relativity theory in well-illustrated, nontechnical volume. Subjects include meaning of time, paradoxes of space travel, gravity and its effect on light, non-Euclidean geometry and curving of space-time, impact of radio astronomy and space-age discoveries, and more. 189 b/w illustrations. xiv+250pp. 8⅜ x 9¼. 0-486-41974-6

STATISTICAL PHYSICS, Gregory H. Wannier. Classic text combines thermodynamics, statistical mechanics and kinetic theory in one unified presentation of thermal physics. Problems with solutions. Bibliography. 532pp. 5⅜ x 8½. 0-486-65401-X

Paperbound unless otherwise indicated. Available at your book dealer, online at **www.doverpublications.com**, or by writing to Dept. GI, Dover Publications, Inc., 31 East 2nd Street, Mineola, NY 11501. For current price information or for free catalogues (please indicate field of interest), write to Dover Publications or log on to **www.doverpublications.com** and see every Dover book in print. Dover publishes more than 500 books each year on science, elementary and advanced mathematics, biology, music, art, literary history, social sciences, and other areas.

CATALOG OF DOVER BOOKS

TENSOR CALCULUS, J.L. Synge and A. Schild. Widely used introductory text covers spaces and tensors, basic operations in Riemannian space, non-Riemannian spaces, etc. 324pp. 5⅜ x 8¼. 0-486-63612-7

ORDINARY DIFFERENTIAL EQUATIONS, Morris Tenenbaum and Harry Pollard. Exhaustive survey of ordinary differential equations for undergraduates in mathematics, engineering, science. Thorough analysis of theorems. Diagrams. Bibliography. Index. 818pp. 5⅜ x 8½. 0-486-64940-7

INTEGRAL EQUATIONS, F. G. Tricomi. Authoritative, well-written treatment of extremely useful mathematical tool with wide applications. Volterra Equations, Fredholm Equations, much more. Advanced undergraduate to graduate level. Exercises. Bibliography. 238pp. 5⅜ x 8½. 0-486-64828-1

FOURIER SERIES, Georgi P. Tolstov. Translated by Richard A. Silverman. A valuable addition to the literature on the subject, moving clearly from subject to subject and theorem to theorem. 107 problems, answers. 336pp. 5⅜ x 8½. 0-486-63317-9

INTRODUCTION TO MATHEMATICAL THINKING, Friedrich Waismann. Examinations of arithmetic, geometry, and theory of integers; rational and natural numbers; complete induction; limit and point of accumulation; remarkable curves; complex and hypercomplex numbers, more. 1959 ed. 27 figures. xii+260pp. 5⅜ x 8½. 0-486-63317-9

POPULAR LECTURES ON MATHEMATICAL LOGIC, Hao Wang. Noted logician's lucid treatment of historical developments, set theory, model theory, recursion theory and constructivism, proof theory, more. 3 appendixes. Bibliography. 1981 edition. ix + 283pp. 5⅜ x 8½. 0-486-67632-3

CALCULUS OF VARIATIONS, Robert Weinstock. Basic introduction covering isoperimetric problems, theory of elasticity, quantum mechanics, electrostatics, etc. Exercises throughout. 326pp. 5⅜ x 8½. 0-486-63069-2

THE CONTINUUM: A CRITICAL EXAMINATION OF THE FOUNDATION OF ANALYSIS, Hermann Weyl. Classic of 20th-century foundational research deals with the conceptual problem posed by the continuum. 156pp. 5⅜ x 8½. 0-486-67982-9

CHALLENGING MATHEMATICAL PROBLEMS WITH ELEMENTARY SOLUTIONS, A. M. Yaglom and I. M. Yaglom. Over 170 challenging problems on probability theory, combinatorial analysis, points and lines, topology, convex polygons, many other topics. Solutions. Total of 445pp. 5⅜ x 8½. Two-vol. set.
Vol. I: 0-486-65536-9   Vol. II: 0-486-65537-7

Paperbound unless otherwise indicated. Available at your book dealer, online at **www.doverpublications.com**, or by writing to Dept. GI, Dover Publications, Inc., 31 East 2nd Street, Mineola, NY 11501. For current price information or for free catalogues (please indicate field of interest), write to Dover Publications or log on to **www.doverpublications.com** and see every Dover book in print. Dover publishes more than 500 books each year on science, elementary and advanced mathematics, biology, music, art, literary history, social sciences, and other areas.